DAVID BEVERIDGE MAIR

EXERCISES IN MATHEMATICS

Elibron Classics
www.elibron.com

Elibron Classics series.

ISBN 1-4021-6144-1 (paperback)
ISBN 1-4021-0500-2 (hardcover)

This Elibron Classics Replica Edition is an unabridged facsimile of the edition published in 1920 by Macmillan and Co., Ltd., London.

EXERCISES IN MATHEMATICS

BY

DAVID BEVERIDGE MAIR, M.A.

SOMETIME FELLOW OF CHRIST'S COLLEGE, CAMBRIDGE

MACMILLAN AND CO., LIMITED

ST. MARTIN'S STREET, LONDON

1920

First Edition 1914.
Reprinted 1915, 1920.

GLASGOW: PRINTED AT THE UNIVERSITY PRESS
BY ROBERT MACLEHOSE AND CO. LTD.

1–17; the following Sections (18–54) cover Geometry and Algebra up to the trigonometrical solution of right-angled triangles; and the text is completed by typical question papers of various examining bodies, whose subject-matter falls within the scope of the book.

The book may be used as a revision course, or each section may be taken with the student's first work upon the branch of which it treats. Few sections are intended to be worked straight through; they contain many questions from which a suitable selection may be made.

Squared paper is required for some questions. It is best to provide two forms of paper in which the units of ruling are 5 mm. and 0·2 inch; but the difference between the two forms is small, and either will in most cases serve in place of the other.

Much trouble has been taken to secure the correctness of the answers. It can scarcely be hoped, however, that all error has been avoided, and any corrections will be gratefully received.

A mathematical line "has length without breadth," but if our diagrams are to be visible their lines must have appreciable breadth; and measurements of the diagram should be taken to the middle of the line.

A figure has sometimes to be transferred from the book. When the position of the figure on the page of the note-book is immaterial it may be pricked through with a pin or compass-point. To place the figure appropriately with regard to the ruling of the note-book, it may be traced upon transparent paper and pricked through from the transparent paper into the note-book.

Acknowledgment of assistance is gratefully made to Mr. J. G. Hamilton for working through the Exercises and making many corrections of results and improvements in wording, and to Prof. R. A. Gregory and Mr. A. T. Simmons for help in many forms.

Most of the questions are taken from public examinations, and

PREFACE

THE motives which give rise to mathematical activities in the human mind are classified by Dr Percy Nunn as the wonder motive, the utility motive, and the systematizing motive. The systematizing motive influences mature minds chiefly and has little concern with school education. In the school the predominant motive is that of utility. This motive must form the basis of a good school course of mathematics; while the wonder motive justifies the inclusion of an occasional striking property which has no utility value.

If we held strictly to these principles we should shun all side streams, the exploration of which occupies time out of proportion to the resulting gain, and confine ourselves to the main course of mathematical development, in order that every pupil—the non-specialist included—may leave school with some real idea of the part played by mathematics in human life.

In practice, however, we are not free to plan an ideal course. The school's freedom to follow the best is limited by a number of restrictions, prominent among which stands the examination syllabus. In consequence we include in this book all the branches which it is customary to teach, and at the head of each Section or Sub-section which a school free to follow the best ought to omit we place a note to that effect.

The book is designed to cover the mathematical studies of the non-specialist up to the close of his school career; the mathematical pupil ought to follow the same course and then proceed to more advanced studies. Arithmetic is treated in Sections

MACMILLAN AND CO., Limited
LONDON · BOMBAY · CALCUTTA · MADRAS
MELBOURNE

THE MACMILLAN COMPANY
NEW YORK · BOSTON · CHICAGO
DALLAS · SAN FRANCISCO

THE MACMILLAN CO. OF CANADA, Ltd.
TORONTO

EXERCISES IN MATHEMATICS

Conversion Relations.	**Less Important Measures.**
1 cm. = 0·394 in.	1 pole or rod = 5·5 yards.
1 m. = 1·09 yd.	1 furlong = 220 yards.
1 km. = 0·621 mile.	1 cable = 608 feet.
1 in. = 2·54 cm.	1 sea mile = 6080 feet.
	10 decimetres = 1 metre.

1 sq. cm. = 0·155 sq. in.	1 sq. rod or pole = 30·25 sq. yards.
1 hectare = 2·47 acres.	1 rood = 40 sq. poles.
1 sq. in. = 6·45 sq. cm.	1 are = 100 sq. metres.

CAPACITY, AND LIQUID MEASURE.

1 c. cm. = 0·0610 c. in.	2 pints = 1 quart.
1 litre = 0·220 gal.	4 quarts = 1 gallon.
1 c. in. = 16·4 c. cm.	8 gallons = 1 bushel.
	8 bushels = 1 quarter.
	100 centilitres = 1 litre.
	100 litres = 1 hectolitre.

1 kg. = 2·20 lb.	7000 grains = 1 pound.
1 tonne = 0·984 ton.	14 pounds = 1 stone.
1 lb. = 454 grammes.	28 pounds = 1 quarter.
	100 pounds = 1 cental.

of πrr, the value of π being 3·1416.
1 gramme.

1. MEASURE OF DISTANCE: or Linear Measure.

Imperial Measures.	Metric Measures.
12 inches = 1 foot.	10 millimetres = 1 centimetre.
3 feet = 1 yard.	100 centimetres = 1 metre.
1760 yards = 1 mile.	1000 metres = 1 kilometre.
100 links = 1 chain = 22 yards.	

2. MEASURE OF AREA: or Square Measure.

144 sq. inches = 1 sq. foot.	100 sq. centimetres = 1 sq. decimetre.
9 sq. feet. = 1 sq. yard.	100 sq. decimetres = 1 sq. metre.
484 sq. yards = 1 sq. chain.	10,000 sq. metres = 1 hectare.
10 sq. chains = 1 acre.	100 hectares = 1 sq. kilometre.
640 acres = 1 sq. mile.	

3. MEASURE OF VOLUME: or Cubic Measure, Measure of

1728 cubic inches = 1 cubic foot.	1000 c. cm. = 1 c. decimetre.
27 cubic feet = 1 cubic yard.	= 1 litre.
1 gallon = 277 cubic inches.	
1 cubic foot = 6·23 gallons.	

4. MEASURE OF WEIGHT.

16 ounces = 1 pound.	1000 grammes = 1 kilogram.
112 pounds = 1 cwt.	1000 kilograms = 1 metric tonne.
20 cwt. = 1 ton.	

5. CONSTANTS.

A circle of radius r has a circumference of length $2\pi r$ and has an area
A gallon of water weighs 10 lb. A cubic centimetre of water weighs
The base e of natural logarithms is 2·718.

are used by the courteous permission of the Controller of His Majesty's Stationery Office, the Commissioners of Intermediate Education in Ireland, the Central Welsh Board, the Oxford and Cambridge Schools Examination Board, the Delegates of the Local Examinations of the University of Oxford, the Local Examinations and Lectures Syndicate of the University of Cambridge, the University of London, the Joint Matriculation Board of the Northern Universities (Manchester, Liverpool, Leeds, Sheffield), the University of Dublin, the University of Birmingham, the University of Bristol, the University of Belfast, the London County Council, the Council of the Institution of Civil Engineers.

DAVID MAIR.

December, 1913.

CONTENTS

PART I. CHIEFLY MENSURATION

PART II. FURTHER ARITHMETIC

PART III. PARALLELS, CONGRUENCE, AND FORMULAS

PART I.

SECTIONS 1–8. CHIEFLY MENSURATION.

Section 1. Weights and Measures.

(i.) MONEY.

1. In the course of a day a tram conductor sold 437 half-penny tickets, 385 penny tickets, 316 penny-halfpenny tickets, and 241 twopenny tickets. How much money did he take altogether?

2. Copy out and complete the following bill:

$3\frac{1}{2}$ yards of flannel at 3s. 3d. a yard.
$2\frac{1}{4}$ yards of ribbon at $5\frac{3}{4}$d. a yard.
6 yards of nun's veiling at 1s. 6d. a yard.
2 dozen pearl buttons at $3\frac{1}{2}$d. a dozen.

The amount of each item is to be given to a farthing, the seller taking the advantage if the amount is not an exact number of farthings.

3. A man earns 12s. 6d. a day and works 240 days in the year. What does he make in the year?

4. One man works 8 hours a day, and is paid 28 shillings for a week of 6 days; another works 9 hours a day, is paid 25 shillings for a week of 6 days, and takes 5 hours to do work the first man does in 4. Which is it more profitable to employ?

5. A labourer, who was earning, wet or fine, £1 a week, took work at 7d. an hour. He worked 8 hours a day and $5\frac{1}{2}$ days a week; but in the course of 52 weeks he lost 15 days' work through rain. What did he gain in the 52 weeks by the change?

6. Express 17s. 9d. as a decimal of £5. 4s. 2d. correct to three significant figures.

7. It has been proposed to divide £1 into 1000 equal parts called mils.

(i) Express the difference between a farthing and a mil as a decimal fraction of 1d.

(ii) Express 4496 mils in pounds, shillings, and pence, correct to the nearest farthing.

8. A side of bacon weighs 63 lb. and costs $9\frac{1}{2}$d. per lb. Show that its value lies between 49 and 50 shillings; and see which of these is nearest the actual value.

9. A publisher produces 10,000 copies of a book at a cost of £1500, and, after giving away 50 presentation copies, sells the rest to booksellers at 4s. 6d. a copy. What profit is made?

10. The tenant of a colliery pays to the landlord a fixed rent of £120, and a royalty of $1\frac{1}{2}$d. for every ton of coal got from the mine. In one year he took 36,000 tons of coal from the mine. What was his total payment to the landlord for that year?

11. A coal shipper contracted to deliver 5500 tons of coal at a foreign port at 21s. 6d. per ton. The coal cost 12s. 9d. per ton in this country, and the freight was at the rate of 7s. $10\frac{1}{2}$d. per ton. What profit did the shipper make on the contract?

12. A publisher agrees to pay an author a fixed sum for a book, and in addition a royalty of 6d. for each copy sold. If the author receives £225 when 5000 copies are sold, what would he receive if 18,000 copies were sold?

13. I buy kerosene and crystal oil in casks of 41 gallons, the cask of kerosene costing 30s., and that of crystal oil 25s. Five shillings is allowed for each cask when returned, and the total cost of carriage on each is 2s. 6d. If half a gallon of each is lost through leakage, etc., what is the net cost in pence per gallon for each oil? Give the results to the nearest farthing, and also to two places of decimals.

14. How many articles, each costing £4. 7s. $9\frac{1}{2}$d., can be bought for £1000? By how much may the price be increased without lessening this number, the price being always an exact number of halfpence?

15. How many bicycles, each costing £6. 7s., can be bought for £300?

By how much must the price be reduced in order that ten more may be bought for the same amount, the price being an exact number of shillings.

16. Write down £3. 13s. $10\frac{1}{4}$d. in pounds and decimals of a pound correct to three places, and shortly explain your method.

17. Do not work the following, but state clearly why the given answer is obviously wrong:

£82. 19s. $4\frac{1}{2}$d. × 133. Answer: £11,034. 19s. 4d.

18. It is required to multiply £4·12678 by 3·14178 so that the answer may be in £. s. d., correct to the nearest penny.

Work this question as shortly as you can.

19. Point out the error of style in the following solution of the question: "How much is the sum of 9d. and half of 5 guineas?"

$\frac{1}{2}$ of £5. 5s. = £2. 12s. 6d. + 9d. = £2. 13s. 3d.

Correct this error without abandoning the equation form.

(ii.) WEIGHT.

20. Find the amount of the following bill:

$5\frac{1}{2}$ lb. of butter at 1s. 3d. per lb.
7 lb. 6 oz. of bacon at $9\frac{1}{2}$d. "
13 lb. 5 oz. of ham at 10d. "
2 dozen eggs at 18 a shilling.

Each item is to be calculated to a halfpenny, the seller taking the advantage when the amount is not an exact number of halfpence.

21. Find the amount of a bill for:

1 lb. 2 oz. rump steak	at	1s. 2d.	a pound.
7 lb. sirloin	"	10d.	"
1 lb. 5 oz. rump steak	"	1s. 2d.	"
1 lb. buttock steak	"	10d.	"
$5\frac{3}{4}$ lb. leg of lamb	"	9d.	"
1 lb. 7 oz. gravy beef	"	8d.	"

Calculate each item to the nearest halfpenny, giving the butcher the benefit of the odd farthing when there is an exact odd farthing.

22. A tobacconist buys 2 hundredweight of tobacco for £22, pays a duty of 3s. 10d. per pound on it, and sells it at 5d. an ounce. What profit does he make?

23. Goods are bought at $7\frac{1}{4}$d. per lb., and the cost of carriage is $1\frac{1}{4}$d. per lb.; they are sold at £4. 10s. per cwt. Find the gain or loss on each cwt.

24. On 16th December the price of wheat was 28s. 5d. per 480 lb., of barley 24s. 5d. per 400 lb., of oats 17s. 11d. per 312 lb. Find, correct to the nearest penny, the price of each per cwt.

25. Coal is sent to a railway station at 21s. 10d. per ton, and there is a further charge for delivery of 1s. per ton per mile. Find the cost of 6 tons 19 cwt. delivered at a place 3 miles from the railway.

26. Find the value to the nearest penny of a bag of sugar containing 48 kilograms at £18 per tonne.

27. What fraction of a ton is 250 lb.? Find the value of 250 lb. of potatoes at 60s. a ton. Let your answer be correct to the nearest penny.

28. Find, to the nearest penny, the cost of 738 tons 11 cwt. of coal at 25s. 9d. a ton.

29. Find the value of a casting weighing 28 tons 19 cwt. 1 qr. at £5. 8s. 6d. a ton.

30. Find the value of 14 tons 6 cwt. 2 qr. 20 lb. of copper at 49s. 6d. a cwt. Apply a rough check to test the accuracy of your result.

31. What is the value of 3 tons 2 cwt. 3 qr. of apples at 6s. 2d. a cwt.?

32. Find the cost of 38 tons 9 cwt. 1 qr. of iron castings at £5. 6s. 6d. per ton.

33. 145 cwt. 3 qr. 21 lb. of tea is bought at £2. 8s. 9d. a cwt. Find the total cost.

34. 40 lb. Troy of standard gold are coined into 1869 sovereigns. Find the value of an ounce of standard gold, giving your answer in sovereigns and a decimal fraction of a sovereign correctly to three places of decimals.

[In Troy weight a lb. = 12 oz.]

35. One ounce (480 grains) of standard gold is worth £3. 17s. 10½d. Find the number of grains in one sovereign correct to four places of decimals.

36. What is the value to the nearest shilling of 100 feet of copper pipe which weighs 8·32 lb. per foot and is worth £48 per ton?

37. The output of coal from a colliery is 750,000 tons per annum. The railway trucks used to distribute this carry on an average 11 tons 2 qr. Assuming that there are 260 working days in the year, how many trucks will on the average be sent out each working day?

38. A set of 80 telephone wires is carried across a house by a post of 50 lb. weight fixed to the roof, the distance spanned on one side of the post being 100 ft. and on the other 140 ft. Find the weight of wire in each span, 6·7 feet of wire weighing a pound. Also find the whole weight borne by the roof, being given that the weight of any span of wire is borne in equal shares by the two posts from which it hangs.

39. A farmer buys 23 pigs at 21s. 6d. each. He keeps them for 13 weeks, the average daily cost to him of each pig being 4·3d. They then average 10 st. 12 lb. if sold by live weight, and 8 stone if sold by dead weight. The farmer has two offers for them, one of 5s. 4d. per stone live weight, and the other of 7s. 2d. per stone dead weight. Which offer should he accept? Also find, to the nearest penny, his profit on each pig, if he closes with the better offer. (A stone = 14 lb.)

(iii.) DISTANCE.

40. In a certain game ten stones are placed at 12 yards interval along a straight line, the nearest being 12 yards from a basket at the end of the line. A boy starts from the basket and fetches the stones one by one to the basket. How far does he walk in doing so?

41. On a railway there is one sleeper in every 2 ft. 9 in. of the track, and there are two chairs to every sleeper. How many chairs will be required in laying down two tracks for 5 miles?

42. A metre of wire weighs 43 grams. Find, to two significant figures, the length of portions of the same wire which weigh

(i) 15 grams, (ii) 1 kilogram.

43. The line AB represents the length of the River Rhine (960 miles); copy AB exactly on to squared paper, and estimate how many miles would be represented by 5 mm. or by 0·2 in. on the same scale.

Fig. for Qn. 43.

Draw a line CD to represent the length of the River Thames (215 miles) on the same scale.

44. How many dual desks, each 3 feet 4 inches long, can be placed across a room 24 feet wide, so as to leave gangways not less than 1 foot 4 inches between each pair of desks, and between the end desks and either wall?

45. 15 balls are placed at equal distances in a straight line with a basket. The first ball is 6 yards from the basket, the second 4 yards from the first, the third 4 yards from the second, and so on. A man, starting from the basket, picks up each ball separately and places it in the basket. How far does he walk in doing so?

46. The thickness of an unbound book of 272 pages is half an inch. Find the thickness of a leaf to the nearest ten-thousandth part of an inch.

47. A railway between two places is 17 miles long. For 4 miles the line is level, for $11\frac{1}{2}$ miles it rises at an average slope of 1 foot in every 500 feet, and for the remainder of the distance descends at an average slope of 1 foot in every 680 feet. What is the difference in level, to the nearest foot, between the two places?

48. The gradient of a line of railway between two places 25 miles apart is 1 in 110 ascending for the first 9 miles, level for the next 10 miles, and 1 in 220 descending for the remaining distance. What is the difference in level between the two places?

49. The cyclometer attached to a bicycle registers 1 mile when the machine has gone 1 mile 23 yards. Find what it will mark, correct to the nearest quarter of a mile, when the machine has been ridden 4529 miles.

50. (i) A watch-chain has 80 circular links joined in the ordinary way. The outside diameter of the link is $\frac{5}{16}$ of an inch; the inside diameter is $\frac{5}{32}$ of an inch. What is the length of the chain?

(ii) When the chain is divided into 4 portions of 20 links each, what is the sum of the lengths of these portions?

(iv.) MEASURES OF AREA AND VOLUME.

51. At a certain place land is selling at 3s. 6d. a square yard. Find how much a landowner would receive if he sold 30 acres of land at this rate.

52. A field containing 7 acres is laid out for building; 3 roods 35 poles is used as a road, and the rest is laid out as far as possible in plots of 520 square yards. How many of these plots are there?

53. What is the tithe rent-charge, to the nearest farthing, on 32 acres 1 rood 14 poles at 4s. 6d. per acre?

54. The inside measurements of a box are 37 in. by 29 in. by 23 in. Find its volume to the nearest tenth of a cubic foot.

55. In digging a trench 578 cub. yd. of material are removed. The trench is then filled with water. How many gallons does it contain? (A gallon is 277 cubic inches.)

56. A steam pump lifts 450 gallons at every stroke. If a stroke is made three times every two minutes, how many minutes will it take to lift 72,000 gallons? Let your answer be correct to the nearest minute.

57. A dressmaker has 3 pieces of material, each $6\frac{1}{2}$ yards long and 54 inches wide, and weighing respectively 4 oz., 5 oz., and 6 oz. to the square yard.

She cuts a dress from each, and, carefully weighing the remnants, finds she has left over 7 oz., $4\frac{1}{2}$ oz., and 11 oz. respectively.

Determine the number of square feet of material used in each dress.

58. Do not work the following questions, but give a reason for thinking that the answers suggested are *obviously* wrong:

(i) Multiply 2459 by 909. Answer: 2230596.

(ii) Find, to the nearest penny, the cost of paving a court 40 metres broad and 100 metres long, at 5s. per square metre. Answer: £997. 6s. 5d.

(iii) £310. 12s. 6d. ÷ £31. 1s. 3d. Answer: 10 shillings.

(iv) What must the width of a field 124 yards long be if it contains the same area as a field measuring 103 yards 1 foot 3 inches by 116 yards? Answer: 107 yards and a fraction of a yard.

(v.) TIME.

59. It was calculated that, in boring the Simplon tunnel, the rock yielded a centimetre of its substance to the drill per minute. The whole length of the tunnel is 19·8 kilometres. How long would it have taken to bore the tunnel if the work had been entirely uninterrupted? Answer to be given to the nearest tenth of a year.

60. A watch ticks five times a second. How long, to the nearest hour, does it take to make a million ticks?

61. The Mont Cenis tunnel is 7·6 miles long and the average number of yards bored per year was 844. How many years did the tunnel take to bore?

How long (roughly) would a Channel tunnel 29 miles long take at twice that rate?

62. A person whose income is £780 a year spends on the average £1. 10s. 6d. a day. How much does he save in a year of 365 days?

63. A balloon voyage was made in November 1906 from Wandsworth, in London, to Vevey, near the Lake of Geneva. The balloon travelled in a direct line and at an even rate of 50 kilometres an hour. Plot on squared paper the relative positions of the places named in the following notes of the journey, taking 5 mm. or 0·2 in. to represent 20 kilometres:

At 3.30 p.m., over Wandsworth.
At 5.15 p.m., over the English coast at Rye.
At 6.45 p.m., over the French coast.
At 10.10 p.m., Paris 92 kilometres to the right.
At 7.30 a.m., over Vevey.

Find also, to the nearest mile, the width of the English Channel where the balloon crossed.

64. Three men A, B, and C are required to fill a tank holding 450 gallons by carrying buckets of water. A pours in $3\frac{1}{2}$ gallons every 4 minutes, B pours in 4 gallons every 5 minutes, and C pours in $4\frac{1}{4}$ gallons every 6 minutes. If each pours in his first instalment at noon, find at what time the tank is filled, how many journeys each man has made, and which poured in last.

65. A man's usual stride measures 32 inches. How many strides will he usually take in walking four miles? If he takes 132 strides a minute, how long will he be in walking the four miles?

66. A difference of 15° in the longitude of two places makes a difference in time of 1 hour. The longitude of New York is 74° West. How much is New York time slow of Greenwich time?

67. A train leaves London at 2.15 p.m. and arrives at Birmingham at 4.35 p.m., and another leaves Birmingham for London at 2.20 p.m. If the two trains travel at the same rate, find at what time they meet.

68. The *Daily Telegraph* newspaper was numbered 15,700 on Friday, August 25th, 1905. What was the day of the week and month when its number was 15,600?

69. If 4 tons of coal were put in the cellar on the 16th September and $\frac{1}{2}$ cwt. were burnt on that day and on every following day, how much coal was left in the cellar on the morning of Christmas Day?

70. A watch, which gains 5 minutes in a week, is set right at 8 a.m. on Monday. What time will it show on Wednesday, when the true time is 9 a.m.?

71. Christmas Day falls in 1914 on a Friday. On what day of the week will it fall in 1963?

72. The 24th of April in 1905 was a Monday. Find what day of the week the 24th April 1967 will be.

73. There are $29\frac{1}{2}$ days between full moon and full moon, and Easter Day is the first Sunday after a full moon. In 1897 it was full moon on Wednesday, the 17th of February, and Easter Day was in April. Find the exact date of Easter Day in 1897.

(vi.) VARIOUS.

Some of the questions at the end of this subsection have little educational value and may well be omitted by classes free to do so.

74. Calculate the following, each to one significant figure:

$$764{\cdot}37 \times 2656{\cdot}84 \times 1{\cdot}35$$
$$5{\cdot}864 \times 0{\cdot}0365 \div 0{\cdot}000743$$

75. Find, correct to three significant figures, the value of:

$$\frac{2{\cdot}73 \times 0{\cdot}0941}{16{\cdot}9 \times 0{\cdot}724}$$

76. A man is required to divide by 3·2. Instead of doing this, he obtains the correct result by multiplying by a certain decimal fraction. Find the decimal fraction.

77. A hollow square of soldiers has 150 men in the front ranks and is 7 men deep. How many men does it contain?

78. Round a penny stamp (still undetached from the sheet) there are 62 perforations, one at each corner, 13 each along the top and bottom, and 16 down each side. Two blocks of six stamps are torn off, consisting of three rows of two stamps and two rows of three stamps respectively. Find the number of perforations in each block, neglecting the broken perforations round the outside of a block.

79. A game is played with cards of three colours, red, green, and blue, a complete set consisting of 40 red cards, 42 green, and 54 blue. If a printer has 100,000 cards of each colour, how many complete sets can he print, and how many cards of each colour will be left over?

If he buys the cards of each colour by the thousand, what is the smallest number of thousands of each colour that he can buy in order that there may be no cards left over after making an exact number of complete sets?

80. A man pays income tax at the rate of 1s. for each pound of his income, which is £831. 10s. How much has he left?

81. The tax on tea at the rate of 6d. on the lb. produced a revenue for the year ending 31 March 1904 of £6,559,705. Find, to the nearest ton, the weight of tea imported.

82. The export of potatoes from Jersey in 1903 was 47,530 tons and realised £475,888. Find, to the nearest penny, the average price realised per stone of 14 lb.

83. Allotments are let by a parish council at the rate of 30s. per acre per annum, but a reduced rate of 26s. 8d. is charged for any rent paid in advance. A tenant of a quarter acre paid at Lady Day 1906 a year's rent from Michaelmas 1905 to Michaelmas 1906. On the amount collected make deductions of 1d. in the shilling for cost of collection and of 2s. 6d. in the £ for rates and taxes, and thus find, to the nearest farthing, the net amount received by the council for the quarter acre allotment.

Lady Day is the March, and Michaelmas the September quarter day.

84. We export £96,000,000 worth of goods to the Colonies in a year. Reckoning that half of this goes in workmen's wages for the manufacture of the goods, and that each workman gets 25 shillings a week, calculate the number of workmen employed in their manufacture.

85. An auctioneer sold a third of an estate to A and $\frac{2}{5}$ of the remainder to B. Forty-eight acres then remained unsold. What was the area of the estate?

86. A man directs in his will that each of his five younger children shall receive one-third of the amount left to his eldest son, and that one-tenth of his whole estate shall be divided amongst his servants. The value of the estate being £54,967, find, to the nearest pound, the amounts received by the children.

87. Two brothers were each entitled to one-third of an estate. The son of one of them inherited two-thirds of his father's share and one-seventh of his uncle's share. What fraction of the whole estate did the son inherit?

If the whole estate is worth £8916. 17s. 3d., what is the value of the son's share?

88. The scale of fees payable on the registration of a company under the Companies Act, 1862, is as follows:

Capital of company, £5000; fee, £5. Capitals above £5000 and up to £100,000, fee 5s. extra per £1000 of capital above £5000; capitals above £100,000, fee 1s. extra per £1000 of capital above £100,000, but there is no higher fee than £50.

What is the smallest capital for which the maximum fee is payable?

89. Demonstrate, in any way you please, that $\frac{3}{4}$ of $\frac{2}{5}$ is equal to $\frac{3}{10}$. [It is not sufficient to quote a rule; some proof must be given.]

90. Justify clearly each step of the reasoning involved in the statements: (i) $2\frac{7}{8}+\frac{4}{3}=4\frac{5}{24}$, (ii) $2\frac{1}{3}\div 3\frac{1}{2}=\frac{2}{3}$.

91. Criticise the following questions which were set to children:

(i) A man can make a box in 7 hours. What fraction of the box can he make in $2\frac{1}{2}$ hours?

(ii) Mary has 13s. 6d. Jane has 0·7 of this amount and Emma has 0·8 of it. How much money have the three girls between them?

(iii) A and B are engaged in digging a trench. A's shovel measures 10 square inches and B's shovel measures 8 square inches. A can dig the whole trench in 3 days. How long will B take to dig it?

92. Two cyclists start abreast at three o'clock to race on a circular track. If one goes round once in 3 min. 12 sec. and the other in 3 min. 30 sec., find at what time they will again be riding abreast, and the number of complete circuits they will then have made.

93. Find the smallest multiple of 81 which is also a multiple of 525, 540, and 616.

94. Express 23,880 and 150,444 as the products of prime factors. Hence write down in factors the H.C.F. and L.C.M. of these numbers.

95. Find the smallest cube which is an exact multiple of 75, 126, and 567, and also the smallest number by which this cube can be divided so that the result may be a square.

96. Assuming that a cubic foot of water weighs 1000 oz., find, in lb. per square inch, the pressure (due to the water) at the bottom of a tank in which water is standing to a depth of 4 feet.

Section 2. Simple Mensuration.

(i.) RECTANGULAR AREAS.

1. Measure up the figure and find its area in sq. cm.

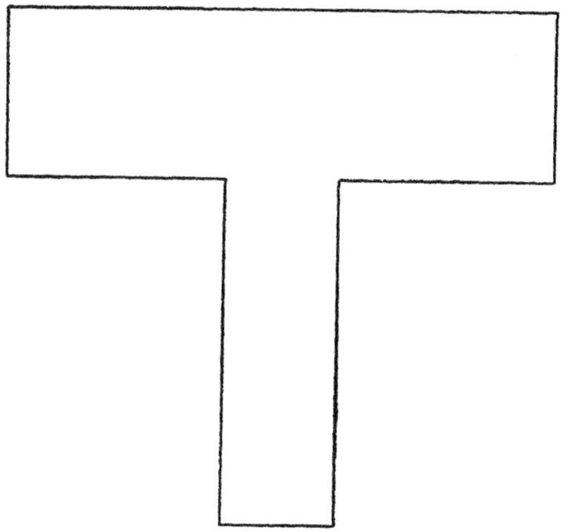

Fig. for Qn. 1.

2. Draw a rectangle which would just enclose all the printing on this page (including the page heading).

Place on the rectangle, in correct position and size, the numbers of the questions, giving (to the nearest millimetre) their distances from the left side and the bottom of the rectangle.

3. Mark out an area of 3 square inches of any shape you like. Mark out also a square measuring 3 inches along each side, and divide it up into squares each of which measures one inch in the side. How many are there, and what is the area of the big square?

4. How many inches go to a foot, and how many square inches to a square foot? With the help of a diagram explain the relation between these numbers.

5. Find the total cost of cleaning the following carpets at 2d. a square foot: one carpet 14 ft. 6 in. by 12 ft., one carpet 13 ft. by 12 ft., and 8 yards of stair carpet 20 inches wide.

6. A square field measures 100 yards in the side. Give its area in acres to the nearest whole number.

7. A square field measures 70 yards in the side. Is its area greater or less than an acre?

8. A railway is 17 miles long and 12 feet broad. Give its area in acres.

9. A street 1520 metres long, and of an average breadth of 23 metres, is being paved with wood blocks at 9s. per square metre. Find the total cost and the cost per hectare.

10. A certain street is 352 metres long, and has a width of 9·7 metres. It has a pavement on each side 1·8 metres broad. Find the area of the roadway in ares. How many stones 60 centimetres square would be required to relay the pavement?

11. The dimensions in centimetres of a rectangular table are: length 47·18, width 39·28. Find the periphery and the area. What is the ratio of the length to the width, as a decimal?

12. A number of square pieces of print, each measuring $6\frac{1}{2}$ inches along the side, are made into a quilt. Before being sewn together $\frac{5}{12}$ of an inch is turned in all round each square or patch. The finished quilt is 24 patches in length, and 18 patches in breadth.

What area does the quilt cover, and what length of tape would be required to bind the edges?

13. A man has to glaze 36 picture frames, each of which requires a piece of glass 17 inches by 9 inches. He cuts these from sheets of glass measuring 18 inches by 30 inches. How many such sheets must he use? If he buys the sheets at the rate of twopence per square foot and sells the glass actually put into the frames at threepence per square foot, what profit per cent. does he make?

14. A cupboard door which is 4 feet high and 3 feet wide is made of wood and glass, the panes of glass being rectangular with their sides parallel to the sides of the cupboard. The wood containing the glass is everywhere $1\frac{1}{2}$ inches in width, and the glass is embedded in it to a depth of $\frac{3}{16}$ inch. There are 12 panes of glass arranged in four horizontal rows with three panes in each row. Find, to the nearest integer, the number of square feet of glass which will be needed to glaze the door.

15. How many cake tins, $6\frac{3}{8}$ inches square at the top and $5\frac{11}{16}$ inches square at the bottom, can be arranged on a rectangular tray measuring $50\frac{3}{4}$ inches by 46 inches, (i) when the tins are placed right side up, (ii) when they are inverted, the tins being so placed that the edges resting on the tray are parallel to the edges of the tray and do not overlap the tray at any point.

(ii.) RECTANGULAR VOLUMES.

16. The figure shows the size of a piece of cardboard which is to be folded along the lines AB, CD, EF, GH, so as to form the outer portion of a match box, the ends beyond AB and GH being pasted together. Find, in square centimetres, the area of the

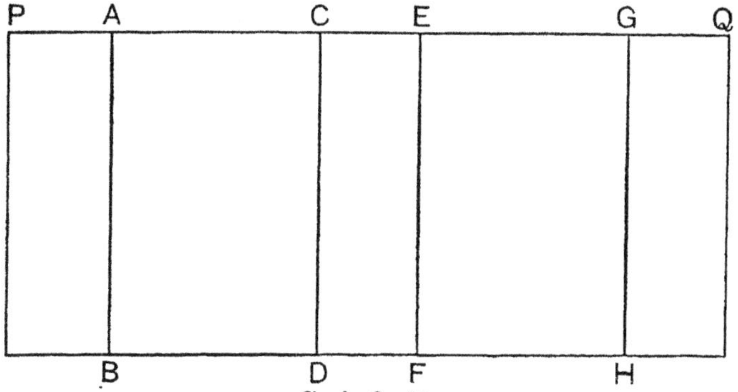

Fig. for Qn. 16.

piece of cardboard, and in cubic centimetres, the volume of the space left for the inner part of the match box. Your measurements should be made to the nearest millimetre.

17. Find the weight of a block, 2 ft. 3 in. by 20 in. by 3 in., of stone which weighs 2 tons 5 cwt. per cubic yard.

18. Find the weight, to the nearest kilogram, of an iron rod of square section, 10 metres long and 2·3 centimetres broad. A cubic centimetre of iron weighs 7·207 grams.

19. If a cubic foot is 6·228 gallons, find the quantity of water contained in a dock 150 yards long, 35 yards wide, and of an average depth of 32 feet.

20. The figure is a section of a bar of angle iron drawn full size. Calculate the weight of a length of 10 feet of this bar, the weight of a cubic foot of the iron being 487 lb.

Fig. for Qr. 20.

21. A tank is 1 metre long, 95 centimetres wide, and 45·5 centimetres deep. How much water will it hold? Give your answer in litres and decimals of a litre.

22. A tank, 3 ft. 9 in. by 3 ft. by 2 ft. 6 in., is filled with water. Find, to the nearest gallon, the volume of water.

23. How many bricks $9'' \times 4\frac{1}{2}'' \times 3''$ can be carried by a truck whose load is 5 tons? The bricks in question weigh 145 pounds per cubic foot. Give your result to two significant figures.

24. Find in pounds the weight of water in a tank 6 feet long, 2 feet 9 inches wide, and 1 foot 8 inches deep, given that a cubic foot of water weighs 1000 ounces.

Illustrate the advantages of the metric system by writing down and working out a parallel question involving the use of the metric weights and measures.

25. Find the volume and weight of a cube of zinc, length of edge 6 in. A cub. in. of zinc weighs 0·26 lb.

If each edge of the cube increases by 0·003 of its length for a rise of temperature of 100° C., what is the increase in the volume when the temperature of the cube has been raised through 100° C.? Find the corresponding increase in the surface of the cube.

26. A jeweller is to make a gold cross, 1 millimetre thick, of the size and shape shown in the figure. The gold used weighs

Fig. for Qn. 26.

16 grams per cubic centimetre. What is the weight of the cross to the nearest gram?

27. A rectangular lawn-tennis court 84 feet long and 42 feet wide is surrounded by a gravel path 4 feet wide. How many cubic yards of gravel would cover the path to a uniform depth of 3 inches, and what would the gravel cost at 4s. 6d. a cubic yard?

28. What will it cost to cover a flat roof $16\frac{1}{2}$ ft. broad, 20 ft. long, with lead $\frac{1}{16}$th inch thick? Lead weighs 712 lb. per cubic foot, and costs $2\frac{3}{4}$d. a pound.

29. How many gallons of tar per mile will be required to give a road 50 feet wide a coating of tar of an average thickness of 0·07 inch? [See the Tables inside the front cover.]

30. A book of gold leaf contains 25 leaves, $3\frac{1}{3}$ inches square and $\frac{1}{282000}$ of an inch thick. If a cubic inch of gold weighs 0·697 lb., and 1 lb. of pure gold is worth £61. 19s., find the value of the gold in the book to the nearest penny.

31. An upright piano is 4′ $9\frac{1}{2}$″ long. The height of the piano is 4′ 2″. When placed with its back against a wall the piano stands out 1′ 2″ from the wall at the base and top, and 1′ 11″ at the key-board. The part containing the key-board measured vertically is 6 inches. Draw a sketch of the piano, and find (to the nearest cubic foot) the volume it occupies.

32. A stone kerb at the side of a road is 8 inches wide and 10 inches deep. If a cubic foot of the stone weighs 168 lb. or 0·075 of a ton, how many tons of stone will be required for edging both sides of a road a mile long?

33. Measure the length and breadth of this page in centimetres. A pile of 1000 leaves of such paper is 6·4 cm. thick, and weighs 1·4 kilograms. Find how many leaves will make a pile 1 cm. thick, and what the paper weighs per cubic centimetre.

34. The base of a tank is 79 cm. by 94 cm., and it is to hold 150 litres. Find its depth to the nearest centimetre.

35. A pane of glass measures 18 inches by 14 inches and weighs a pound. What is its thickness if the glass weighs 160 pounds per cubic foot?

36. If a square foot of steel $\frac{3}{4}$ inch thick weighs 30·5 lb., what is the weight of a plate 6 ft. 6 in. long, 3 ft. 6 in. wide, and $3\frac{1}{2}$ in. thick made from the same steel, and how much does the steel weigh per cubic foot?

37. A book containing 325 leaves is 1·7 inches thick; allowing 0·09 inch as the thickness of each board of the cover, find to four decimal places the thickness of the paper.

38. A field is in the form of a rectangle 187 yards long by 160 broad. Find the number of gallons of water falling upon it during a rainfall of 0·154 inch. [See the Tables inside the front cover.]

(iii.) VARIOUS.

39. Trace the figure on transparent paper, place your tracing on squared paper, and find its area in sq. cm. by counting squares.

Also draw on your tracing a rectangle that you judge by eye to be of the same area, and find the area in this way.

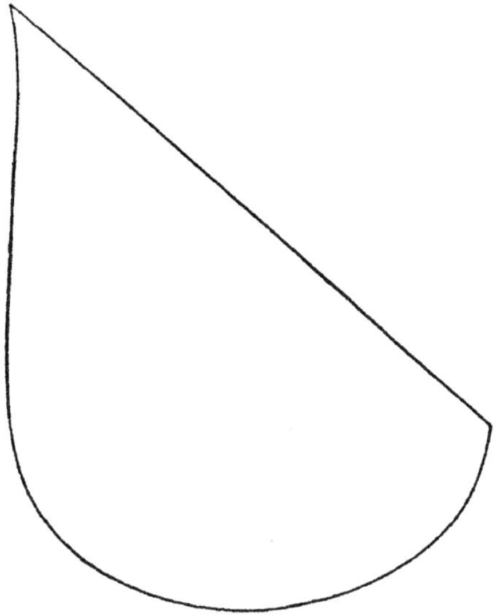

Fig. for Qn. 39.

40. A piece of platinum 17 cubic inches in volume weighs 12 pounds avoirdupois, while a piece of gold 77 cubic inches in volume weighs 54 pounds; find to the nearest ounce the difference between the weight of a cubical block of platinum measuring 5 inches each way and the weight of an equally large block of gold.

41. The outside measurements of a rectangular box are 1 foot 10 inches, 1 foot 2 inches, and 1 foot respectively. If it is wrapped completely in a sheet of brown paper measuring 60 inches by 42 inches, how many square inches of paper are folded so as to cover parts of the box with more than one thickness?

Find also the length of cord required to go once round the long way (*i.e.* along the length and *across* the end) of the box and twice round the short way (*i.e.* across the four longest faces), adding 10 per cent. for knots and waste.

42. A pond whose sides are vertical has an area of 0·764 hectares, and in the centre of it is a square tower whose side measures 5·5 metres. How many litres of water must be pumped out that the surface may sink 50 centimetres?

43. A street improvement scheme involves purchasing the houses on one side of the street, and the land on which they stand, along a frontage of 352 feet, and to an average depth of 53 feet. The houses cost £87,700, and the land costs £3. 7s. 0d. per square foot. The roadway is then widened 9 feet, and re-making this extra piece costs 8s. per square yard. The remaining land is sold again at £5. 13s. 0d. per square foot. Find the net cost of the improvement.

44. A rectangular stone trough is 3 ft. 9 in. long, 2 ft. 7 in. broad, and 1 ft. 2 in. deep externally, and weighs 874 pounds when empty. The interior surface is irregular, but it is found by trial that the trough holds 38 gallons. Find, to one place of decimals, how many times the stone is as heavy as water, bulk for bulk. [See the data on the front cover.]

45. Two cans are exactly alike and contain $2\frac{1}{2}$ gallons of water and $2\frac{1}{2}$ gallons of oil respectively. The oil weighs $3\frac{3}{4}$ lb. less than the water. Find the weight of a cubic foot of the oil.

46. A cistern which is 9 feet long, 6 feet wide, and 6 feet deep contains water to a depth of 4 feet. Porous bricks 9 inches long, $4\frac{1}{2}$ inches wide, and 3 inches thick are immersed in the cistern, and each brick absorbs 0·05 of its own volume of water. Find the largest number of bricks which can be immersed without the water overflowing.

47. A bucket filled to the brim with water is suspended from one arm of a balance and weighs 67 lb. 11 oz. A non-porous brick whose edges are 9 inches, $4\frac{1}{2}$ inches, 3 inches in length, is then placed in the bucket. The bucket with its contents then weighs $75\frac{1}{2}$ lb. Find the weight of the brick.

48. During a storm 1 inch of rain fell on a 1 acre field. Given that a cubic foot of water weighs 1000 ounces, calculate the total weight of water which fell on the field. Express your answer in tons.

49. Using the data on the front cover, find to the nearest ton the weight of water represented by an inch of rainfall over a field of 10 acres. Would your answer be affected by using the value given in the preceding exercise for the weight of water?

50. The amount of water running under a bridge in each second is 15·36 gallons. If the stream is entirely supplied from a lake a square mile in extent, how much does the surface of the lake sink in 24 hours, supposing that the lake has no means of replenishing itself?

51. The surface areas of two reservoirs are 2000 sq. ft. and 3500 sq. ft. respectively. If the first discharges into the second at the rate of 500 gallons a minute, how many minutes will it be before the difference in the two levels is diminished by 1 foot?

52. Snow falls uniformly to a depth of six inches over an island 35,700 acres in extent. If water is 3·5 times as heavy as this snow, and a cubic foot of water weighs 1000 oz., find the weight in tons (to the nearest million) of the snow that has fallen.

53. After a certain snowfall of 8 inches an analysis of the melted snow is said to have given an average of 30 grains of solid matter to every gallon of water. Taking snow a foot deep as equivalent to 0·95 inch of rain, find to the nearest ton the weight of solid matter deposited over an area of 16 square miles by a snowfall 8 inches deep.

54. The length, breadth, and height of a rectangular room are k, b, and h yards respectively. Find the formulae for (i) V, the volume of the room; (ii) A, the total area of walls, ceiling, and floor. Evaluate these when $k=8$, $b=6$, $h=4$.

If v and a are the volume and area of a similar room, each dimension of which is half the corresponding dimension of the large room, state the relations between v and V and between a and A.

Show that a rectangular block, whose edges are whole numbers of centimetres, may be divided up into cubes measuring a centimetre in the edge, and extend the proposition and proof to blocks whose dimensions are given as decimals.

55. Lake Tsana has a surface area of 30,000,000 ares, and discharges annually a quantity of water that would raise its level by $1\frac{1}{4}$ metres. It is proposed to erect a barrage at the outlet to control the discharge, with sluice-gates that allow the passage of 170 cubic metres of water per second. If this is done, and the annual discharge is to remain what it now is, during how many days per year (to the nearest integer) may the sluice-gates be open?

56. A town has a population of 400,000 and a water supply of 30 gallons per head per day is required. A reservoir is provided to hold six months' supply [say, July to December inclusive], with a mean depth of 20 feet. What must the area of the reservoir be? Find the area of the catchment basin, if 20 inches of rainfall in the year, one third of which runs into the reservoir, will yield sufficient water to supply the town.

57. A town contains 730,000 inhabitants, and every inhabitant uses 42 gallons of water a day. Give in cubic feet the quantity of water they use in a day.

The water supply is drawn from a lake the area of whose surface is 4000 acres. How many cubic feet of water are there in a layer 1 foot deep on the surface of the lake?

During a period of 21 days no rain fell. Find how much the surface of the lake fell during these 21 days, assuming that the total influx of water into the lake during that period was a quantity sufficient to restore the loss caused by surface evaporation.

Draw a rectangle whose area will represent, on the scale of 1 inch to a mile, the area of the surface of the lake.

Section 3. Mensuration by means of Scale-drawing.

(i.) AREAS.

1. The figure is the plan of an estate. Trace it on to transparent paper, and on your tracing draw a rectangle having the same area as nearly as you can judge by eye. Measure the length and breadth of the rectangle, and use them to determine the acreage of the estate.

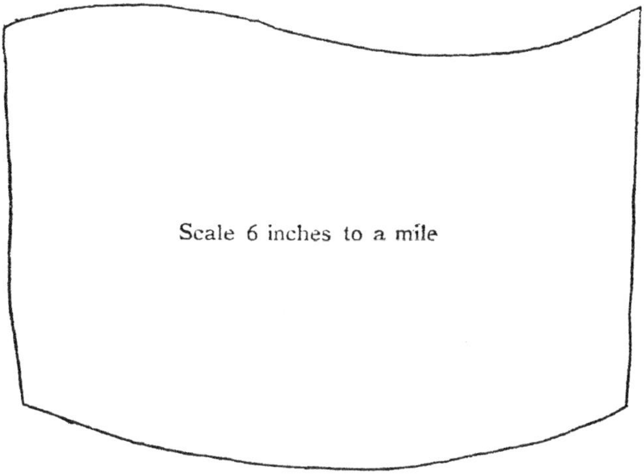

Fig. for Qn. 1.

2. The accompanying sketch-map of Lake Geneva is drawn on a scale of 1 in 1,000,000.

Calculate, in kilometres, the distance as the crow flies from Geneva to Lausanne. Also, trace the outline of the lake upon transparent paper, place your tracing on metric squared paper, count the squares enclosed, and calculate the area of the lake in hectares.

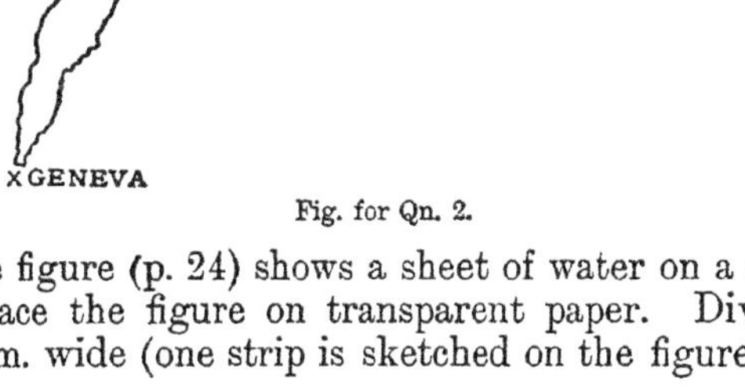

Fig. for Qn. 2.

3. The figure (p. 24) shows a sheet of water on a scale of 1 to 2000. Trace the figure on transparent paper. Divide it into strips 5 mm. wide (one strip is sketched on the figure). Replace each strip of the figure by a rectangular strip equal in area as near as you can judge. Measure up and give the area of the figure in sq. cm., and the area of the sheet of water in hectares.

Or, instead of drawing the 5 millimetre strips, you may pin the tracing on square ruled paper, and make use of its ruling.

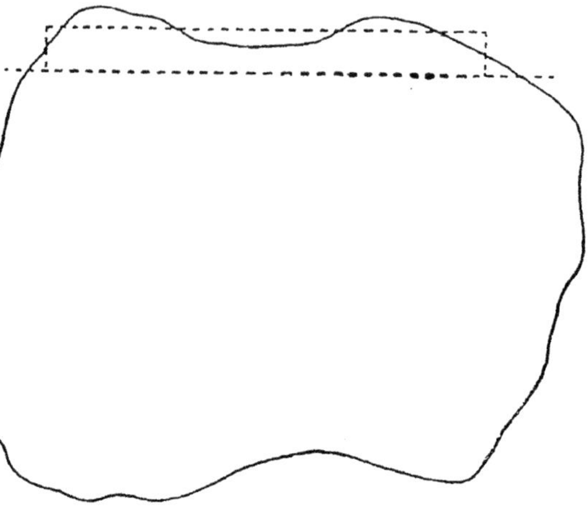

Fig. for Qn. 3.

4. What area of land is represented by 0·36 sq. in. on a map whose scale is 20 miles to 1 in.?

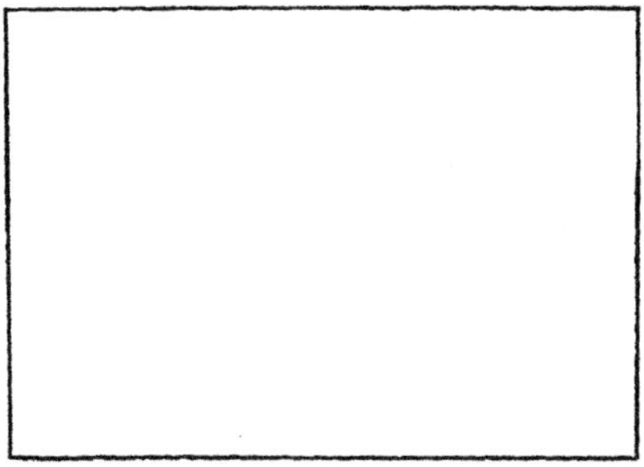

Fig. for Qn. 5.

5. The figure represents, to the scale of $\frac{1}{16}$th of an inch to the foot, a courtyard which is to be paved at 3s. 6d. per square yard. Find, to the nearest pound, what the expense will be.

6. The dimensions of a rectangular carpet being those of the room shown in the figure, how many square yards must be cut away from it to make it fit the projection, the plan being on a scale of 1 inch to 2 yards?

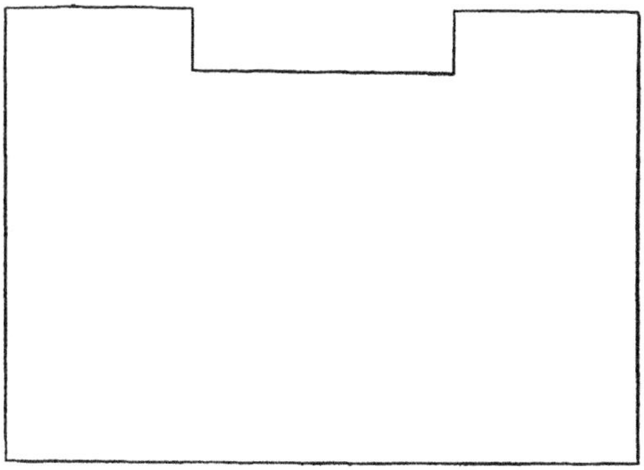

Fig. for Qn. 6.

7. The figure represents a plot of land on a scale of 1 to 1000, the shaded portion representing a house built on it. Find, to the nearest pound, the cost of gravelling the portion not built upon at 6d. a square metre.

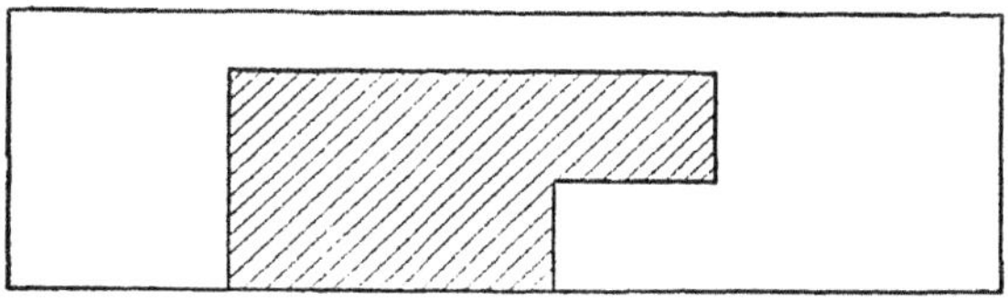

Fig. for Qn. 7.

8. If a small square of squared paper ruled at intervals of 5 mm. represents an area of 1 hectare, so that each side of the square represents 100 metres, draw a rectangle whose area represents a square kilometre, and also one whose area represents a square mile, taking a mile as 1·6 kilometres.

From your figures find how many hectares there are in a square kilometre and how many in a square mile.

9. The figure (p. 26) is the plan of an open tank 70 cm. deep; it is drawn to a scale of 1/20. What area of lead sheeting

would be required to line the tank? What would the lead cost at 7s. 1½d. a sq. metre?

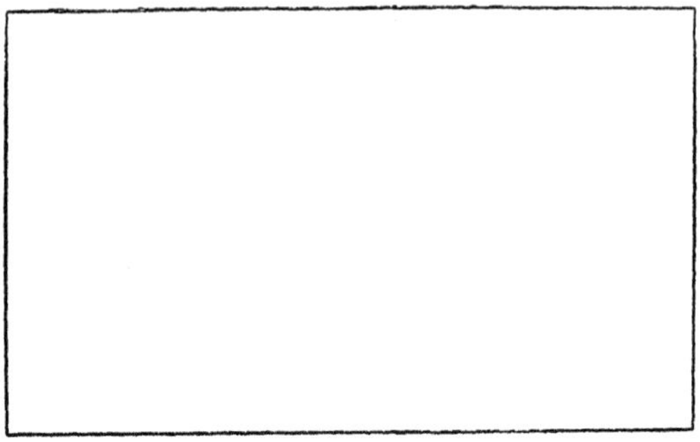

Fig. for Qn. 9.

10. The figure represents a piece of country of 17 square miles. What area does a square inch of the figure represent?

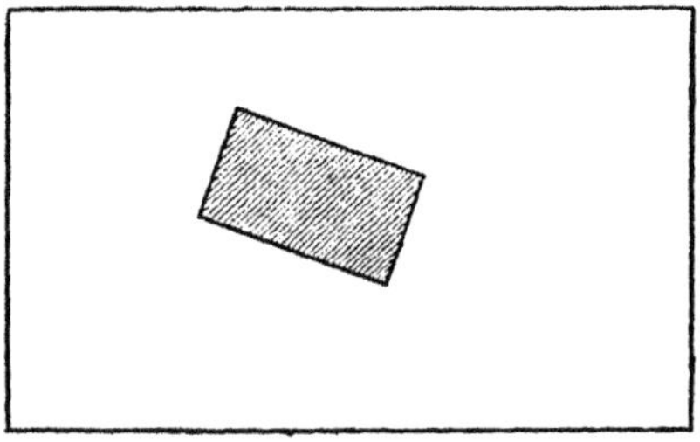

Fig. for Qn. 10.

The shaded portion of the figure is an estate valued at an average price of 70s. per acre. What is its total value?

(ii.) VOLUMES.

11. The counter of a shop, as in the figure (p. 27), is made of oak planking 5 cm. thick, weighing 860 grams to the cubic decimetre. Find its weight. The scale of the diagram is one thirtieth, that is, the length and breadth are each 1/30 of the actual length and breadth.

Fig. for Qn. 11.

12. The figure is the plan of the floor of a coal-cellar to a scale of 1/60; that is, the length and breadth of the figure are 1/60 of the actual length and breadth; and the depth of the

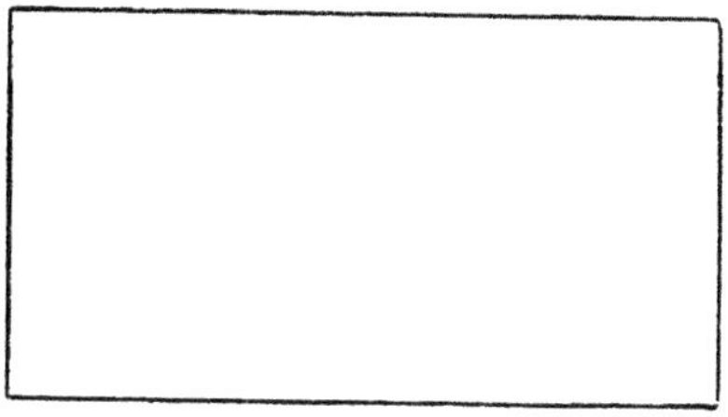

Fig. for Qn. 12.

cellar is 6 ft. 6 in. Find the cubic contents of the cellar. Find also, to the nearest cwt., how much coal the cellar holds, a cubic foot of the coal weighing 70 lb.

13. The figure shows the floor of a church, on a scale of 1 to 1000. Give the area of the floor in square metres, and, taking

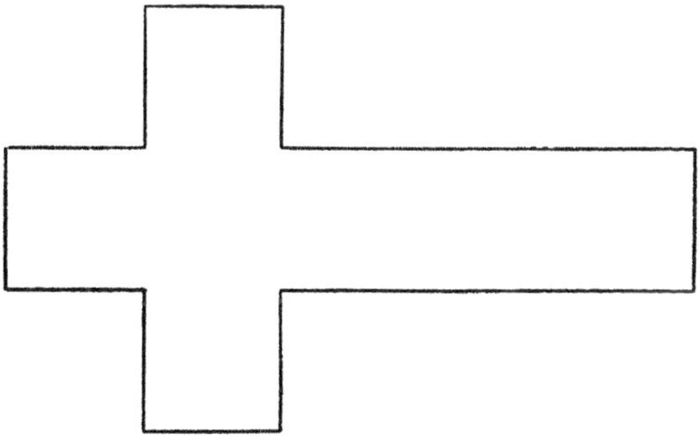

Fig. for Qn. 13.

the average height as 30 metres, give the air space in cubic metres.

14. The figure is the plan of the floor of a building to a scale of 1 inch to 40 feet. If the height of the building is 20 feet,

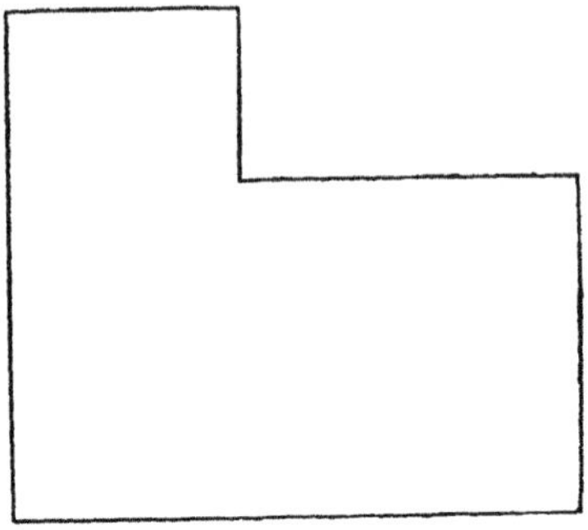

Fig. for Qn. 14.

what is the air space, and how many people may be employed in it at the rate of 250 cubic feet per head?

15. The figure shows, quarter size, the top and side of a wood paving block. Give its length, breadth, and thickness in inches to the nearest tenth. Calculate the volume of the block.

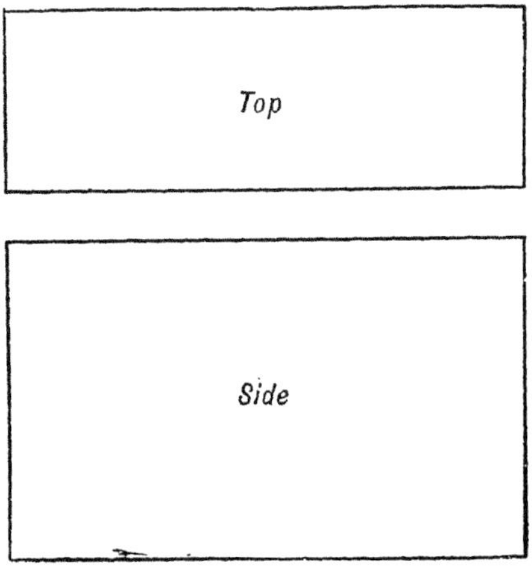

Fig. for Qn. 15.

The block is made of wood that weighs 63 lb. per cubic foot. Find the weight of the block.

How many blocks will it take to pave 100 yards of a street 60 feet wide? And what will these cost at 2s. 1d. per cubic foot?

16. This page represents, on a scale of 1 inch to 10 yards, a courtyard which is to be paved with blocks of wood, 6 inches long, 3 inches broad, and 4 inches deep. How many blocks are used, and how many cartloads of $1\frac{1}{2}$ tons each will they make, if the wood is 0·9 times as heavy as water? And what will they cost at 8s. 6d. a ton?

17. The figure represents the section of a cast-iron girder, 20 feet long, scale 1/4. Calculate the weight of the girder. Cast-iron is, bulk for bulk, 7·3 times as heavy as water.

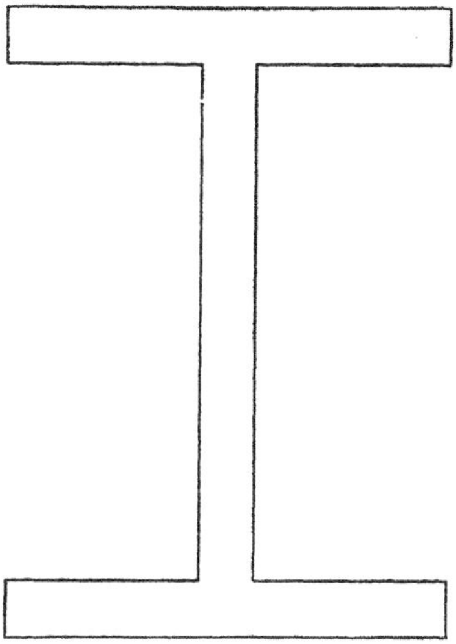

Fig. for Qn. 17.

18. The figure (p. 30) represents the lead lining of a water cistern, when spread out flat. Find the length, breadth, and depth of the cistern, and find either in litres or in gallons how much water the cistern would hold. Any length measured on the plan is meant to be one-twentieth part of the corresponding length measured in the cistern, but the drawing is not absolutely accurate, and you had better measure the height of each side and each end, and use the average of these four measurements. [See the Tables inside the front cover.]

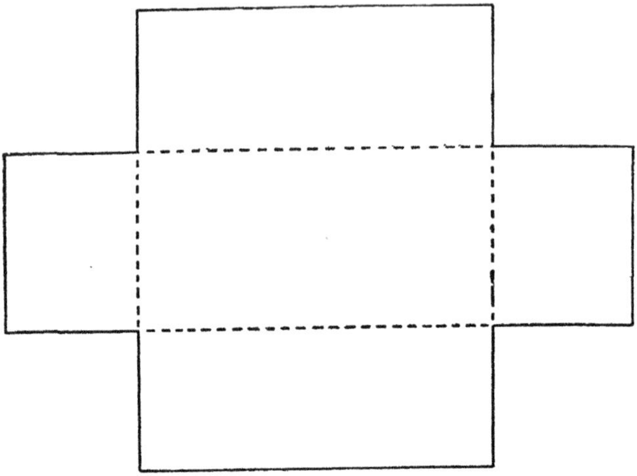

Fig. for Qn. 18.

19. An open box 10 centimetres deep is to be made from the sheet of cardboard shown in the figure to a scale of 1/10 of actual size, by cutting squares of the proper size from the corners and

Fig. for Qn. 19.

turning up the projecting pieces left. What will the dimensions of the box be, and how much will it hold?

20. The diagram shows, on the scale of $\frac{3}{4}$ inch to the foot, the breadth of a mill-course and the depth of water in it. Find the number of gallons of water which pass any point in a minute when the stream is running at 3 miles per hour.

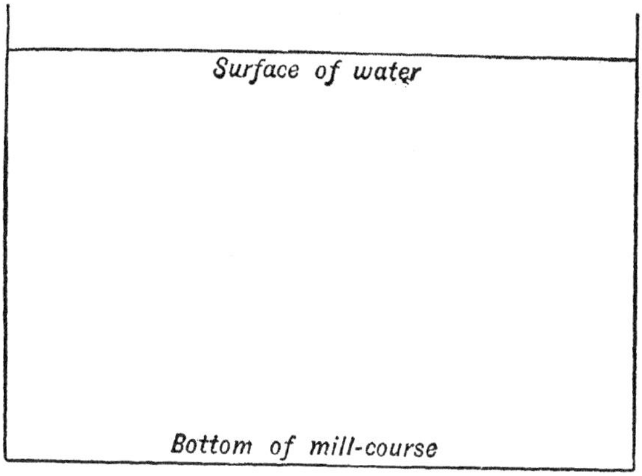

Fig. for Qn. 20.

Also, take the scale as 1 to 20, the speed of the water 5 kilometres an hour, and give the result in litres per minute.

(iii.) VARIOUS.

21. The area of the United Kingdom is 120,000 sq. miles, that of the German Empire 210,000 sq. miles, and that of Switzerland 16,000 sq. miles. Taking 5 mm. or 1/5 inch as unit of length, represent the area of the United Kingdom by a rectangle measuring 10 units by 12 units, and draw two other rectangles, each with one side 10 units long, to represent the areas of Germany and Switzerland. Mark on each rectangle the number of units in its second side.

22. Mark out a rectangular area measuring 14 cm. by 9 cm. The length of England from north to south being 645 kilometres, and the breadth from east to west 520 kilometres, what is the scale of the largest map of England that will go into this rectangle? State the scale as a fraction with unity for numerator, and give the denominator to three figures.

Shade an area of any shape you like to represent on this scale the area of England, 131,680 sq. kilometres.

23. The figure shows, on a scale of 1 to 400, a level field ABCD, containing a pond E, whose average depth is 60 cm. below the level of the field. It is proposed to remove the water from E and fill up with earth taken from the surface of the rest of the field. Find, by pricking the diagram through on to squared

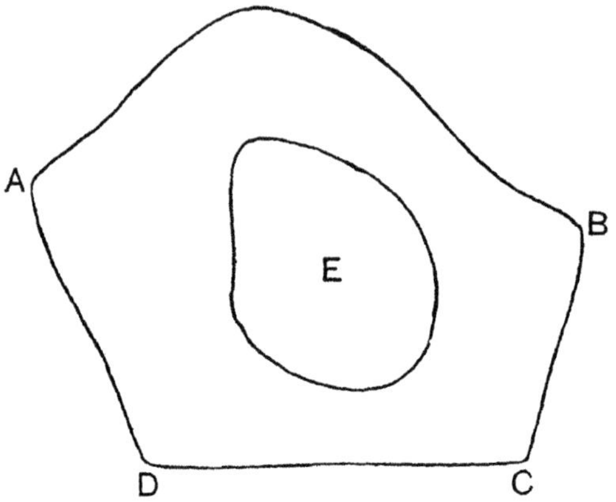

Fig. for Qn. 23.

paper and counting squares, what fraction of the whole field is covered by the pond; and hence calculate how many centimetres the surface of the field will have to be lowered in order to fill up the pond.

If the work costs 2s. per cubic metre moved, what is the total cost?

24. The British Isles lie between latitudes 50° and 58° and between longitudes 1·4° East and 10·7° West. A degree of latitude measures 111 kilometres and a degree of longitude may be taken as 71 kilometres for the British Isles. Would a page of this book hold a map of the British Isles on a scale of 1 to 5,000,000?

Section 4. Mensuration of the Triangle.

(i.) THE TRIANGLE ITSELF.

1. The figure is drawn on a scale 10 feet to the inch, and represents the end view and side view of a shed, open at the ends, the roof and walls of which are made of corrugated iron. Find the cost of the roof and walls at $1\frac{3}{4}$d. per square foot.

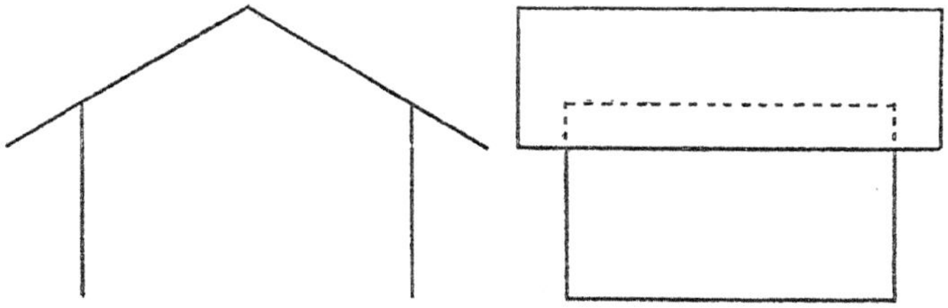

Fig. for Qn. 1.

2. The sketch (not drawn to scale) shows the longitudinal section of a punt, with dimensions marked. The sides are vertical and the breadth is 3 feet. The slope is the same at the two ends. By drawing find the second dimension of the slant end—one dimension you know to be 3 feet.

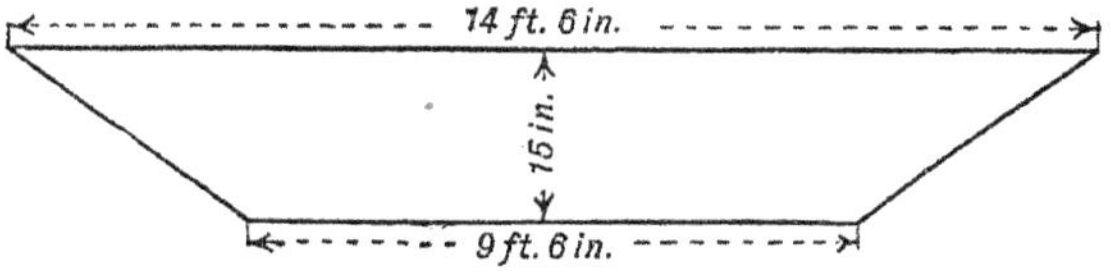

Fig. for Qn. 2.

The punt is made of wood, a square foot of which weighs 2·9 lb.; the thickness of the wood you may neglect. Find the weight of the punt. Find also the difference between that weight and the weight of water it would displace if depressed until the water-line on its side measured 12 feet and half the height was immersed. [See the Tables inside the front cover.]

3. Find the area of the figure (p. 34) by transferring it to squared paper and counting squares, or in any other way you please. Give a description of the method you adopt.

The figure represents the cross-section of a canal, on a scale of $\frac{1}{100}$. Find how many thousand cubic metres of soil would be excavated in making a kilometre of this canal.

Fig. for Qn. 3.

4. The figure is the plan of a coal-hole drawn on a scale of 1 inch to 4 feet. When this coal-hole is filled with coal to an average depth of 3 feet, how much coal is there in it? A ton of coal occupies 20 cubic feet. The area of the plan is the average of the two sides marked a and b multiplied by the side marked c, area and lengths being expressed in corresponding units.

Give the value of the coal at 22 shillings a ton.

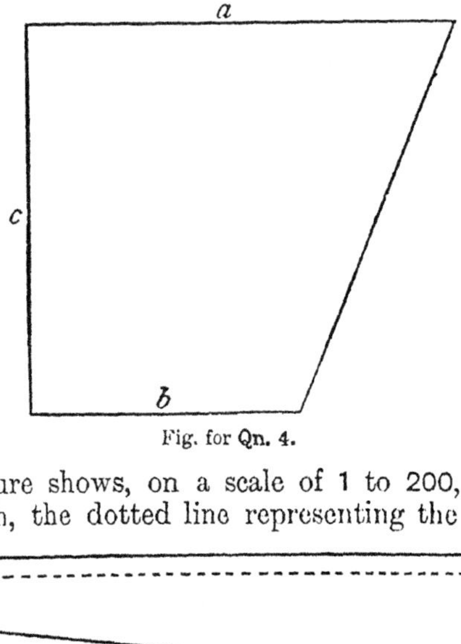

Fig. for Qn. 4.

5. The figure shows, on a scale of 1 to 200, the side of a swimming bath, the dotted line representing the surface of the

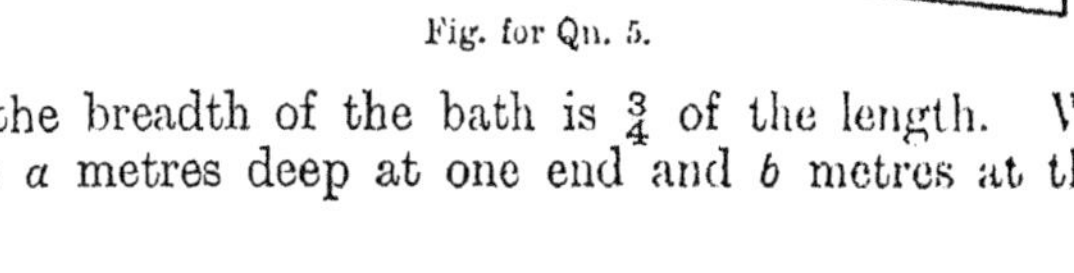

Fig. for Qn. 5.

water; the breadth of the bath is $\frac{3}{4}$ of the length. When the water is a metres deep at one end and b metres at the other, the length of the bath being c metres and the breadth d metres, the volume is given in cubic metres by the formula $\frac{a+b}{2} \times c \times d$. Find how many litres of water it takes to fill the bath to the level shown, and the cost at 2d. per thousand litres.

6. The figure gives the plan and elevation of a work-shed on a scale of 1 inch to 20 feet. Fifteen per cent. of the space in the work-shed is occupied by machinery and fittings. If each workman is to have 400 cubic feet of air-space, how many men may

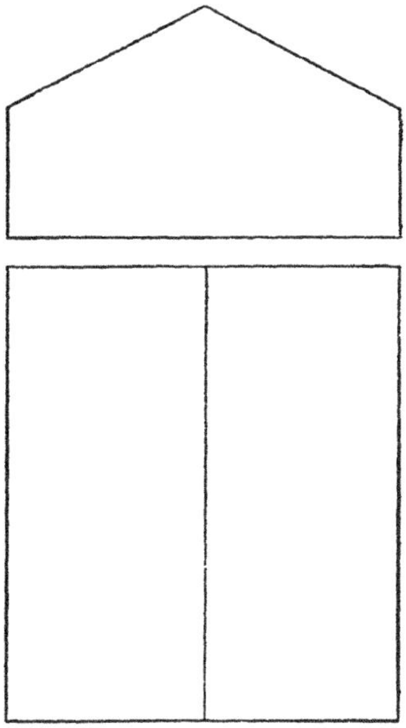

Fig. for Qn. 6.

work in the shed? [The area of a triangle is half the product of the base and the height.]

Suppose a model of the dimensions of the figure made of the shed. Give in decimal form the ratios of the height, floor area, and air space of the model to the corresponding quantities of the shed; each to one significant figure.

(ii.) RECTILINEAR AREAS DEPENDING ON THE TRIANGLE.

7. The figure represents, half the real size, a section of picture moulding. Find, to the nearest lb., the weight of a 10-foot length

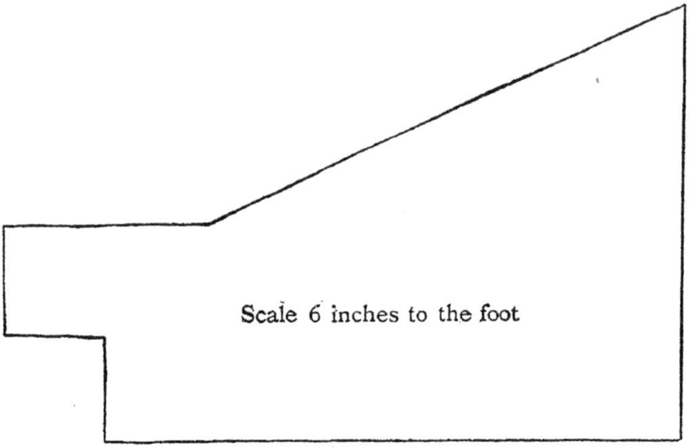

Fig. for Qn 7.

of this moulding, if a cubic foot of the wood it is made of weighs 47 lb.

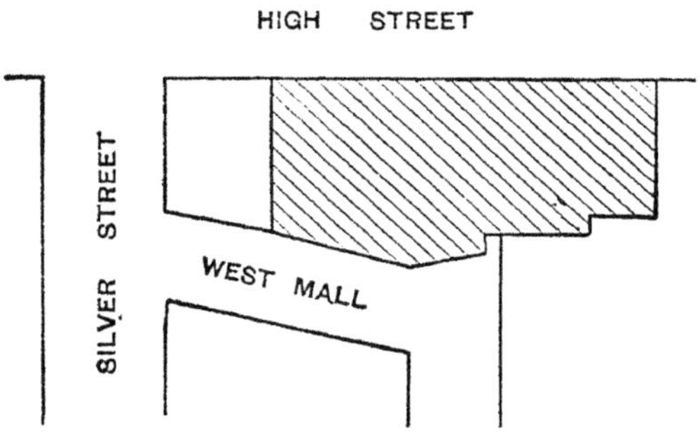

Fig. for Qn. 8.

8. The shaded part of the plan (scale, 1 inch to 80 feet) was advertised for sale. Find the lengths of the frontage to High Street and of the frontage to West Mall up to the first angle, and the number of square feet of area in the shaded part.

Check this last result by tracing the figure on transparent paper, laying your tracing on squared paper, and counting the number of squares enclosed.

9. The figure represents a field on a scale of an inch to 100 yards. What weight of gravel, to the nearest ton, would be

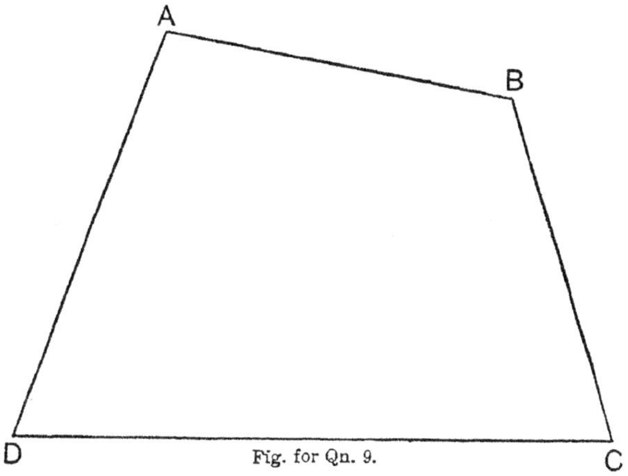

Fig. for Qn. 9.

required to cover a path 3 feet wide from A to C to a depth of 3 inches? A cubic foot of gravel weighs 110 lb.

What is the area, to the nearest acre, of the field ABCD? The area of a triangle is half the product of the base and the height. Trace the figure on transparent paper, and judge by eye the perpendicularity of any line you draw.

10. Find, in acres, the area of a field which is represented, on a scale of 10 inches to the mile, by the figure ABCDEA; *or*, if you prefer, take the scale as 1 to 6000, and give the area in hectares. If you want to draw perpendiculars, judge their positions by eye.

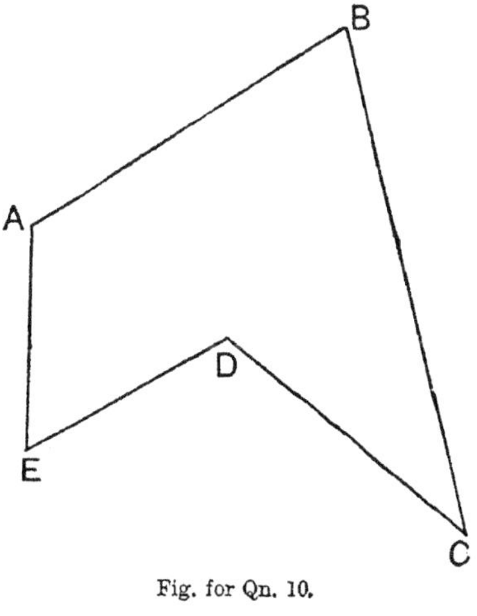

Fig. for Qn. 10.

(iii.) SIMPSON'S RULE, ETC.

11. The figure shows a metal plate. Find the area of the plate in sq. cm., taking the area of each strip to be the breadth multiplied by half the sum of the two sides; you may assume the strips to be equal in breadth.

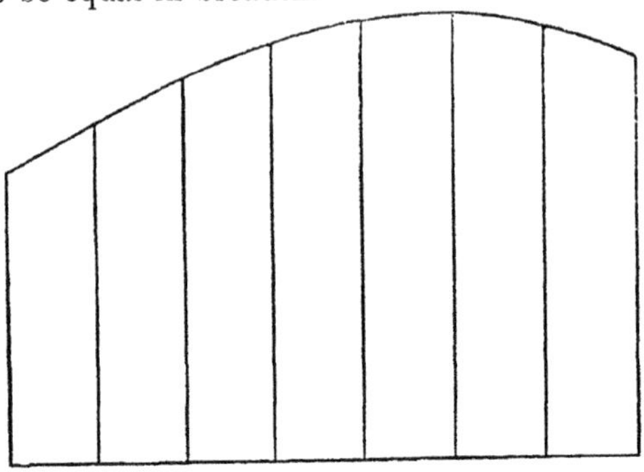

Fig. for Qn. 11.

If the plate is worth 1s. 2d. per square centimetre, how much is the whole worth?

12. The figure is a drawing of a sheet of metal on a scale of an inch to 2 feet. To determine the area perpendiculars to the

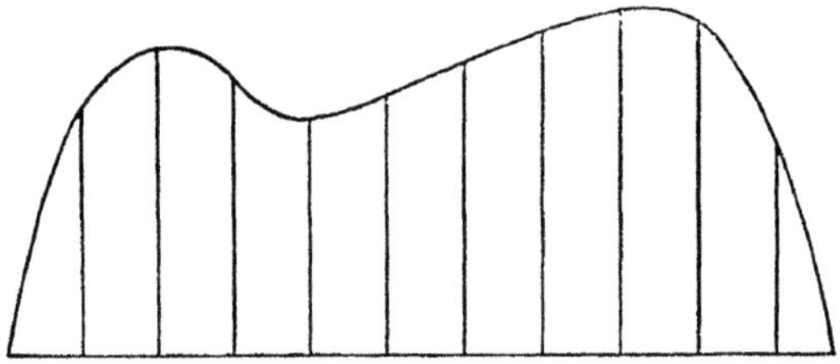

Fig. for Qn. 12.

straight side have been drawn at equal intervals, starting from the left. Measure the necessary measurements, and find the area of the drawing and of the sheet of metal.

13. State Simpson's, or any other, rule for finding areas. A ditch 100 yards long filled with water is to be drained and filled with earth. The ditch is 30 yards wide, and soundings at the bank and at every third yard across its width give depths in yards of 2, 7·2, 10, 12, 11·5, 9, 7·5, 7·3, 6·4, 3, 0; the bottom is to be taken to be the same all along the ditch. Find the quantity of earth required.

14. The depth of a river channel at a number of points situated in a straight line is given in the following table:

Distance from one bank in yards	0	40	80	100	112	120	140	160	188	200
Depth in feet - -	0	16	24	21	20	20	28	32	20	0

Assuming that there are no sudden holes, show the shape of the section of the river on squared paper, and find its area.

Section 5. Mensuration of the Circle,

INCLUDING PROBLEMS ON THE CYLINDER AND THE SPHERE.

1. I go to an ironmonger and ask for 18 yards of a certain wire. The wire is in a coil; the ironmonger measures the distance across the coil (which is 20 inches) and gets the right length by counting off a certain number of rounds from the coil. If the length round is 3·14 times the distance across, how many rounds must the ironmonger cut off?

2. Express the difference between $3\frac{1}{7}$ and $\frac{355}{113}$ as a vulgar fraction.

If you were to calculate the area of a circle, using in turn for π each of these approximations, show, by trial with a radius of 28 cm., that the radius must be greater than 28 cm. for the difference in the results to amount to 1 sq. cm.

3. The length of a window blind is 45 inches and the diameter of the roller is 1·5 inches. How many turns of the roller must be made in order to roll the blind up completely if the thickness of the blind is so small that it may be neglected? This blind is drawn up by means of an endless cord passing over two pulleys, which are of the same diameter as the roller, and are 3 feet apart. How far would a knot in the cord travel in rolling up the blind?

4. The two wheels of a bicycle are 28 in. and 30 in. in diameter, and the spokes which initially point vertically downwards are marked. How far must the bicycle travel before the two marked spokes are at right angles, and what time will elapse before they are again at right angles if the bicycle is moving at the rate of 12 miles an hour? [See the Tables inside the front cover.]

5. A rectangular board measuring 6 decimetres by half a metre is just large enough to display 120 circular medals, all of the same size, arranged edge to edge in 12 rows. Find, to two significant figures, (i) the area of a face of one medal; (ii) the area of the board that is left uncovered by the medals.

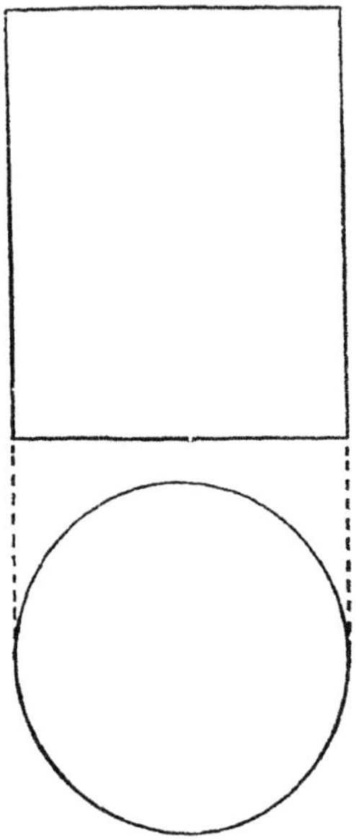

Fig. for Qn. 7.

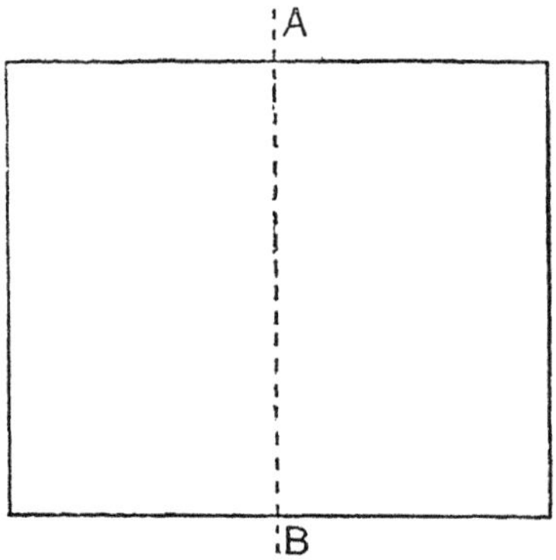

Fig. for Qn. 8.

6. The diameter of a coin is 3 cm.; the united thickness of three of the coins placed together is 4 mm.; and three coins together weigh an ounce. Find, to two significant figures, the number of cubic millimetres in an ounce of the metal of which the coins are made. The volume of a coin a cm. across and b cm. thick, may be calculated from the formula:

$$\text{Volume in c. cm.} = 0{\cdot}785 \times a \times a \times b.$$

7. The figure (p. 40) shows the plan and elevation of a cylindrical tank to a scale of $\frac{1}{20}$ of natural size. Find, in gallons, how much water it will hold, and, in inches, to what depth it would be filled by 40 gallons of water. Mark the depth of the water by a dotted line on a copy of the elevation.

The volume of a cylinder is 3·142 × the square of the radius of the base × the height, volume and lengths being in corresponding units. [See the Tables inside the front cover.]

8. The figure (p. 40) shows the section of a stone for a circular pillar on a scale of an inch to 2 feet; AB is the axis. Measure the figure and calculate the weight of the stone, taking the weight of a cubic foot as 140 lb. The volume of a cylinder is $\pi r^2 h$, where π is 3·1416, r is the radius of the circular base, and h is the length of the cylinder.

9. The volume of a cylinder of diameter d inches and length h inches is given in cubic inches by the formula $0{\cdot}7854 \times h \times d \times d$. How many cubes having an edge an inch in length can be cast from a cylinder of lead whose diameter is 3·4 in. and length 10 in.?

10. A garden roller, 13 inches radius and 3 feet long, is $\frac{4}{5}$ filled with water to make it heavier. Find the weight of water, given that the internal volume of a roller of r inches radius and h inches long is $\pi r^2 h$ cubic inches.

11. The diameter of some galvanised wire is 1·4 millimetres, and a cubic decimetre of the material weighs 6·65 kilograms. Calculate, to the nearest metre, the length of wire in a coil which weighs a kilogram. A piece of wire d cm. in diameter and h cm. long has a volume of $0{\cdot}785 \times d \times d \times h$ c. cm.

12. A coil of wire has 36 turns; the average diameter of the turns is 10 inches and the thickness of the wire is $\frac{1}{10}$ inch. Find (i) the length of the wire, (ii) the volume of wire in the coil.

13. Buttons are made with lead discs 0·8 of an inch across and 0·09 of an inch thick. The volume of a disc, a inches across and b inches thick, may be calculated from the formula:

$$\text{Volume in cubic inches} = 0{\cdot}785 \times a \times a \times b,$$

and lead weighs 711 lb. per cubic foot. How much lead, to the nearest pound, is wanted for a thousand buttons?

14. A boy measures up a sheet of lead to see how many bullets he can cast from it. Its thickness lies between 2·5 and 3 millimetres, and its area is shown in the figure on a scale of $\frac{1}{5}$. He

Fig. for Qn. 14.

means to cast bullets 8 mm. in diameter, and knows that the volume of each will be $0{\cdot}52 \times 8^3$ c. mm. How many can he cast?

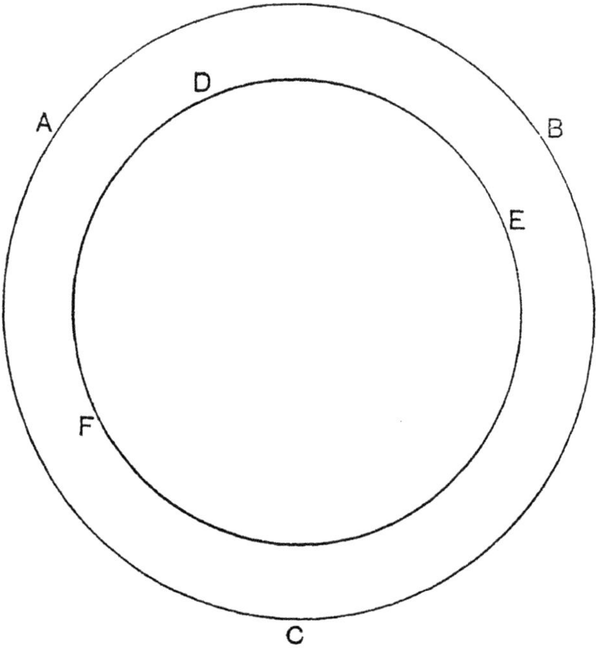

Fig. for Qn. 15.

15. In the figure (p. 42) ABC represents the circumference of a new grindstone weighing 102 lb. Find its weight when it is worn down to DEF. Ignore the hole through which the axle passes.

16. The length of a cylindrical boiler with flat ends is 26 feet and the diameter of each end is 62 inches. Find its weight in tons correct to the nearest tenth, assuming the thickness of the boiler plates to be $\frac{3}{8}$ inch, and the weight of a cubic foot of iron to be 480 lb.

17. A circular mirror of radius 1 ft. 3 in. is surrounded by a circular gilt frame of width 3 in. If resilvering the glass costs $\frac{3}{4}$d. per sq. in. and regilding the frame 4d. per sq. in., what is the cost (to the nearest shilling) of doing up the mirror? The area of a circle is $\pi \times (\text{radius})^2$.

18. Find the weight of 45 metres of garden hose from the following data: Volume of a hollow cylinder $= \pi h(R^2 - r^2)$, where h is the length of the cylinder, R and r are the external and internal radii. External diameter of hose = 3 5 cm. Thickness of rubber of which the hose is made, 0·5 cm. The rubber is 1·57 times as heavy as water, bulk for bulk.

19. Find the weight per foot length of a cast-iron pipe, internal diameter $5\frac{1}{2}$ in., thickness $\frac{1}{2}$ in. If the internal diameter were increased by 4 per cent., the thickness remaining unchanged, what would then be its weight, and the percentage change in its weight? 1 cub. in. of cast iron weighs 0·26 lb.

20. A brass tube in a surface condenser is $\frac{3}{4}$ inch external diameter and 8 ft. long. What is the external surface of such a tube? What is the total surface in square feet if there are 500 such tubes? If the thickness of each tube is $\frac{1}{20}$ inch, what is the total weight of the tubes? 1 cub. in. of brass weighs 0·29 lb.

21. The diameter of the collecting funnel of a rain gauge is 5 inches at the top, and the diameter of the measuring glass (a cylinder) is 1·65 inches. Find the distance between the graduations, each representing one-hundredth of an inch in rainfall, on the measuring glass.

22. The figure shows the full-size section across a hollow bar; the outside is 8-sided, and the dotted lines show how much it wants of being a square. Give in c. cm. the volume per metre length of the bar.

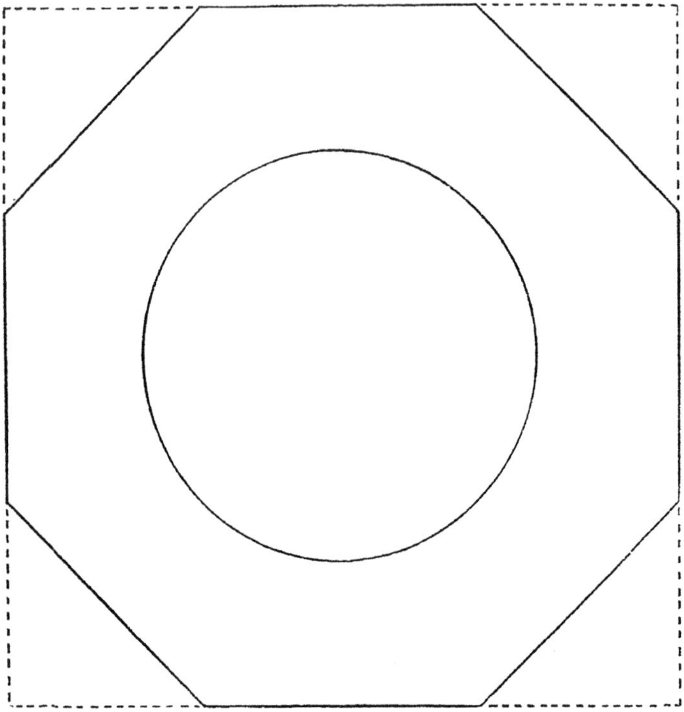

Fig. for Qn. 22.

23. Steel wire is made to bear a strain of 215,000 lb. per square inch of section. Find, to the nearest cwt., the weight that can be suspended from such a steel wire of diameter 0·104 of an inch.

The area of the section is $0{\cdot}785 \times d \times d$, where d is the diameter.

24. The area of the surface of a ball in square inches may be calculated from the formula $3{\cdot}1416 \times d \times d$, where d is the number of inches in the diameter of the ball. Find, to the nearest shilling, the cost of covering with gold leaf a wooden ball whose diameter is 6 feet 5 inches, the cost of material and work being at the rate of 4s. 6d. per square foot.

What would be the cost for another ball of half the diameter?

25. A hollow ball of external diameter 6 centimetres, and of thickness half a centimetre, weighs 162 grams when it is filled with a certain liquid. Find, to the nearest hundredth of a gram, the weight of a cubic centimetre of the liquid, a cubic centimetre of the substance of the ball weighing 2 grams. The volume of a sphere of radius r is $\frac{4}{3}\pi rrr$.

26. A tank consists of a cylinder with hemispherical ends. If the length of the cylinder (without the ends) be 10 feet, and the diameter of its cross section 5 feet, find its volume in cubic feet to the nearest foot.

27. A cylindrical vessel has an internal diameter of 2 cm., and is filled with water to 1 cm. from the top. It is found that 58 steel pellets can be dropped in before the vessel is quite full. Find in c. mm. the volume of a pellet; that of a cylinder of diameter d cm., and height h cm., being $\frac{1}{4}\pi ddh$ c. cm.

28. A vertical boiler is a cylinder of height 6 ft., its base being a circular disc of diameter 3 ft. 3 in. If it is refilled to a depth of 5 feet as soon as the depth of water falls to 3 ft., and if this occurs every 35 minutes, how many hundred gallons of water are required for a working day of 10 hours? The boiler must be filled in the morning to the depth of 5 ft., the water left at night being drawn off and not used again.

Section 6. Mensuration by Formula.

1. To find the weight that a steel wire rope will safely lift, take the number of inches in the circumference, square this number, and then multiply by 0·45; you thus get the number of tons the rope will lift with safety. Use this formula to find (i) what load may be lifted by a rope $2\frac{1}{4}$ inches in circumference, and (ii) if it is safe to lift 5 tons with a rope $3\frac{1}{2}$ inches in circumference.

2. A railway embankment H feet high, A feet wide at the top, and B feet wide at the bottom, contains $\frac{A+B}{2} \times H \times L$ cubic feet of earth in a length of L feet. Find the volume of earth in 200 feet of an embankment 20 feet high, 14 feet wide at the top, and 37 feet wide at the bottom.

If 18 cubic feet of the earth weigh a ton, what is the weight of this length of embankment to the nearest hundred tons?

3. An oil drum is 18 inches high and 15 inches in diameter. The capacity C of a drum is given in gallons by the formula:

$$C = 4{\cdot}89 \times L \times D \times D,$$

where L is the height and D the diameter of the drum, both in feet. Use this to find the capacity of the above drum to the nearest tenth of a gallon.

What would be the value of the oil which the drum would hold at 6½d. per gallon?

4. The weight W of copper wire in kilograms per kilometre and the diameter D of the wire in centimetres are connected by the relation $W = 698 \times D \times D$. Find the weight of a kilometre of wire 0·8 mm. in diameter, and its value at 1s. 2d. a kilogram.

5. Find, to within a kilogram, the weight of a piece of lead piping which is 7 metres long, and has an inside diameter of 8·6 cm. and a thickness of 1·15 cm. The weight W in kilograms of a piece of this piping, 1 metre in length, inside diameter A cm., thickness B cm., is found from the formula:

$$W = 3{\cdot}523 \times B \times (A + B).$$

What would this piece of lead piping cost at 4d. a kilogram?

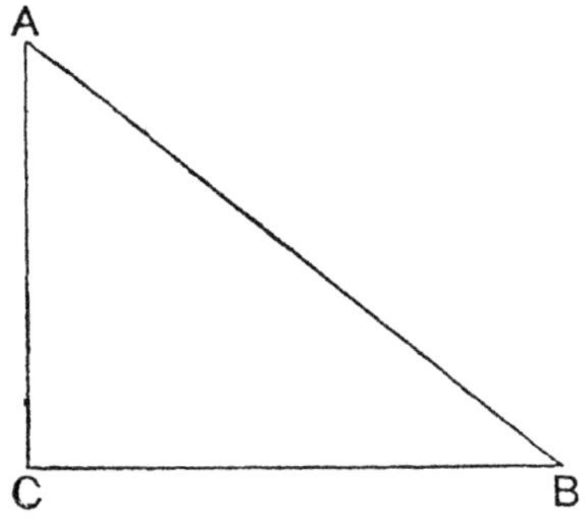

Fig. for Qn. 6.

6. What is the area, in square inches, to within a tenth, of the triangle ABC in the figure? If the triangle is turned once completely round AC as an axis, what is the volume of the cone swept out; that is, of a cone of which AC is the height and BC the radius of the base? The volume of the cone is $\frac{1}{3} \times 3{\cdot}14 \times AC \times BC \times BC$ cubic inches, AC and BC denoting the numbers of inches in the lengths of those lines.

7. A regular pyramid on a square base is 10 inches high, 12 inches edge of base, and weighs 80 lb. It is not solid, but contains a cavity. If the metal of which it is made weighs 4 ounces per cubic inch, find the size of the cavity. The volume in cubic inches of a pyramid is one third of the product of the height in inches by the area of the base in square inches.

8. A cast-iron weight is made in the form of the frustum of a square pyramid, altitude 7·5 in., side of base 5·7 in., side of upper face 5 in. Find its volume and weight.

The side of the base being a in., of the upper face b in., and the altitude c in., the volume in cubic inches is $\frac{1}{3}c(aa + ab + bb)$.

(A cub. in. of cast iron weighs 0·26 lb.)

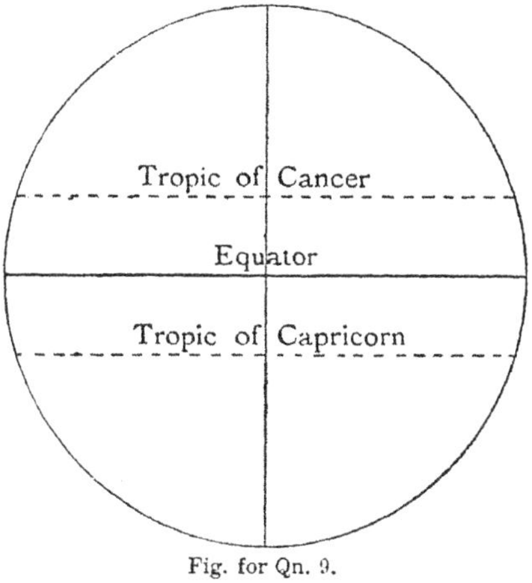

Fig. for Qn. 9.

9. The radius of the earth being R miles, and the distance between the planes of two parallels of latitude h miles, the area of the earth's surface that lies between these parallels is $2 \times \pi \times R \times h$ square miles. The figure is a section through the earth's axis, drawn to a certain scale. From measurements of the figure along the axis, express this scale as so many miles to the inch, find the area of the Tropics, and find how many times this is as great as that of the British Empire (11,400,000 square miles). Take the earth's radius as 4000 miles.

10. A measuring vessel is of such a shape that when liquid is poured into it to a depth d inches the volume of the liquid contained is $0{\cdot}065 \times d \times d \times d$ cubic inches. How many pints does the vessel hold when the depth of the liquid is 14 inches?

11. If in a map x ft. is represented by 1 inch, then the side of the square of area 1 acre is represented approximately by $200/x$ inches. Examine the truth of this statement for a map drawn on a scale of 30 ft. to 1 inch, and find by how much per cent. the result obtained by the rule differs from the true result.

12. The pressure of the atmosphere is 15 pounds per square inch. When a bicycle tyre, of capacity A cubic inches, is inflated by a pump of capacity B cubic inches, the pressure of air in the tyre, after n strokes of the pump, exceeds the atmospheric pressure by $n \times \mathrm{B} \div \mathrm{A}$ times the atmospheric pressure. In a certain case the pump is ten inches long and $\frac{3}{4}$ inch in diameter; the tyre may be treated as a cylinder 88 inches long and $1\frac{3}{4}$ inches in diameter; and the capacity of a cylinder a inches long and b inches in diameter is $0{\cdot}79 \times a \times b \times b$ cubic inches.

Find the capacity of the tyre and of the pump, and the pressure of the air in the tyre after 40 strokes.

Section 7. Mensuration Questions involving Models.

1. Estimate the length, breadth, and height of the room you are sitting in, and find the number of cubic feet of air space. Count how many people are in the room, and find how many cubic feet of air there are per head.

2. Measure the rectangular block given to you, and find how many cubic centimetres of wood there are in it.

3. From measurements of the model supplied (rectangular block pierced with large square hole), find the number of cubic centimetres of wood in it.

Weigh the model, and calculate the weight of a cubic centimetre of the wood.

4. Measure, to the nearest tenth of a centimetre, the diameter of the coin supplied, and thence find the circumference of the coin.

Also find the distance passed over by the coin while rolling along a straight line until it has made three complete revolutions. If this when divided by 3 does not agree with the previous result, give the probable reason.

5. Measure in centimetres the length and diameter of the cylindrical block supplied. If the length of such a block is h cm. and the diameter d cm., it is known that the volume is $0{\cdot}79 \times h \times d \times d$ cubic centimetres. Use this to find the volume of the block.

6. The volume of a pencil h cm. long, d cm. in diameter, and g cm. in girth may be expressed as $0{\cdot}785 \times h \times d \times d$ c. cm., or as $0{\cdot}0796 \times h \times g \times g$ c. cm. Measure the length and diameter of the pencil supplied, and use the first formula to find the volume.

Also find the girth by trying what length of cotton it needs to go 20 times round the pencil, and use the second formula to find the volume. Do the two results differ? if so, explain why.

7. Draw a square which will have as nearly as possible the same area as the surface of the lawn-tennis ball supplied to you. State what measurements you take, and indicate clearly how you make the square from them.

As a check strip off the covering, spread it on squared paper, and so find its area.

Section 8. Various Mensuration Questions.

1. A piece of netting measures 15 ft. by 6 ft. 3 in. and the meshes are squares of side $\frac{1}{4}$ inch. Find to the nearest yard the length of string in the netting, ignoring the length consumed in knots.

2. In a piece of fine linen there are 180 parallel threads to the inch. What is the total length of thread used to make a square yard of such linen?

3. It is found that 5800 yards of linen thread go to the weaving of a square yard of certain linen cloth, and as many threads run lengthwise as crosswise in the square yard. How many run each way in a square inch?

4. A piece of netting measures 3 ft. by 2 ft., and each mesh is a square of side $\frac{3}{4}$ in. How many meshes are there, and what is the length of string used, excluding the amount required for knots?

If 6 % extra is used for knots, what is the actual length of string to the nearest foot?

5. The value of a certain set of ornamental silver bowls of the same shape varies with the weight of silver used, and this varies as the square of the diameter.

The diameters of the bowls are 5, $5\frac{1}{2}$, and 6 inches respectively, and the price of the smallest is £3. 10s. 0d. Calculate the price of each of the other two to the nearest shilling.

6. An order is given for a marble statue 10 feet high. A plaster model is first made, which is 2 feet 8 inches high and weighs 40 lb. What will be the weight of the marble statue to the nearest pound, marble being 1·7 times as heavy as plaster.

Do you see any objection to stating the result to the nearest pound?

7. Find the weight that a spherical balloon of 10 metres diameter, made of thin material weighing 30 grams per square metre, will lift when filled with coal gas, specific gravity compared with air 0·44. Assume 1 litre of air to weigh 1·28 grams.

PART II.

SECTIONS 9–17. FURTHER ARITHMETIC.

Section 9. Conversion

FROM ONE SYSTEM OF WEIGHTS AND MEASURES TO ANOTHER.

Every student should have some rough idea of the value in imperial units of a quantity expressed in metric units, and vice versa; he should know that a pound is about half a kilogram and that 20 *kilometres is* 12 *or* 13 *miles. But the time required for accurate conversion (such as most of the questions in this section require) ought to be saved, to be spent upon more valuable portions of mathematics. Consequently this section should be omitted whenever possible. (See page* v.)

1. Give to the nearest penny the sum wanted to complete the following letter:—"Dear Sir, we hand you herewith London draft for   , being the equivalent in English money of 15 dollars, at the rate of 4·875 dollars to the pound sterling."

2. A man in India wishes to send his wife in England £100 a year in monthly instalments. How many rupees worth 1s. 4d. each must he transmit monthly?

3. Show that 1s. = £0·05, 6d. = £0·025, and $\frac{1}{4}$d. = £0·001 nearly. Convert £7·652 to pounds shillings and pence (to the nearest farthing), and express £3. 4s. 8$\frac{1}{2}$d. as a decimal (to three places). Show how you arrive at your results.

4. An Indian official, whose income is 2300 rupees a calendar month, has to send half his salary for his family in England. The rupee is worth 1s. 4d. What are the annual expenses of his family in English money?

5. In the Liverpool corn market the unit weight is a cental (100 lb.), and quotations of the price of a cental are given to the nearest eighth of a penny. Form such quotations from the following data:

Wheat - - -	33s. 0d. per quarter (480 lb.).
Barley - - -	22s. 10d. ,, ,,
Oats - - -	18s. 3d. ,, ,,

6. A novel sold in France for 3·50 francs is sold in England for 2s. 8d. By how much, to the nearest farthing, is this less than the exact equivalent, if 25·20 francs be taken as equal to £1?

7. A book costs 18s. 9d. in England and 23·85 fr. in France. If one shilling exchanges for 1·25 fr., is the book dearer or cheaper in France, and by how much per cent.?

8. What sum in English money would a Frenchman get in London for 16,345 francs 27 centimes, if £1 is worth 25 francs 47 centimes? A franc is equal to 100 centimes.

9. If the price of an article in England is 3s. 0d., find, to the nearest 5 cents, what the price of it would be in American money, assuming that 10 dollars (=1000 cents) exchange for 41s. $1\frac{1}{4}$d.

10. I was weighed in France and England. In France I was 51 kilograms, in England 8 stone. Which is the greater? By how much (in imperial measures)? (See the tables inside the front cover.)

11. The 10-franc piece weighs 3·2258 grams and contains 90% of pure gold, the sovereign weighs 123·274 grains and contains $\frac{11}{12}$ of pure gold. Taking the value of the coins to be equal to that of the gold they contain, find how many 10-franc pieces are equal in value to 100 sovereigns.

12. An Egyptian pound (£E.) is worth £1. 0s. $6\frac{1}{4}$d. Use the information supplied in the following table to make out a new table in which the sums are expressed in pounds sterling, to the nearest integer:

	Merchandise.	
	Imports.	Exports.
Year.	£E.	£E.
1904 - -	20,559,588	20,811,040
1905 - -	21,564,076	20,360,285

13. Find the cost per hundred in francs, in dollars, and in marks, of exercise books priced at 10s. 5d. per gross; each to two decimal places.

£1 = 25·26 francs = 4·84 dollars = 20·51 marks.

14. A certain article is sold in England at 7½d. per lb., and in France at 2 fr. 50 centimes per kilo. Find the difference in shillings between the cost in England and France of that quantity of the article which can be bought for 100 francs in France.

See tables inside front cover. 25 francs = £1.

15. A new railway in Holland, 110 kilometres long, will cost 6,441,000 florins. Find the whole cost, and the cost per mile, in English money to the nearest thousand pounds.

Exchange, £1 = 12 florins 11 cents. 1 florin = 100 cents.

16. The duty on tea in France is 2 francs 68 centimes per kilogram. Give the equivalent in pence per lb. to the nearest penny.

1 franc = 9½d. = 100 centimes.

17. In 1902 France produced 115,530,692 hectolitres of wheat on 6,563,711 hectares. Find the yield in bushels per acre to the nearest whole number.

18. In 1901 Denmark, with a total area of 14,840 sq. miles, produced 332,000 hectolitres of wheat on an area of 23,655 hectares, and Great Britain, with a total area of 56,875,000 acres, produced 52,458 thousand bushels of wheat on 1,700,828 acres. Find, in each case, the percentage of land under wheat and the yield of wheat-land in bushels per acre. Give each result to the nearest tenth.

19. I buy wheat in France at 14·2 francs per hectolitre. At what price, to an exact penny, must I sell it in England per bushel, so as not to lose? Freight is 1 franc per 100 hectolitres, commission 2½ per cent., insurance 1 per cent., and exchange is 25·30 francs for £1.

20. A wine-merchant buys a cask of claret in Bordeaux, and the freight to London is £1. It contains 208 litres of wine at 75 centimes the litre, and for the cask itself he has to pay 12 francs. He has to pay a duty of 1s. 3d. per gallon, and he bottles the wine at a cost of 2d. a bottle. If he sells it at 18d. a bottle, how much of this price is profit?

£1 = 25 francs = 2500 centimes. 6 bottles = 1 gallon.

21. A piece of land in France is worth 3 francs a square metre. If £1 = 25·22 francs, and 4047 square metres = 4840 square yards, what is the land worth per acre in English money to the nearest pound?

22. Find, from the data inside the front cover, a multiplier for converting hectolitres per hectare into bushels per acre.

23. A certain mixture should be made by dissolving 12 grams of bluestone and 8 grams of lime in 1 litre of water. The mixture is actually made by dissolving 2 oz. of bluestone and $1\frac{1}{4}$ oz. of lime in 1 gallon of water. On this plan how many grams of bluestone and of lime are there to the litre of water?

24. Find, to the nearest integer, the velocity in cm. per second of a boat travelling at 15 knots.

1 knot = 1·1515 miles per hour.

25. Find, to the nearest shilling, the cost of filling a tank 37 cm. by 15 cm. by 29 cm. with liquid which costs 17s. 4d. per gallon.

26. A sheet of metal weighs $4\frac{1}{2}$ lb. to the square foot. How many kilograms to the square metre is this? Give your answer to the nearest kilogram.

27. An acre of land in England costs £92. 15s., and a hectare of land in France costs 2150 francs. Express the cost of land in France as a percentage of the cost in England, giving the result to one decimal place.

£1 = 25 francs.

28. The cost of travelling third class on the Swiss railways is 5·2 centimes per kilometre, and luggage is charged for at the rate of 5 centimes per 100 kilograms per kilometre. Find the greatest number of lb. of luggage a traveller can take with him so that his total expenditure shall not exceed 1d. per mile.

25 francs = £1.

29. The average cost of travelling in England is 0·85 pence per mile, and the charge for excess luggage is 6d. per fifty pounds for every hundred miles travelled. Express these statements in (i) centimes per kilometre and (ii) centimes per 50 kilograms for every hundred kilometres travelled, giving the results correct to two significant figures.

100 centimes = 9·525 pence.

30. The rainfall in a certain district is 5·76 cm. How many litres per hectare is this?

Taking an inch as $2\frac{1}{2}$ cm. and a cubic foot of water as weighing $62\frac{1}{2}$ lb., how many tons per acre is this? Give your answer to the nearest hundred tons.

31. The weight in ounces of a leaden bullet is 3·43 times the cube of its diameter in inches. Express this in terms of grammes and centimetres.

Bullets of a certain size cost 17s. per 1000; express this in francs.

£1 = 25 fr.

32. The formulas for the components A and B of a photographic developer given in English measures are as follows:

A.	Quinol	80 grains.
	Sodium sulphite	1 oz.
	Water up to a total volume of	10 oz.
B.	Caustic soda	40 grains.
	Potassium bromide	30 ,,
	Water up to a total volume of	10 oz.

Calculate the corresponding quantities in the metric system required to make 100 cubic centimetres of each solution. In the above formula the weight of sodium sulphite is given in avoirdupois ounces, and the volume of water in fluid ounces of which twenty go to a pint.

Section 10. Rates

OF VARIOUS KINDS, SUCH AS DENSITY OF POPULATION, PERCENTAGE OF INCOME OR EXPENDITURE DUE TO A PARTICULAR ITEM, RATE PER HEAD OF THE PRODUCTION OR CONSUMPTION OF A COMMODITY.

1. The following statistics are given in the Census of 1901:

District.	Number of Acres.	Population.
London	74,839	4,536,541
Westmoreland	505,330	64,409
England	32,611,033	30,829,695

Find, in each district, the population per acre correct to three decimal places, that is, to the nearest thousandth. And calculate to two significant figures the ratio of the density of London to that of England and the ratio of the density of Westmoreland to that of England.

2. The area and the estimated population in 1905 of the United Kingdom and of India are shown in the following table:

	Area in Square Miles.	Estimated Population.
United Kingdom -	121,000	43,200,000
India - - - -	1,900,000	300,000,000

From these data calculate in each case the density of population, that is, the number of inhabitants per square mile. Give your answers to the nearest whole number.

3. The following table shows the areas and populations of Belgium and China. Find, to the nearest whole number, how many persons there are to the square mile in each country; and, using the results of the preceding question, state whether each is more or less densely populated than the United Kingdom.

	Area in Square Miles.	Population.
Belgium - - - -	11,400	7,000,000
China - - - -	1,500,000	390,600,000

4. From the data of the two preceding questions, find what the population of the United Kingdom would be, to the nearest million, if it were as thickly populated as Belgium?

5. The county of Surrey has an area of 707 square miles and a population of 519,654. If the people were spread evenly over the land, how many would there be on each square mile? Is Surrey more or less densely populated than the United Kingdom considered as a whole? (See Question 2.)

6. From the data below, find, (i) for the City of London and (ii) for the County of London, the number of people there are to an acre, and the number of people to an inhabited house, and the number of inhabited houses to the acre; each result to the nearest integer:

	City of London.	County of London.
Area in acres - - - -	673	74,144
Population - - - -	26,923	4,509,506
Number of inhabited houses -	3,865	567,903

7. The area of the counties of Wales is given in the following table:

County.	No. of Acres.
Anglesey	120,417
Brecon	473,087
Cardigan	595,285
Carmarthen	464,587
Carnarvon	322,742
Denbigh	378,309
Flint	74,023
Glamorgan	576,537
Merioneth	523,708
Montgomery	591,973
Pembroke	357,118
Radnor	238,660

What percentage of the whole area is furnished by the largest, and what by the smallest county? The answers are to be given to the nearest integer.

8. The population of London at the middle of 1903 was estimated at 4,613,812. Taking the annual birth-rate as 28·4 per 1000 and the annual death-rate as 15·7 per 1000, and assuming the influx and efflux of population to balance one another, estimate the population of London at the middle of 1904.

9. The population of London at a certain date was 4,796,000, the birth and death rates per annum respectively 25·2 and 13·8 per thousand, and the area 74,820 acres. Neglecting the movement of population, find how many births and deaths take place annually on an area of 10 acres, and what the annual increase of population on that area is.

10. From the table given below, find (for 1906) what percentage of the population were parliamentary electors in England and Wales, in Scotland, and in Ireland. Neglect the changes in population between 1901 and 1906.

Find also what population was on the average represented by a Member of Parliament in England and Wales, in Scotland, and in Ireland.

	Population in 1901.	No. of Electors in 1906.	No. of Members of Parliament.
England and Wales	32,527,843	5,824,884	485
Scotland	4,472,103	750,401	72
Ireland	4,458,775	691,423	103

11. The following statistics refer to England and Wales:

Year.		Population.	Males.	Females.	Average School Attendance.
1903	-	33,373,338	16,133,344	17,244,994	5,030,219
1904	-	33,763,434	16,316,647	17,446,787	5,142,774

Find, to the nearest integer, the percentage of the population attending school each year. If the proportion of boys to girls in the schools in any year is the same as that of males to females in the whole population for that year, find, to the nearest thousand, the average attendance of boys in each year.

12. The following table gives the destination of emigrants from the United Kingdom for the year 1904:

United States - - - - -	146,797
British North America - - -	69,744
Australasia - - - - - -	13,762
Cape Colony and Natal - - -	26,681
Other places - - - - - -	14,673

Find, to the nearest whole number, the percentage of the whole number of emigrants under each head.

13. The table below shows the quantity of wheat grown in the United Kingdom and its average price for the years 1903 and 1904. Find the total value of the wheat in each year to the nearest thousand pounds, and the percentage increase or decrease, to the nearest whole number, in the value in 1904 as compared with 1903.

Year.	Amount of Wheat in Quarters.	Average Price per Quarter. s.	d.
1903 - - -	2,386,010	26	$5\frac{1}{4}$
1904 - - -	2,129,430	27	$2\frac{1}{4}$

14. In 1903 the United Kingdom imported 88,131,000 cwt. of wheat. Of this 24,196,000 cwt. came from the United States, whilst 28,490,000 cwt. came from British possessions, India sending 17,058,000 cwt., and Canada 10,804,000 cwt. What percentages of the total import came from the United States and what from British possessions? What percentage of the imports from British possessions came from India and Canada together? Give your answers to the nearest tenth.

15. In the year 1903 there were in Ireland 37,595 acres of land under wheat, which produced 1,175,972 bushels. Find the average yield per acre to the nearest hundredth of a bushel, and the total value of the wheat, to the nearest pound, at 26s. 9d. per quarter.

16. The number of quarters of corn that were sold in a certain week in 1898 was 16,302, and the average price was 30s. 7d. per qr.; in the corresponding week of 1904 the number of quarters was 19,200 and the price 28s. 8d. Find, to the nearest pound, the difference in the total values of the sales in the two weeks.

17. Below are given the population of the United Kingdom in 1903 and the quantity and value of sugar and tea imported in that year. Ascertain, to the nearest pound, what quantity of tea and of sugar we consumed per head, and, to the nearest penny, how much per head we paid for each.

	Quantity.	Value in £.
Sugar imported - -	31,230,317 cwt.	15,458,857
Tea ,, - -	255,365,953 lb.	9,666,790

Population, 42,372,556.

18. From the following table, find, to the nearest shilling, the value per cwt. of coffee imported in 1903, in 1904, and in 1905. Find also the percentage falling off in 1905 in quantity and in total value as compared with 1904, each percentage to the nearest integer and reckoned on the 1904 numbers:

Year.	Quantity Imported. Cwt.	Value in £.
1903 - - -	1,143,526	3,134,924
1904 - - -	1,055,866	3,329,598
1905 - - -	956,226	2,707,169

19. From the following data, find, to the nearest whole number, how many gallons of beer were drunk per head, and, to the nearest shilling, how much per head was contributed to the revenue in this way:

Year.	Population.	Barrels of Beer drunk. 1 Barrel=36 Gallons.	Revenue from Beer in £.
1904-5 -	43,331,000	35,862,068	13,123,679

20. The table below shows how many gallons of beer per head the inhabitants of New Zealand drank in 1884 and in 1895. Express the diminution in 1895 as compared with 1884 as a percentage, to the nearest whole number, of the amount for 1884.

The table gives the data for wine and spirits also. Express the diminution for each in a similar way.

Find whether the total consumption of beer increased or diminished in this period.

New Zealand. Consumption per Head in Gallons.

Year.	Population.	Beer.	Wine.	Spirits.
1884	543,000	8·77	0·27	1·00
1895	688,000	7·42	0·13	0·63

21. It is estimated that a reservoir, having a surface of 50 acres and an average depth of 20 feet, will hold a week's supply of water for 100,000 persons. How many gallons a day, to the nearest ten gallons, does this estimate allow to each person?

22. It has been proposed to construct a reservoir with an area of 100 acres and a depth of 40 feet to supplement the London water supply. It is said that in the event of other sources being cut off, the reservoir when full would hold sufficient water to supply London for six days. From these data ascertain the number of gallons consumed in London daily.

23. The value of woollen yarn exported from this country in certain years was:

Year.	Value in £.
1902	23,307,987
1903	25,385,941
1904	27,474,649
1905	29,916,807

Calculate the increase in value in 1903 as compared with 1902 as a percentage of the value in 1902, to the nearest integer. In the same way give the increase in 1904 as a percentage of the 1903 value, and the increase in 1905 as a percentage of the 1904 value.

24. In New South Wales the returns for 1903 showed that the number of sheep in the country was 26,656,000, and the value of wool *produced* £8,593,000. In New Zealand similar returns showed that the number of sheep was 18,955,000, and the value of the wool *exported* £6,041,000. Assuming that the value of wool produced per head of sheep was the same in the two States, calculate the percentage of the total wool production of New Zealand which was exported.

25. The table below gives the value of the wool exported in various forms from Canada, and of the wool exported in each form to the United Kingdom. Find, to the nearest whole number, the percentage of each form that is sent to the United Kingdom.

	Total exported.	Exported to United Kingdom.
Deals - - - - -	£11,087,715	£9,860,320
Planks and boards - -	12,568,991	1,623,292
Shingles - - - - -	1,525,386	0
Wood for pulping - -	1,315,038	120,445
Wood pulp - - - -	2,046,398	818,580

26. The area of Ireland, under flax and potatoes, in 1855 and in 1901 is given below. Find, to the nearest whole number, the decrease of area under each crop as a percentage of the area in 1855.

Year.	Acres under Flax.	Acres under Potatoes.
1855 - - - - -	97,075	982,301
1901 - - - - -	55,442	635,321

27. The following returns were made by a group of four mines in July 1906:

	Tons of Ore treated.	Yield of Gold in Oz.	Value in £.
Wolhuter - - -	18,700	6,260	26,338
Abosso - - - -	4,034	2,868	11,539
Rice Hamilton - -	5,035	1,456	4,560
Theta - - - -	1,325	949	3,479

Find, to the nearest integer, the average number of ounces of gold per hundred tons of ore for the group, and, to the nearest shilling, the average value of an ounce of the gold.

28. The traffic receipts from 52 English railways for the week ending 23 August 1908, were £2,401,244, and the number of miles of railway open was 21,201. For the corresponding week in 1907 the figures were £2,444,178 and 21,186. Find the decrease in the receipts per mile of railway correct to the nearest shilling.

29. The following table shows the number of persons engaged in certain important industries in the United Kingdom for the years 1891 and 1901:

	No. in 1891.	No. in 1901.
Agriculture - - - -	2,266,000	2,054,000
Building - - - - -	836,000	1,133,000
Mining - - - - -	707,000	839,000
Cotton - - - - -	625,000	606,000

The population of the United Kingdom was 38,104,975 in 1891, and 41,976,827 in 1901. Find, to the nearest integer, the number of persons per 10,000 of the population engaged in each of the above four industries in 1891 and in 1901; and express each of these numbers for 1901 (taken accurately, and not to the nearest integer) as a percentage, to the nearest integer, of the corresponding number for 1891.

30. The following table shows the value, in thousands of pounds, of merchandise exported from the United Kingdom in the years 1885 and 1904 respectively, the foreign trade being distinguished from the colonial trade:

	Exports, in Thousands of Pounds.	
Year.	To Foreign Countries.	To Colonies.
1885 - - -	185,985	85,490
1904 - - -	250,232	120,783

Find the increase per cent. during the period in question in the exports, first to foreign countries, and secondly to the Colonies. Also find by what per cent. the total exports increased in the same period. In all cases give your answer to the nearest tenth.

31. In the year 1902 British exports amounted to £283,423,966, and were distributed as shown in the table below:

Destination.	Value in £.
British Possessions (excluding India)	109,028,611
United States - - - - - -	23,760,913
India - - - - - - -	32,681,979
France - - - - - - -	15,587,300

Find, to the nearest whole number, the value of British exports to each of the regions named as a percentage of British exports to the whole world.

32. From the following data find, to the nearest pound, the imports per head in 1902; and find the same for 1903 on the supposition that the population has increased by 1 per cent.

Population in 1902.	Imports in 1902.	Imports in 1903.
41,976,827	£528,391,274	£542,600,289

33. Below are given the population and the value of the exports of the United Kingdom for certain years. Calculate, to the nearest shilling, the value of the exports per head of population for each year.

Year.	Population.	Value of Exports.
1881 -	- 35,241,482	£297,082,775
1891 -	- 38,104,975	309,113,718
1901 -	- 41,605,323	347,864,268

34. The Canadian Minister of Agriculture said some years ago: "During the past year, Great Britain proved the best customer for our farm products, taking 90 per cent. of our total exports. An ample margin remained, however, for the further extension of our trade in the United Kingdom, as for every hundred dollars' worth of farm products supplied to the mother country by Canada, other countries furnished her with about four hundred and forty-seven dollars' worth."

Taking the value of Canada's exports of farm products to Britain as 13·1 million pounds sterling, calculate approximately the value of farm products exported by Canada to all countries, and the value of farm products imported by Britain from all countries.

35. The income-tax on earned income is 9d. in the pound, on unearned income it is 1s. 2d. in the pound. If a man's gross income is £840, five-sixths of which is earned, what is his net income after paying income-tax?

36. Income-tax was at one time charged at the rate of 1s. in the pound; with the conditions that an income of less than £160 paid nothing, an income between £160 and £400 had the first £160 exempt and paid only on the excess above £160, an income between £400 and £500 had the first £150 exempt, between £500 and £600 had £120 exempt, between £600 and £700 had £70 exempt. Thus small incomes paid less than 1s. in the pound if the total tax they paid was reckoned on the whole income. Find, reckoned in this way, the rate paid by incomes of £170, £340, £510, £680, £850; giving each to the nearest tenth of a penny in the pound.

37. A man whose income is £490 pays income-tax only on the excess over £150, whilst another, whose income is £610, pays on the excess over £70. If the tax is at 1s. in the £, what is the difference in the virtual rate of tax each pays on his whole income?

38. In a certain district a man pays yearly an education rate of 1s. 4d. for every pound of his yearly rent. How much a year, and how much a week does a man pay for education who pays a weekly rent of 13s.?

39. The following figures are for the year 1908:

	County of London.	Westmoreland.
No. of acres	74,839	505,330
Population	4,536,541	64,409
Rateable value	£44,308,721	£580,893
Poor rate	£3,836,949	£13,302
No. of paupers	148,956	1,357

For each county calculate the following, each to two significant figures, and show your results in a table:

(i) The percentage of paupers to the whole population.
(ii) The percentage of the poor rate to the rateable value.
(iii) The average population per acre.

40. The table below shows the area and also the population and valuation in 1901 of the Borough of Poplar and the City of Westminster:

	Area.	Population.	Valuation.
Poplar	2,333 acres	168,822	£800,675
Westminster	2,555 ,,	183,011	5,516,725

Use the table to find for each of these places (i) the number of persons per acre, to the nearest integer, (ii) the valuation per head of population, to the nearest shilling.

41. A German workman earning £2 a week has to pay, per annum, income tax 36s., church tax 9s., and town tax 45s. How much, to the nearest farthing, does the total of these taxes amount to per pound of his income? The year may be taken as 52 weeks.

42. The national income for the year 1901–1902 was £143,581,542, and the main items were:

Income tax	£35,378,700
Excise	31,597,962
Customs	31,046,869

Express each of these items as a percentage of the whole, to the nearest whole number.

43. In a certain year the total revenue was £144,000,000, and the tax on tea produced £5,790,000. What per cent. of the whole revenue, to the nearest whole number, was produced by the tea tax?

44. For the year 1903–4, of the gross National Revenue, England and Wales contributed £121,658,500; Scotland, £15,526,000; and Ireland, £9,748,500. The following sums were spent on local services (civil government, tax-collecting, etc.), in the three countries: In England and Wales, £38,850,000; in Scotland, £5,377,000; in Ireland, £7,548,000.

Find the net balance available for Imperial Services, and the percentage of this net balance derived from each of the three countries.

Taking the population of England and Wales to be 34,000,000; of Scotland, 4,500,000; and of Ireland, 4,400,000; find, to the nearest shilling, the gross amount of revenue per head for each country.

Section 11. Averages.

1. The weights of the Oxford and Cambridge crews one year were as follows:

	Oxford.		Cambridge.	
	st.	lb.	st.	lb.
Bow - - - -	10	9	10	9
No. 2 - - - -	11	5½	12	9
,, 3 - - - -	12	0	12	4
,, 4 - - - -	12	0	12	4½
,, 5 - - - -	13	10½	13	1
,, 6 - - - -	13	2	12	7
,, 7 - - - -	11	4	12	3
Stroke - - - -	11	1½	10	4

Which was the heavier crew, and what, to the nearest half pound, was the average weight of each? (See the Tables.)

2. In 1903 there were, at a certain place, 196 days on which no rain fell, and the total rainfall recorded during the year was 47·12 inches. Find, to two places of decimals, the average fall on the days on which there was rain.

3. The figure shows a metal plate full size. Find, in inches, the average width of the plate by adding together the lengths of the lines *aa*, *bb*, *cc*, ... *kk*, and dividing the sum by the number of the lines. Find also the area, in square inches, of the plate by multiplying the average width by the length.

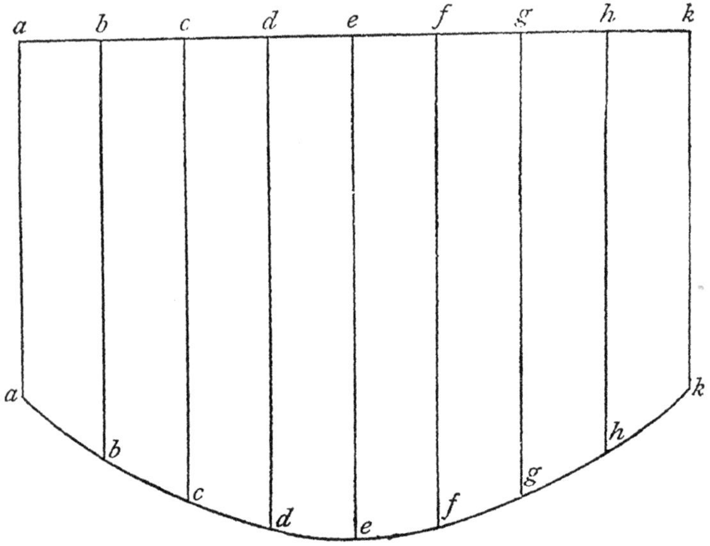

Fig. for Qn. 3.

4. The table below shows the performance of two bowlers in a certain cricket match. Find for each bowler how many runs a wicket cost him in the first innings, in the second innings, and in the match as a whole. Which is the better bowler judged by the first innings, by the second innings, and by the match as a whole?

		Number of Wickets taken.	Number of Runs made off him.
1st innings	Odell	4	84
	Thompson	2	45
2nd innings	Odell	6	54
	Thompson	6	59

5. The following table gives the results of an inquiry into the average weekly incomes of workmen's families in London and suburbs:

No. of Returns.	Average Weekly Income. s.	d.
53	21	5
49	26	$7\frac{3}{4}$
80	32	$1\frac{1}{4}$
69	36	$1\frac{1}{4}$
96	52	3

What was the average weekly income for all the families taken together?

6. From a return as to the wages of railway men, it appears that:

		£	s.	d.
111,086	earn an average weekly wage of	0	17	6
76,770	,, ,,	1	2	6
38,441	,, ,,	1	7	6
15,045	,, ,,	1	12	6
6,598	,, ,,	1	17	6
9,891	,, ,,	8	0	0

What is the average weekly wage of the whole lot?

7. From 1 April 1908, to 31 March 1909, the number of letters, etc., delivered by the Post Office were estimated as follows:

England and Wales	$2{,}482\frac{6}{10}$ millions.
Scotland	$256\frac{9}{10}$,,
Ireland	$167\frac{9}{10}$,,

and the average numbers of letters, etc., delivered per head of the population as:

England and Wales	70
Scotland	53
Ireland	38

From the data estimate the number of letters delivered per head of the whole population of Great Britain and Ireland combined.

8. The following table gives the population and birth-rate per 1000 for the three greatest cities of the United Kingdom:

	Population.	Birth-rate.
London - - -	4,613,812	28·4
Glasgow - - -	786,897	31·9
Liverpool - - -	716,810	33·4

Find the birth-rate for the three cities taken together.

9. The following table gives the number of persons engaged in textile trades whose rates of wages were increased or decreased during each of the four years 1894–1897:

	Increases.			Decreases.		
Year.	Number of Persons affected.	Total Rise of Weekly Wages.		Number of Persons affected.	Total Fall in Weekly Wages.	
		£	s.		£	s.
1894 - -	8,662	585	7	3,936	250	17
1895 - -	10,192	572	14	5,396	420	18
1896 - -	7,122	334	17	2,834	288	19
1897 - -	3,894	266	6	4,115	189	13

From the above data find the average net increase of weekly wages *per head* of those affected for each of the four years. Give your results to the nearest farthing.

10. A batsman has an average of 34·31 for 26 completed innings. In the next match he plays two complete innings and scores 99 and 12. What is his average on the 28 innings?

11. A man's average annual expenditure for the five years, 1900 to 1904, was £462. 6s.; he spent £335. 10s. in 1900, and £589 in 1905. What was his average expenditure for the five years, 1901 to 1905?

12. The temperature averaged 63° F. for the first six days of a week, for the Saturday alone it averaged 69° F. Find what it averaged for the whole week.

Section 12. Percentages.

1. A tradesman's takings for a week are £25; what is his net profit if he marks his goods 20 per cent. above cost price and pays for rent, labour, etc., 10 per cent. of his takings?

2. The population of a certain village is 7516 and it increases at the rate of 5 per cent. per annum. What will the population be, to the nearest thousand, at the end of 4 years?

3. The population of a town is increasing uniformly, *i.e.* the increase in each period of ten years is a constant fraction of the population at the beginning of that period. If the population was 172,000 in 1891 and 193,000 in 1901, what was it in 1911 and what was it in 1871?

4. A man firing at a target makes 11 per cent. misses; 12 per cent. outers, which score 3 points; and 13 per cent. inners, which score 4 points; and places the rest of his shots on the bull's eye, scoring 5 points each. What percentage of the greatest possible number of points does he score?

5. The price of flour is 28s. 6d. per sack of 280 lb. In making bread 2 parts by weight of flour are mixed with 1 part of water to make dough, and during the baking some of the water evaporates, so that 117 lb. of dough make only 100 lb. of bread. If rent, labour, firing, etc., come to 30 per cent. on the cost of the flour, and a 2 lb. loaf is sold for $2\frac{3}{4}$d., what percentage of this $2\frac{3}{4}$d. is profit?

6. An ore from which lead and silver are extracted contains 85 per cent. of these metals. How many hundredweight of metal will be obtained from 500 tons of ore, supposing that 90 per cent. of the metal in the ore is extracted?

If the metal extracted contains 4·5 per cent. of silver, and if no silver (but only lead) was left in the ore, what percentage of silver existed in the ore?

7. Explain clearly what is meant by saying that A is 5 per cent. greater than B.

The population of a country increases by 14·5 per cent., while the expenditure to be met by taxation increases 23·2 per cent. Find, to the nearest integer, the increase per cent. in the average taxation per head of the population.

8. A buys a horse and sells it to B, losing 10 per cent. by the sale. B resells the horse to A, and makes a profit of 5 per cent. in doing so. A then sells again at the price paid to B. Express A's gain or loss on the whole transaction as a percentage of the price he first paid.

9. The receipts and expenditure of a railway for 1907 have increased by 5 per cent. and 2 per cent. respectively, when compared with 1906. In 1907 the expenses were 62 per cent. of the receipts. What was the corresponding percentage in 1906?

10. 55 per cent. of the workers in a factory are men getting an average wage of 35s. a week each, and the rest are women getting an average wage of 22s. 6d. a week each. Find, to the nearest integer in each case, what percentage of the total wages earned is paid to (i) men, (ii) women.

11. Below are tabulated the total values of certain commodities imported during corresponding months of the years 1907 and 1908, together with the percentage increase or decrease in the quantities imported:

	1907.	1908.	Percentage Increase or Decrease in Quantity as compared with 1907
Bacon - -	£1,164,121	£1,236,807	11·0 increase.
Butter - -	2,384,943	1,997,546	19·9 decrease.
Wheat - -	2,268,151	3,145,264	5·8 increase.
Oats - - -	273,891	277,755	3·0 decrease.

Calculate *to two significant figures* for each commodity:

(i) The percentage increase or decrease in the total value imported.

(ii) The total value for 1908 reckoned at 1907 prices.

12. An English manufacturer makes an article, the catalogue price of which is £10. 10s. An American firm buys these articles at a discount of $33\frac{1}{3}$ per cent. from the catalogue price, and has to pay an import duty of 40 per cent. on the *net* cost of the article to them. Their ideal being to make a profit of 20 per cent. on their total outlay, what should they charge for the article?

As a matter of fact, they charge the manufacturer's catalogue price plus the import duty thereon; what percentage of profit on their total outlay does this give them?

13. A manufacturer gives a shopkeeper a discount of 25 per cent. from the ordinary prices, and then makes a profit of 15·7 per cent. Find, to the same degree of accuracy, his profit when selling direct to a customer who is allowed no discount.

14. A manufacturer increases the cost price of his goods by 60 per cent., and sells them at the price thus obtained to a retail dealer, allowing him 5 per cent. discount off this price for cash. The dealer puts a profit of 40 per cent. on his net outlay, and allows his customer 10 per cent. discount for cash. By how much per cent. is this selling price greater than the cost price of manufacture?

15. One cubic centimetre of concentrated sulphuric acid weighs 1·842 grams. If 70 parts by volume of this acid are mixed with 30 parts of water, it is found that one cubic centimetre of the mixture weighs 1·615 grams. Find how much per cent. the mixture has shrunk. (See the Tables inside the front cover).

16. Barley weighs 56 lb. a bushel, and when made into malt loses 15 per cent. of its weight and increases 3 per cent. in bulk. What is the bulk of 1 cwt. of malt?

17. In a certain year 56,000 tons of ore were taken from a gold mine and yielded 14,000 ounces of gold. If the next year the output of ore was increased by 20 per cent. and the yield of gold per ton of ore was also raised by 2 per cent., what number of ounces of gold would be obtained?

18. A rectangular box is increased 20 per cent. in length and 24 per cent. in breadth, but diminished 16 per cent. in height. Find by how much per cent. its volume is increased or diminished.

19. An employer finds that one result of reducing the hours of labour from 9 hours a day to 8 is to increase the amount of work done per hour by 8 per cent. How should the rate of wages per day be increased or decreased in order that the employer may pay the same sum for a given amount of work done?

20. In an arithmetic examination 75 per cent. of the candidates are boys and the remainder girls; 40 per cent. of the boys and 44 per cent. of the girls are successful. If 177 candidates failed, what was the total number of candidates who entered?

Section 13. Approximation.

1. Evaluate

(i) $26{,}607 \times 1{,}493{,}721$ to the nearest million;

(ii) $\dfrac{1}{266{\cdot}07 \times 0{\cdot}1493721}$ to the nearest thousandth;

(iii) $\dfrac{1}{0{\cdot}26607} + \dfrac{1}{14{\cdot}93721}$ to three significant figures.

2. For finding the volume in cubic feet of an oil drum the following rules are used:

(i) Multiply the square of the radius by the height, both in feet, and the product by $3\frac{1}{7}$.

(ii) Multiply the square of the diameter by the height, both in inches, and divide by 2200.

(iii) Multiply the square of the diameter by the height, both in feet, and the product by 0·7854.

Apply these rules to the case in which the diameter and height are each 1 foot, and compare with the result given by the Tables inside the front cover to find which of the rules is the most accurate.

3. Evaluate as shortly as you can the three most important figures in the product of 4680·13576 and 286423·5791, stating shortly the reasons for your procedure.

4. Work the following questions as shortly as you can:

(i) Find the cost of 84,127 books at 1s. 8d. each.

(ii) Simplify $55{\cdot}23 \times 1\frac{1}{8} + 8{\cdot}77 \times 1{\cdot}125$.

(iii) Evaluate 7·69569 of 6s. 8d. correct to the nearest shilling.

(iv) Find the total cost of 38 lb. of coffee at 1s. $2\frac{1}{2}$d. per lb. and 76 lb. of chicory at $7\frac{3}{4}$d. per lb.

5. The five left-hand significant figures in the product of 0·013057 and 23·7814 are given in a table of products as 31051. How many more significant figures would be required to give the exact product? Write down the product correct to three places of decimals.

Find, to two decimal places, the remainder when 4·37214 is divided by 0·257819.

6. The diameter of a circular brass disc, cut out of a sheet 5 millimetres thick, is known to lie between 3·48 and 3·52 centimetres. If its volume is calculated on the assumption that the diameter is 3·5 centimetres, what is the greatest possible error in the result?

7. The straight lines A and B give the length and breadth of a rectangle; measure these lines correct to the nearest millimetre, and find the area of the rectangle in square centimetres.

If your measurements do not differ from the true measurements by more than half a millimetre, show that the error in your result is not more than about one square centimetre.

Fig. for Qn. 7.

8. The length and breadth of a room, *measured to the nearest decimetre*, are 17·4 and 7·8 metres. Between what limits do the actual length and breadth lie? Calculate the area, and state the degree of accuracy of your result.

9. The figure gives the end and side view of an oil drum on a scale of $\frac{1}{10}$. How many gallons will the drum hold? (See the Tables.)

Assuming that each of your measurements may be wrong by $\frac{1}{50}$ of an inch, find how much per cent. your result may be wrong.

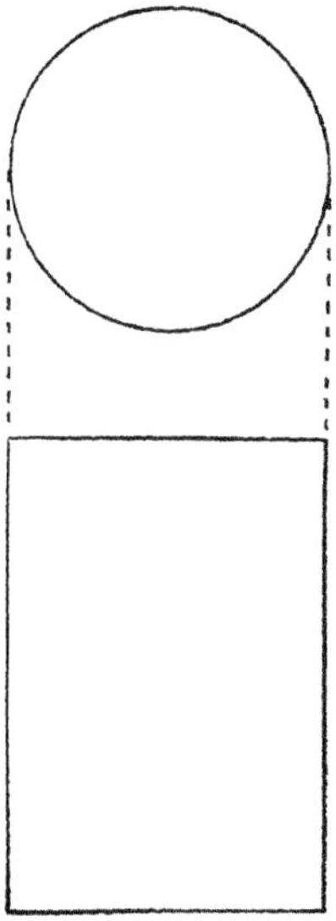

Fig. for Qn. 9.

10. In order to determine the volume of a cube and the area of one face, a boy measures the edge of the cube and finds it to be 5·6 cm. His measurement may be as much as 0·1 cm. too great or too small. Find the greatest error that might arise in (i) the area and (ii) the volume, calculated from his measurement. In view of this possible error, to what degree of accuracy should (i) the area, (ii) the volume, be given?

11. When a circular piece of a certain paper, of diameter $18\frac{1}{2}$ cm., is burnt, the weight of the ash is 0·00605 gram. Find the weight of the ash of a circular sheet of the same thickness and material whose diameter is $29\frac{1}{2}$ cm. The areas of circles vary as the squares of their diameters.

If the weight of ash from the latter is catalogued as 0·01355 gram, find by how much per cent. the calculated result differs from that given in the catalogue.

12. The following rules for conversion of units are sometimes employed: Having given that 1 kilogram is 2·204 pounds, find the percentage error in each case:

(i) To convert ounces to grams multiply by $\frac{85}{3}$.

(ii) To convert pounds to kilograms multiply by $\frac{5}{11}$.

13. A man measures the length and breadth of a room by means of a walking-stick and estimates them at 15 feet and $12\frac{1}{2}$ feet. In making the calculation he assumes that this stick is 3 feet long. The actual length is 2 ft. 9 in. Find the errors in the measurements of the room arising from this mistake, and the error in the area to the nearest square foot.

14. A man measures the length and breadth of a rectangular field by pacing, and takes the length of his pace as 30 inches. He makes the area 2 acres 3 roods. His pace is really 32 inches. Find the true area.

15. The area of the cross-section of a piece of wire is to be determined by dividing its volume by its length, these quantities having been measured. It is known that the error of measurement of the volume is not more than 5 per cent. of its true value, and that of the length not more than 3 per cent. of its true value. Find, to one decimal place, by how much per cent. of its true value the calculated value of the area may be in error (i) in excess, (ii) in defect.

16. Between 3 and 4 per cent. of milk is fat. Of a mixture of milk and water 2·9 per cent. was fat. Within what limits did the percentage of milk in the mixture lie?

17. A motor car was timed over a measured quarter mile, and the time given as 34 seconds. The error made in taking the time may have been 1 second. Within what limits does the speed, in miles an hour, lie?

18. The New York Department of Agriculture estimated the total farm wealth for the year 1905 at $6,415,000,000. This total is stated to be 4 per cent. higher than in 1904, 8 per cent. higher than in 1902, and 36 per cent. higher than in 1899, each of these percentages being reckoned on the earlier year concerned. Calculate the total farm wealth for each of those years.

Assuming that the percentage increase on 1904, as stated above, is correct to the nearest integer only, what are the limits of error in your result for that year?

19. The total amount of tea imported into the United Kingdom in a certain year was 216,568,503 lb. The percentage from China was 13, to the nearest whole number; and the percentage from India 53, to the nearest whole number. The remainder came from Ceylon. Give the amount from Ceylon in pounds to as many figures as are reliable.

20. Calculate the values of the crops of wheat, barley, and oats grown in Great Britain in 1904 from the table:

	Crop in Bushels (8 to a quarter).	Average Value per Quarter to the nearest Penny.
Wheat - - -	36,880,246	30s. 5d.
Barley - -	57,193,067	24s. 4d.
Oats - - -	127,407,848	16s. 1d.

21. The following table gives the mileage and annual cost of certain urban main roads in the County of Bedfordshire:

	Miles.	Fur.	Ch.	Cost.
Bedford - - - - -	9	5	6	£3,000
Biggleswade - - -	10	6	5	1,575
Dunstable - - - - -	2	0	1	353
Leighton Buzzard - -	8	1	4	1,225
Luton - - - - - -	16	2	6	3,300

Calculate the average cost per mile in each case. You may assume that the distances are correct to the nearest chain.

22. A formula in use by enginners is that "the weight of a rail in pounds per yard $\times 1{\cdot}571$ = the weight of rails for a mile of single line of railway in tons." Find out whether this is accurate.

23. For every sovereign which a money-lender lends to his clients over a fixed period he reckons that he receives at the end of this time £1·2155062. What sum, to the nearest pound, will repay fully a loan of £1534 for this period?

Test the accuracy of your result by a rough check.

Have you any criticism to offer on the first number, £1·2155062?

24. The volume of a segment h cm. high of a sphere of r cm. radius is $\frac{\pi h}{3}\{3rh - h^2\}$ c. cm. How many litres of milk will fill to a depth of 12 cm. a hemispherical bowl of 30 cm. diameter?

The following rule for finding approximately the volume of a sphere is sometimes given: "Take half the volume of the cube on the diameter and add 5 per cent. to it." Show that this rule is unreliable when more than two significant figures are required in your answer.

25. Having to find the value of 0·539 of £184. 7s. 6d. to the nearest pound, a clerk assumed that he would get the right result by finding the value of 0·54 of £184 to the nearest pound. Show that he was right, and explain why he was.

26. A and B, in the figure, are supposed to be two places in a map whose scale is $\frac{1}{10000}$. Find, to the nearest minute, by measurement and calculation, the time that would be taken to walk straight from A to B at 5 kilometres an hour. Also find whether the area represented by the whole figure is more or less than 1 sq. km.

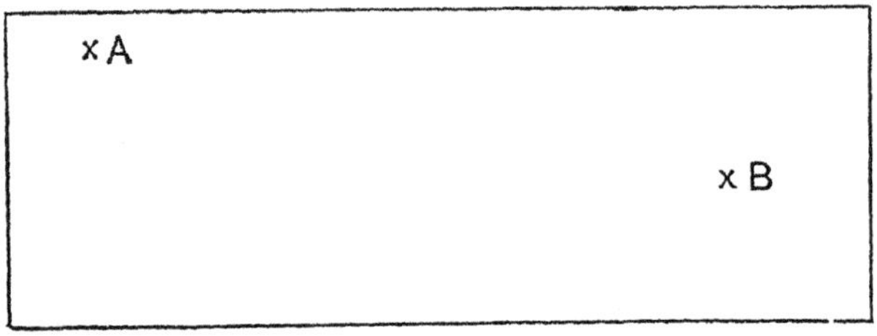

Fig. for Qn. 26.

27. Do not work the following questions, but give a reason for thinking that the answers suggested are *obviously* wrong:

(i) Multiply 573 by 963. *Ans.* 550,799.

(ii) Multiply $9\frac{31}{57}$ by $6\frac{12}{17}$. *Ans.* 69.

(iii) £310. 12. 6d. ÷ £31. 1s. 3d. *Ans.* 10s.

(iv) What must the width of a field, 124 yards long, be if it contains the same area as a field measuring 103 yards 1 foot 3 inches by 116 yards? *Ans.* 107 yards and a fraction of a yard.

(v) How much tea at 1s. 7d. a lb. must be mixed with a cwt. of tea at 2s. 6d. a lb. in order that the mixture may be worth 1s. 11½d. a lb? *Ans.* 49 lb.

(vi) Which is the greater—26 cubic feet, or a cubic metre? *Ans.* 26 cubic feet.

(vii) The average age of the 169 boys in a school is 10 years 1 month, while the average for the 374 boys and girls is 9 years 3 months. Find the average age for the girls. *Ans.* 9 years 7 months.

28. I wish to form a rough estimate of the acreage of a rectangular field by stepping along two adjacent sides. If my steps average 33 inches, and the two sides are 57 and 159 steps respectively, show that the field contains about 1½ acres.

29. Find, with as little work as possible, whether a stack of coal, 74 feet long, 24 feet wide, and 12 feet high, contains as much as 300 tons of coal. A ton of coal occupies 43 cubic feet.

30. The figure shows a lawn on a scale of 1 inch to 40 feet. Find whether there is room for a lawn tennis court, reckoning

Fig. for Qn. 30.

that for the court and space round it a length of 100 feet and a breadth of 44 feet are necessary.

31. A journey between two places can be taken either by land or by water. Travelling by land at an average speed of 35 miles an hour it takes 18½ hours; travelling by water at an average speed of 20 knots it takes 22 hours. A knot is a speed of

1·15 miles an hour. Find, with as little work as possible, whether the distance is greater by land or by water.

32. Find out, with as little work as possible, whether a 2-inch pipe through which water is running at 4 miles an hour is large enough to fill a bath holding 15,000 gallons within 10 hours. A 2-inch pipe has, roughly, a cross-sectional area of $3\frac{1}{7}$ square inches, and a cubic foot is 6·23 gallons.

Section 14. Arithmetical Graphs.

1. The figure shows the yearly importation of frozen beef for the years 1895–1904. The length of the rectangle for any year shows the amount for that year, 1 cm. representing 50,000 quarters of beef. Find the amount imported in the whole period, and express the least and greatest amounts imported in a year as decimals of the amount imported in the whole period.

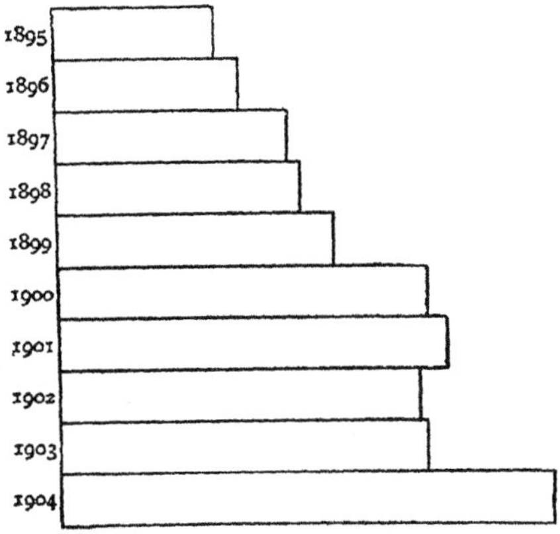

Fig. for Qn. 1.

2. The figure (p. 79) gives in shillings per ton the price of a certain iron ore during 1904 and 1905. What were the highest and lowest prices within this period, and when were they reached? At what other times did the price (i) cease to rise and begin to fall, (ii) cease to fall and begin to rise? When was the price rising fastest?

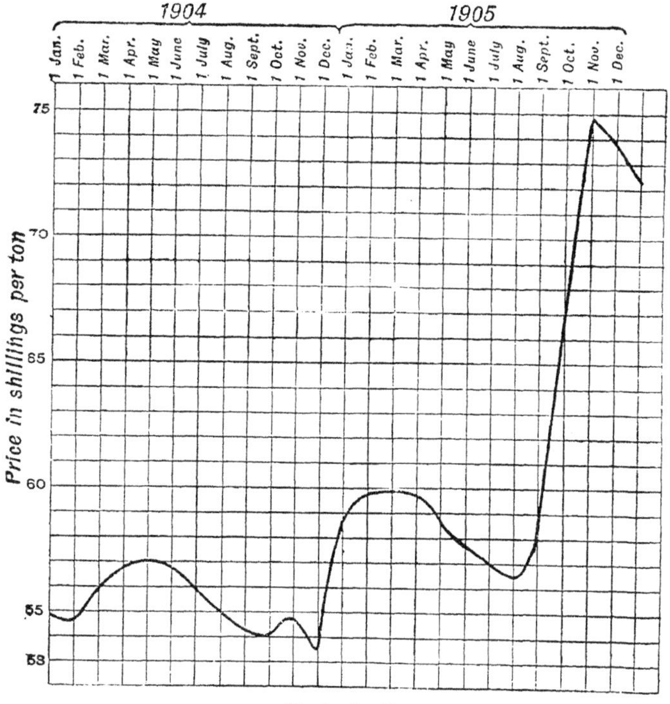

Fig. for Qn. 2.

3. From the corners of a square of cardboard measuring 30 cm. in the side, four equal small squares (shown shaded in the sketch) are to be cut away, and the remainder is to be made into a lidless box by folding along the dotted lines. The volume of the box will depend on the size of the corner squares cut away; calculate the volume, taking in succession the length of the side of a corner square as 2, 4, 6, 8, 10, 12, 14 cm.

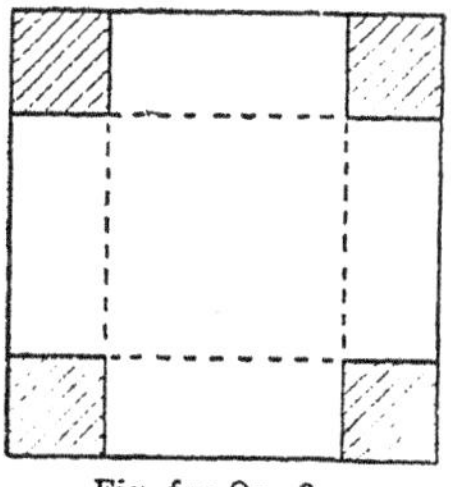

Fig. for Qn. 3.

Mark off along the bottom line of a page of squared paper, distances 2, 4, 6, ... cm. At the distance 2 cm. measure off up the page a line to represent the volume of the box when the side of the squares cut away is 2 cm., using 1 cm. to represent a volume of 100 c. cm. Do the like for the distances 4, 6, ... cm.; and through the ends of these lines measured up the page draw a smooth curve.

This curve shows the volume of the box for any size of corner square. From it find what size of corner square gives the box the greatest volume. Find also what this volume is.

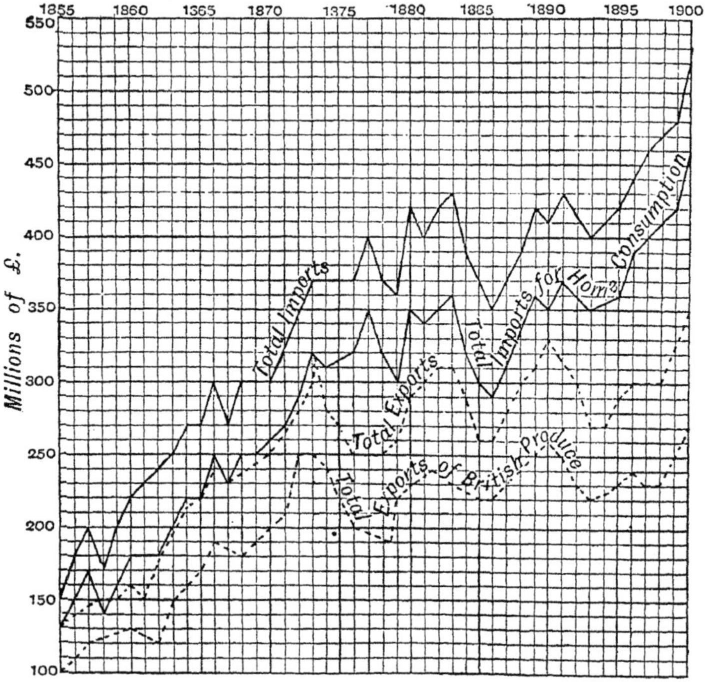

Fig. for Qn. 4.

4. From the diagram, giving the value of imports and exports in various years, find, for the year 1875, what proportion of

imports were for home consumption, and what proportion of exports were British produce. Find also whether, in the period 1855–1900, the exports of British produce increased in a greater or in a less ratio than the exports of goods that had first been imported.

5. The figure gives curves showing (i) the population of England and Wales at any time during the years 1899 to 1903,

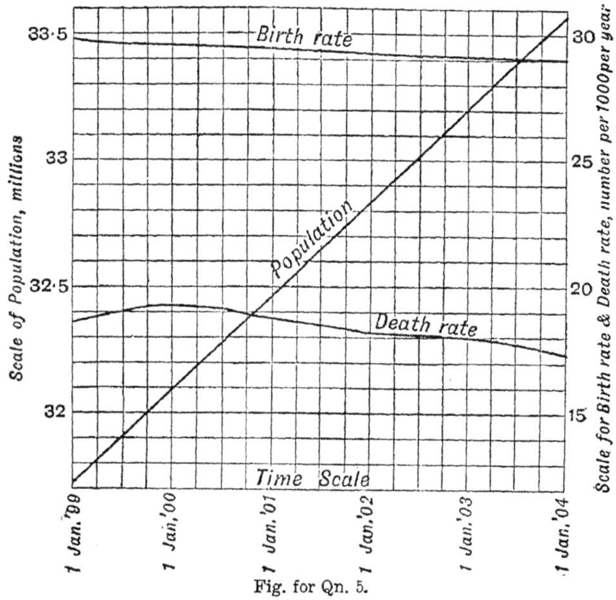

Fig. for Qn. 5.

(ii) the birth-rate per 1000 of population per year for the same period, and (iii) the death-rate per 1000 of population per year for the same period. Read off the population on 1st January 1902 and on 1st January 1903. Read off the birth-rate at each of those dates, and supposing the birth-rate for the year 1902 midway between these, find the number of children born in that year. Similarly, find how many people died in that year. Do the births and deaths account for the change of population in that year?

6. A sum of money is borrowed, to be repaid by a number of equal annual instalments. The amount of the annual instalment depends on the rate of interest and on the number of years over which instalments will be spread. The curve in the figure enables you to find the amount to repay a loan of £100 when interest is reckoned at $3\frac{1}{2}$ per cent. It is used in the following way: Trace the figure on to transparent paper. Divide OX in twenty-fifths of an inch and mark it in years, so that, for example, a point Q 24/25 of an inch from O is marked 24 years. From Q draw QP at right angles to OX and meeting the curve in P. Find how many fifths of an inch QP contains; this number is the number of pounds in the annual instalment when payments are to go on for 24 years; and so for any number of years.

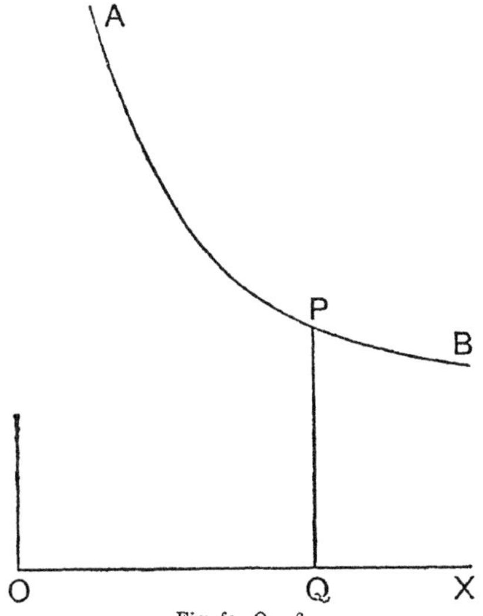

Fig. for Qn. 6.

What is the amount of the instalment if payments are to go on 30 years? If for 40 years? And for how many years must a payment of £640 go on to pay off a loan of £10,000?

7. By the following method determine which of the following express runs is the fastest: (1) 36 miles in 39 minutes; (2) $25\frac{1}{4}$ miles in 28 minutes; (3) $48\frac{1}{2}$ miles in 55 minutes; (4) $34\frac{1}{8}$ miles in 36 minutes.

Take a straight line OX parallel to the long edge of a page. Mark off from this line Op to represent 39 minutes; at p draw a line pq perpendicular to OX of a length to represent 36 miles. Join Oq. Treat each of the other runs in the same way.

Scales: 1 cm. to represent 4 minutes.
1 cm. " 4 miles.

8. A train leaves Paddington at 4.15 p.m. and arrives at Bath at 6.6 p.m. Another train leaves Bath at 4.45 p.m. and arrives at Paddington at 6.50 p.m. At what time will they pass each other? Assume that each train travels uniformly. A graphical solution is permissible.

9. Three places A, B, C on a railway are passed by a train at 9, 9.40, and 10.32; another train passes A and B at 9.50 and 10.16; both trains travelling uniformly. At what time should the second train pass C? If the first is travelling 24 miles an hour, what is the rate of the second? You may work by squared paper if you choose.

10. A thief steals a motor car at 7 p.m. and drives off at 25 miles an hour, followed 10 minutes later by the owner in another car; the owner travels at 24 miles an hour while driving, but spends 15 per cent. of his time in making enquiries. The thief, after driving 15 miles, transfers the car to an accomplice, who drives 6 miles at 36 miles an hour and then has a breakdown. After spending 15 minutes in repairing the machine, he finds that he can drive it at only 16 miles an hour. At what time will he be overtaken? This question may be solved graphically or otherwise.

Section 15. Square Root and Logarithms.

SUBSECTION (i). *The questions in this Subsection should be answered without the use of logarithms.*

1. Find the square root of 9,042,049.

2. Find, to three significant figures, the value of $\sqrt{879}$, and also find, to two significant figures, by how much per cent. the square of this approximation differs from 879.

3. Find the value of

$$\frac{\sqrt{13}-\sqrt{11}}{\sqrt{13}+\sqrt{11}-6}$$

correct to five significant figures.

4. It has been suggested that the income tax should be proportional to the square root of the cube of the income, in such a way that a man whose income is one pound per week should pay twopence per week tax. Find, to the nearest tenth of a penny, how much tax would be paid per pound per annum on incomes of £300 and £1200 per annum. Take a year as 52 weeks.

5. Work the following correct to two significant figures, using contracted methods if possible:

(i) $\frac{1}{16} \times 3{\cdot}142$,

(ii) $(0{\cdot}0625)^2 \times 0{\cdot}7854$,

(iii) If $d=\sqrt{0{\cdot}198 \div 0{\cdot}7854}$, find the numerical value of d.

6. A length of 10 feet of a water pipe, $\frac{3}{4}$-inch in diameter, freezes. In freezing any volume of water expands to 1·087 times that volume. Sometimes this freezing bursts the pipe, sometimes the ice occupies an additional length of pipe. What additional length of pipe (in inches) is necessary in the present case? If no additional length of pipe is available, to what diameter does the water expand in freezing?

7. It is proposed to make an ornamental lake of $5\frac{1}{2}$ acres of the shape shown in the figure (p. 85), and an average depth of 4 feet. The cost of excavation and carting is estimated at 1s. per cubic yard. Find the cost of making the lake.

A streamlet, 15 inches across and 2 inches deep, running at one mile an hour, is to be diverted to run through the lake. How many gallons of water per day will enter the lake for every thousand gallons in the lake?

Find, approximately, in any way you please, the scale of the figure in chains to the inch. An acre is 10 square chains.

8. A cubical block of metal, whose edge is 25 centimetres long, weighs 8 grammes per cubic centimetre. A hole of square sectional area is to be cut completely through the metal perpendicular to a face of the cube in order that the weight of the metal left may be 100 kilograms. Find, to the nearest millimetre, the side of the square section.

9. The time of oscillation of a pendulum is proportional to the square root of its length. If the length of a seconds pendulum (whose time of oscillation is 1 second) is 39 inches, find the time of an oscillation when the length is decreased by 3 inches.

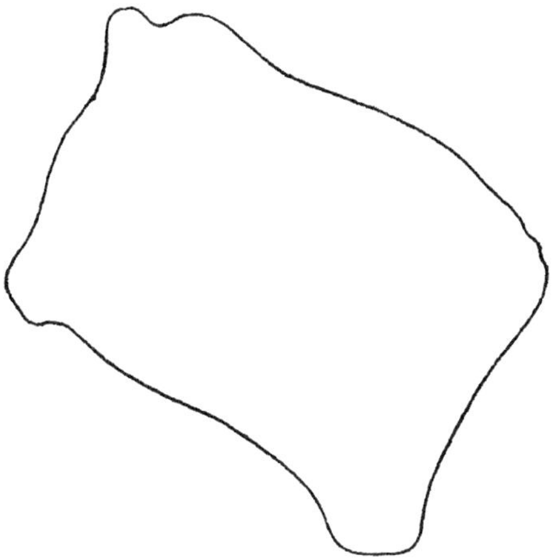

Fig. for Qn. 7.

10. A space in the shape of a broad ring is bounded by two concentric circles, radii 2 cm. and 10·5 cm. Find the radius of the smaller of the two circles which will divide it into 3 parts of equal area? (Nearest tenth of a millimetre.)

SUBSECTION (ii). *In this Subsection the use of logarithms is optional, except where otherwise specified.*

11. Light travels 186,000 miles per second; find how long the sun's light takes to reach the earth if the sun is 93,165,000 miles away. Discuss the accuracy of your answer if the data are only just reliable to the last significant figure.

12. Find from your tables $\log 3167$ and $\log 82{\cdot}4$. Hence find $\log(3167 \times 82{\cdot}4)$ and $\log(82{\cdot}4/3167)$, and then by means of your tables state the values of $3167 \times 82{\cdot}4$ and $82{\cdot}4/3167$ to 3 significant figures.

13. The population of England and Wales has increased in the successive decennial periods 1801–11, 1811–21, etc., by the following percentages: 14·00, 18·06, 15·80, 14·48, 12·89, 11·90, 13·21, 14·36, 11·65, 12·17, each percentage being reckoned on the population at the beginning of the decennial period concerned. The population was 32,527,843 in 1901. Find the population in 1851 and in 1801 to the nearest hundred-thousand.

14. It is known that $\sqrt{(a+b+c)(b+c-a)(c+a-b)(a+b-c)}$ is either twice or four times the number of square centimetres in the area of a triangle whose sides are a, b, and c centimetres long. Find whether it is twice or four times by applying it to the triangle BCD in the figure below.

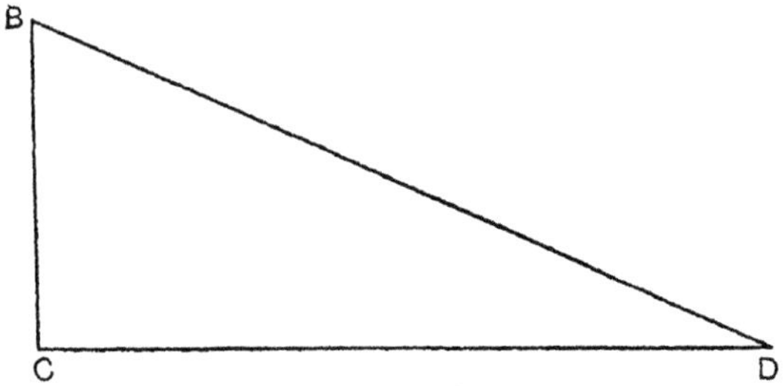

Fig. for Qn. 14.

15. The values of a, b, c are known approximately, viz.: $a = 3·741$, $b = 53·92$, $c = 0·04168$. Calculate, as accurately as the data permit, the value of $ab/\sqrt{c}$ and c/ab.

16. The area of a triangle is given by the formula

$$\sqrt{s(s-a)(s-b)(s-c)}$$

where s is half the perimeter of the triangle and a, b, c are the lengths of the sides. By the aid of this formula and a table of logarithms, find, to the $\frac{1}{100}$ part of an inch, the length of the side of a square equal in area to a triangle whose sides are 3·7, 2·4, and 1·9 inches.

17. Logarithms should be used in both the calculations of this question; good answers obtained without their use will be valued on a somewhat lower scale.

If h is the length and d the diameter of a piece of wire, both in inches, the volume of the wire in cubic inches is given by the formula $\frac{1}{4}\pi d^2 h$, in which π denotes 3·1416. Find the number of feet of wire 0·04 of an inch in diameter that can be made from a cubic foot of metal.

Also find the diameter of a wire of which a length of one mile can be made from a cubic foot of metal.

18. Find the diameter of the solid sphere that could be made out of 10,000 tons of wrought iron. The weight of a cubic foot of wrought iron is 480 lb. The volume of a sphere of diameter d is $\frac{1}{6}\pi d^3$.

19. Find by the aid of logarithms the edge of a cube equal in volume to a right circular cylinder whose height is 13·72 cm. and the radius of whose base is 3·94 cm.

20. The weight of the projectile of a gun whose calibre (internal diameter) is d inches is kd^3 lb., where k is constant.

Calculate, by your tables, or in any other way, the calibre of a gun whose projectile is twice as heavy as that of a gun whose calibre is 10 inches.

21. For an open rectangular cistern of given internal surface area the volume is greatest when the bottom measures each way twice the height. How many gallons does a cistern of this shape and 2 ft. 3 in. high hold, and what must be the height of one that will hold 1000 gallons?

22. If a water-pipe is L yards long, d inches in diameter, and one end is H feet higher than the other, then $\sqrt{(3d)^5 \times H \div L}$ gallons of water will flow through the pipe in a minute. Use this formula to find how many gallons per minute will flow through a pipe a mile long, $4\frac{1}{4}$ inches in diameter, one end being 38 feet higher than the other.

23. A circular drain whose fall is 1 in 100 and whose diameter is D feet can carry off $\dfrac{A \cdot D^3}{C + \sqrt{D}}$ cubic feet of water per minute, where $A = 454{\cdot}3$ and $C = 1{\cdot}005$. Calculate the volume carried off by drains of diameter 6 inches, 12 inches, 18 inches, and 24 inches. Use these values to draw a graph showing the volume carried off by drains of different diameters, and determine by means of your graph the diameter of a drain which will carry off (i) 100 cubic feet per minute, (ii) 500 cubic feet per minute, (iii) 1000 cubic feet per minute.

24. In finding the volume of a stream, it is customary to erect a temporary weir and measure the depth of the water above the sill of the weir, and the velocity of the water approaching the sill.

The following formula may be used: Cubic feet discharged per minute over each foot width of the sill is $214\sqrt{H^3 + 0\cdot035V^2H^2}$, where H is the depth in feet of the water above the sill and V the velocity in feet per second.

Calculate the quantity of water discharged per minute over a sill 6 feet wide when the depth of the water is 3·1 in. and the velocity 1700 feet per hour.

25. A triangular field ABC is to be levelled. The side BC is level and A is 2·3 metres higher than BC. Find how much the field must be lowered at A, and how much it must be raised along BC.

Section 16. Speed and Flow.

1. Find the average rate per hour in the following non-stop railway journey: London to Crewe (L.N.W.R.), distance 158 miles. Train leaves Euston 12.10 p.m., reaching Crewe 3.5 p.m.

2. Motor cars are allowed in this country to travel at the rate of 20 miles an hour. Express this rate in feet per second (to the nearest foot).

3. In certain motor-car trials one car travelled a kilometre in $21\frac{2}{5}$ seconds. What was the average speed of the car in kilometres an hour? Give your answer to the nearest whole number.

In later trials another car travelled the same distance in $20\frac{3}{5}$ seconds. Express the increase in the speed as a percentage, to the nearest whole number, of the speed of the first car.

4. Which goes the faster, a train travelling at 40 miles an hour, or a turbine steamer travelling at 33 knots? A knot is a speed of 6080 feet per hour.

5. Two railway companies announce non-stop runs of 164 miles in 2 hours 53 minutes, and 147 miles in 2 hours 38 minutes respectively. Which of these is the faster run, and what, to the nearest tenth of a mile an hour, is the difference of the average speeds of the two runs?

6. An express leaves Geneva at 8.5 p.m. What must be its average speed in kilometres per hour (to the nearest integer) that it may reach Paris, 625 kilometres away, at 6.45 a.m. the following morning?

If its average speed is 2 kilometres per hour short of this, how much will the train be late on arrival?

7. The following table gives the scheduled times of a certain train:

Station A	-	depart 5.35
,, B, 48 miles from A	-	arrive 6.30
,, B	-	depart 6.32
,, C, 50 miles from B	-	arrive 7.28

Find the average value of the scheduled speed of running for each section of the journey and for the whole journey. The average speed for the whole journey is the quotient of the distance by the difference between times of departure and arrival.

In an actual journey the train leaves A 5 minutes behind time, arrives at B 10 minutes behind time, stops 3 minutes at B, and arrives at C 4 minutes behind time. Find the average value of the speed of running for each section, and for the whole journey, as before.

8. Sound travels at the rate of 1100 feet in a second. If a big gun is fired 50 miles away, how long will the sound of it take to reach you?

9. Find, to three significant figures, a multiplier that will convert a speed expressed in kilometres per hour into a speed expressed in centimetres per second.

10. Correct, justify, or complete the following statements:

(i) 0·07 is greater than $\frac{1}{15}$ and less than $\frac{1}{14}$.

(ii) 30 inches = ... centimetres (to nearest centimetre).

(iii) 15 miles per hour = ... feet per second.

(iv) 17s. $6\frac{1}{2}$d. = £0·877.

(v) In the specification of wire of a certain material, 35·6 feet per pound is equivalent to ... pounds per mile.

11. Work the following, (i) to (iv), in the most intelligent way you can:

(i) 30 miles an hour is how many feet per second?

(ii) If 15 per cent. of a certain amount is £27. 10s. 6d., what would $16\frac{1}{2}$ per cent. of the same amount be?

(iii) $3{\cdot}142(7{\cdot}86^2 - 5{\cdot}93^2)$.

(iv) $(1{\cdot}23456)^2$ to 5 significant figures.

12. An aviator covers 117 kilometres in 1 hour 7 minutes. Find his speed in miles per hour to one decimal place. See the Tables inside the front cover.

13. A man runs after a tram-car which is 20 yards ahead of him and going at 9 miles an hour. If he is willing to run 100 yards at any pace up to 12 miles an hour, will he catch the car?

14. A, who cycles to business every morning at 10 miles an hour, regularly passes at the same spot a man B, who is walking 4 miles an hour. One morning A passes B a quarter of a mile further along the road than usual. Supposing B is punctual, how many minutes late is A?

15. An up train, travelling 45 miles an hour, always passes a down train at the same spot when both are punctual, the down train travelling at 30 miles an hour. One day the up train is 4 minutes, and the down train 3 minutes late. Where will they pass each other?

16. On a single line railway, 55 miles long, there are stations at every 5 miles, where it is possible for two trains to pass one another. A fast train starts at noon from one terminus, running at the rate of 50 miles an hour, and stops 2 minutes at the fifth station out. A slow train starts 5 minutes later from the other terminus, running between stations at the rate of 20 miles an hour, and stops 2 minutes at every station. At what station must the slow train lie by for the fast train to pass it? The slow train must be out of the way at least a minute before the fast train is due.

17. On the railway, 30 miles long, from London to X fast and slow trains run side by side. A slow train starts from London every 5 minutes, and runs 20 miles an hour. A fast train starts two minutes after a slow train and runs 44 miles an hour. How many slow trains will it pass before reaching X?

18. PQ is a straight path running along the edge of a shingly beach, A and D two points on the shingle, 10 yards from the path and 75 yards from one another. A boy who wishes to get as quickly as possible from A to D can go 2 yards a second on the shingle and 3 yards a second on the path. Find the time taken from A to D direct; and also the time taken by getting off the shingle, doing 75 yards on the path and then getting over the shingle to D.

What would be the difference in time between the two routes if the distance AD were 76 yards? And if it were 77, 78, or 79 yards? For what distance between A and D would the two routes take the same time?

19. On a tramway route cars are running both ways every five minutes, at the rate of ten miles an hour. A man travels two miles in a car. How many cars, travelling in the opposite direction, may he meet?

20. I cycle at 12 miles an hour along a road where tramcars are running both ways every five minutes at 10 miles an hour. How far do I run between meeting one car and meeting the next, and what time elapses between my overtaking one car and overtaking the next?

21. On a certain road there is a bus every ten minutes going at 6 miles an hour. If a man walks at $3\frac{3}{4}$ miles an hour, what is (i) the greatest, (ii) the least, number of buses that may overtake him in walking 3 miles along the road?

22. The Massachusetts law requires that a class-room should be supplied with 1800 cubic feet of air per child per hour. To avoid draught, the incoming air must not be admitted at a velocity greater than six feet per second. Find the area of the smallest inlets that will allow of these conditions being satisfied in a class-room containing 50 children.

(Assume that the number of cubic feet passing through any inlet per second = velocity measured in feet per second × area of inlet in square feet.)

23. Water for the irrigation of a field of 18 acres is brought by a pipe 8 inches in diameter. Calculate the amount of water in 40 feet of the pipe; the area of a section across a pipe of diameter D inches being $0{\cdot}785 \times D \times D$ square inches.

When the pipe is running full, and at a speed of 40 feet per minute, how long will it take to deliver an inch of water over the whole field?

24. A town has a population of 20,000, and a daily water supply of 30 gallons per head is required. It is estimated that water will flow along the pipe, which is 15 inches in diameter, from the reservoir at 3 miles per hour. For how many hours a day must the water be allowed to flow along the pipe so as to bring the town its daily supply? A pipe of diameter d inches has a cross-section of area $0{\cdot}785 \times d \times d$ sq. inches.

Section 17. Loans and Investments.

Subsection (i). INTEREST.

1. Find the simple interest on £437. 8s. 6d. for three years at $4\frac{1}{2}$ per cent. per annum, to the nearest penny.

2. Find, to the nearest penny, the sum required to pay interest at $2\frac{3}{4}$ per cent. on £768,443,386.

3. Find the simple interest upon £3285 for 87 days at $7\frac{1}{2}$ per cent. per annum.

4. Find, to the nearest shilling, the interest upon £218. 6s. 0d. for 78 days at 5 per cent. per annum.

5. What will be the amount payable for interest on a sum of £250 which is lent at the rate of $4\frac{1}{2}$ per cent. per annum on 24 Dec. 1905, and is repaid on 5 May following?

6. A man receives £2. 10s. per annum as interest on a sum of £81. 6s. which he has lent, and £3 per annum on another loan of £97. Find the rate per cent. per annum in each case to one place of decimals.

7. Do not work the following questions, but give a reason for thinking that the answers suggested are *obviously* wrong:

(i) Multiply £8654. 7s. 11d. by 240.

Ans. £2,077,055. 5s. 6d.

(ii) Find the remainder, after making the greatest possible number of packets, each containing 0·03 lb., out of 217 lb. *Ans.* $\frac{1}{10}$ lb.

(iii) Find the simple interest on £4220 for 1 year at $4\frac{11}{12}$ per cent. *Ans.* £211. 8s. 4d.

(iv) Find the income derived from the investment of £5101 at 3 per cent. after paying income-tax at the rate of $7\frac{1}{2}$d. in the £. *Ans.* £153. 16s. $8\frac{1}{2}$d.

(v) Find the cost of paving a courtyard 23 yards 1 foot long and 24 yards 2 feet wide at 6d. per square foot.

Ans. £150. 10s. 6d.

8. £100 is lent, and $2\frac{1}{2}$ per cent. per year is charged for the use of the money. But this sum, instead of being paid, is every year added to the original loan: thus the loan increases year by year. The curved line of the figure can be used to find to what amount the loan has increased in any number of years: thus, if the amount in 23 years is required, draw the upright line AB from the 23rd division, read off the number of divisions in AB,

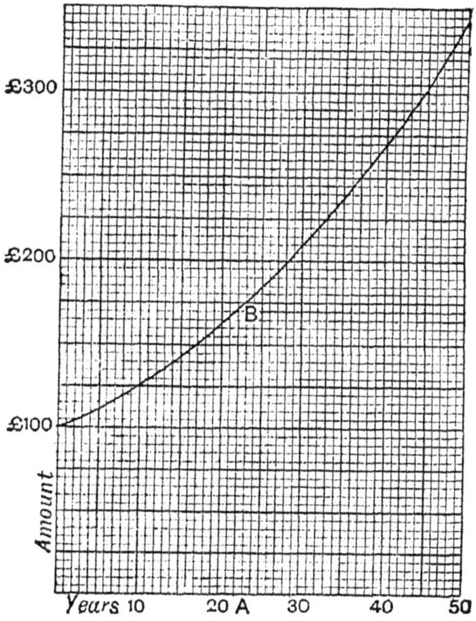

Fig. for Qu. 8.

and then calculate what amount AB would represent, supposing, as here, that each division represents £5. In this case it is obvious that the amount sought is about £175.

Similarly use this curve to find the amounts to which £100 will increase in 17, 31, and 45 years respectively, and also determine how many years must pass before £100 increases to £260.

9. A customer has a deposit account at a bank, and his balances are as follows, during the several months mentioned: January, £374. 10s. 6d.; February, £563. 11s. 4d.; March, £672. 6s. 3d.; April, £864. 10s. 6d.; May, £462. 7s. 6d.; June, £343. 2s. 6d. The bank allows him interest on his monthly balance at the rate of $2\frac{1}{2}$ per cent. per annum, or at 1 per cent. below "bank-rate," at his option. "Bank-rate" being as follows: January, February, and March, 3 per cent.; April and May,

4 per cent.; June, 6 per cent.; calculate the difference of the interest for the six months that he will receive on the two systems, correct to the nearest shilling.

10. Find, approximately, the compound interest on £625 for 17 years at 5 per cent. per annum; the error in the answer should not exceed 10 shillings. Log 1·05 = 0·0211893.

11. If p pounds accumulate for n years at r per cent. compound interest payable annually, find an expression for the amount at the end of that time.

As an example, find, to the nearest hundred pounds, the amount due after £5000 has accumulated for 25 years at $4\frac{1}{2}$ per cent. Log 1·045 = 0·0191163.

12. Find, by how much compound interest on £100 at 5 per cent. per annum for three years exceeds simple interest for the same period and rate.

13. Find the dividend, at the rate of 12s. $10\frac{1}{2}$d. in the £, on

(i) £475. 12s. 6d.;
(ii) 45 francs 37 centimes;
(iii) 453 dollars 70 cents.

[1 franc = 100 centimes; 1 dollar = 100 cents.]

14. What sum will amount to £390. 12s. in 2 years at 4 per cent. simple interest?

15. A man borrows £60 from a moneylender, and agrees to repay £72 in four quarterly instalments of £18 each, the first instalment to be paid at the end of one quarter from the date of the loan. The moneylender invests each repayment at 6 per cent. simple interest as soon as he receives it. Calculate how much he has received by the end of the year in return for his loan.

16. A man borrowed £20 and agreed to pay 1s. a month for the use of the money. At what rate per cent. per annum did he pay interest?

17. A landlord receives rent amounting to £180 per annum in half-yearly instalments paid on January 1 and July 1. He always immediately after receiving a payment places the money in a bank which allows $2\frac{1}{2}$ per cent. interest, the interest being added to the total amount at the end of each year. Find the amount to his credit in the bank after he has paid in the money received on 1 January 1907, the first payment having been on 1 January 1905

18. A man proposes to invest £200 at $2\frac{1}{2}$ per cent. and to reinvest the interest at the same rate each year. If he continues this process to the end of the thirtieth year, by how much will his original £200 have multiplied itself?

19. A company of capital £5000 makes 4000 articles at a cost of 6s. each, and sells them for 12s. 6d. each. Out of the resulting profit £750 is appropriated for rent and salaries, and 5 per cent. of the remainder for the reserve fund. If a dividend of $7\frac{1}{2}$ per cent. is then paid, what balance remains?

20. A buys a house for £500. He pays £300 in cash, and 6 per cent. interest upon the balance. If he pays £7. 10s. as ground rent and 2s. 6d. in the pound upon a valuation of £36, as rates, find the lowest rent which will yield him 8 per cent. upon his investment.

21. What sum of money must be paid now to provide for a first payment of £100 three years hence and a second payment of £100 four years hence, reckoning compound interest at 4 per cent. per annum? Log 1·04 = 0·0170333.

22. A man borrows £12,000 and undertakes to repay it with compound interest in 2 equal payments at the end of one and two years. If the rate of interest be 5 per cent. payable annually, find the amount of each payment to the nearest shilling.

23. A man has contracted to pay £25. 5s. a year hence. Instead he pays two equal sums, one at the end of six months and the other at the end of the year. What ought to be the amount of each payment if the rate of interest is 4 per cent. per annum?

24. A person may make a payment of £516. 13s. 4d. to-day, or of £526. 7s. 1d. in half a year's time. What rate per annum of simple interest is being charged on the £516. 13s. 4d.?

Subsection (ii). STOCKS AND SHARES.

The transactions concerned in the questions of this Subsection are too involved for ready comprehension. The students' time can be more advantageously employed otherwise, and the omission of this subsection is recommended. *See page* v.

25. On January 1st A started in business with a capital of £8000. On May 1st B joined him as a partner, contributing £7000 on the understanding that A received 20 per cent. of the profits, in lieu of salary as manager, in addition to his share as partner. If the profits on 31st December (accrued since 1st May) were £1330, how much should each receive?

26. Three partners A B C agree to share the profits of their business thus: In the first place each is to receive £500. In the second place the remaining profit is to be divided among them in the ratio 3 : 2 : 1 up to the point at which C's total share, including his former £500, amounts to £1000. Lastly, any profit still remaining is to be divided between A and B in the ratio 2 : 1. How should they divide profits of (i) £2397, (ii) £5176?

27. Three men A, B, C enter into partnership, bringing respectively capital to the amounts of £1300, £4200, and £7800. They agree that only one of them need work at any one time, and it is arranged that the number of days worked by each shall be inversely proportional to the amount of capital subscribed by him. Find on how many of the 305 working days in the year each should attend.

28. The profits of a business are £4875. One of the partners sells 0·24 of his share of the profits for £780. What fraction of the profits belongs to him?

29. A company of traders have contributed a total capital of £110,000, and realise an annual profit of 10 per cent. thereon. They subsequently admit more capitalists into the company, who subscribe altogether an additional £10,000. The profits of the company thereupon in the first year increase by £600, and are divided evenly over the whole capital subscribed. Find the dividend per cent. for each investor in this year.

30. Of a company's capital 16 per cent. was invested in a farm, 28 per cent. in ships, 36 per cent. in a railway, and the rest in minerals. The ships rose $22\frac{1}{2}$ per cent. in value, the railway rose 30 per cent., and the farm fell in value $12\frac{1}{2}$ per cent. The company was just as rich as before. How much had the minerals fallen in value?

31. The claims against a bankrupt company are £5000 claimed by the bank for an overdraft, £4000 claimed by A, and £11,000 by other creditors. Moreover A has guaranteed the bank against loss. The available assets are £4000. The case is taken into court, where A's claim against the company is reduced to £2000, and the other claims are allowed in full. How much more is A out of pocket after discharging his guarantee than he would have been if the court had not reduced his claim?

32. A company has issued the following stocks and shares:

£20,000 $3\frac{1}{2}$ per cent. debenture stock,
£25,000 $4\frac{1}{2}$ per cent. preference stock,
100,000 ordinary shares of £2 each.

If the net profits in a certain year be £12,825, find what rate per cent. of dividend can be paid on the ordinary shares.

33. How much money, invested in India 3 per cents. at 83, will yield £50 in annual interest?

34. In December last a man invested £2475 in $2\frac{1}{2}$ per cent. Consols at $82\frac{1}{2}$ (including brokerage). Find (i) his total income from this investment and (ii) his yield per cent. on his money invested.

35. A bankrupt owes £12,000, and his assets consist of £6325 stock standing at $92\frac{1}{2}$ and a bill for £3200 due at the end of six months. If the latter is realised at once, money being worth 5 per cent. per annum, how much in the £ can he pay?

36. French 3 Per Cent. Rentes are quoted at 97 francs $22\frac{1}{2}$ centimes per 100 francs stock. Find the equivalent price of £100 stock in English money, in pounds and fractions of a pound, to the nearest sixteenth.

Find also the yield on £100 sterling invested, giving your answer to the nearest threepence.

37. A man invests £13,050 in a 3 per cent. stock which stands at 87, and has to pay income-tax at 1s. 2d. in the pound on the income therefrom. What is the net income therefrom, and what percentage is this net income of the sum invested? (Neglect brokerage, and give the percentage to one decimal place.)

38. A person invests $\frac{2}{7}$ of his capital in a $2\frac{1}{4}$ per cent. stock at 90, $\frac{1}{3}$ of his capital in a $3\frac{1}{2}$ per cent. stock at 98, and the remainder in a $4\frac{1}{2}$ per cent. stock at 108. Find the average rate of interest he receives on his capital.

39. A man buys railway shares at £118. At the end of a year he receives a dividend of $4\frac{1}{4}$ per cent. on his stock, and sells out at £117. 10s. Express his profit as a percentage of the sum paid out in buying. At each transaction he pays brokerage at the rate of 2s. 6d. on each £100 of stock.

40. A man invests £6000 in 3 per cents. at $98\frac{3}{4}$. How much must he invest in 6 per cents. at £$102\frac{7}{8}$ that the rate of interest on his whole outlay may be 4 per cent. for the first year after paying brokerage of $\frac{1}{4}$ per cent. in each case?

41. A man dies possessing £22,500 Consols. His heir is required by law to pay estate duty amounting to 4 per cent. of the equivalent in cash of the investment. How much was paid in duty, and what was the heir's income if he sold just sufficient of the Consols to pay the estate duty? Consols pay $2\frac{1}{2}$ per cent. interest and were then at $87\frac{1}{2}$.

42. A man invested a sum of money in a $3\frac{1}{2}$ per cent. stock when it stood at £12 above its par value (100), and sold it 10 years later at £12 below par, receiving £600 less than he paid for it. Find the amount of stock he bought, and the annual income he received from it. (Disregard brokerage and other expenses of the buying and selling.)

43. How much 3 per cent. stock does a man hold if, by selling out at 80 and investing the proceeds in a 5 per cent. stock at 85, he increases his income by £203 per annum? (Neglect brokerage.)

PART III.

SECTIONS 18–26. PARALLELS, CONGRUENCE, AND FORMULAS.

Section 18. Drawing;

AND THE SIMPLEST PROPERTIES OF FIGURES.

1. Draw an angle of 112° with your protractor. Bisect it without using your protractor. State how you bisected the angle.

2. Draw a triangle ABC, given AB = 74 mm., BC = 42 mm., CA = 88 mm. Bisect the angle ACB by a line CD meeting AB in D.

3. Draw a straight line AB, $2\frac{1}{2}$ inches long, and at A and B erect, on the same side of AB, perpendiculars AC and BD, 3 inches and 1 inch long respectively. With centre C and radius CD describe a circle cutting AB in E. Measure the length of AE.

4. A man walks due east for a distance of 2 miles, then 2 miles due north, then 1000 yards in the south-east direction. Find his final distance from his starting-point by drawing a plan and measuring the distance.

5. Draw any line AB, then construct and show clearly the boundaries of the area within which a point P must lie so that the triangle PAB may have APB for its greatest angle and PBA for its least angle, proving that your construction is correct.

6. Prove that the sum of the lengths of two straight lines drawn from the extremities of the base of a triangle to any point within the triangle is less than the sum of the lengths of the two sides of the triangle.

ABC is a triangle and F is any point in the base AB. Prove that CF is less than half the perimeter of the triangle.

7. AB BC CD DE are four straight lines placed so that the angles at B C D are equal. If the lines bisecting these angles meet in a point O, show that OB OC OD are all equal.

8. Show that the straight lines which bisect two adjacent angles, made by one straight line meeting another, are at right angles.

9. The sum of the sides of any quadrilateral is greater than the sum of the diagonals.

10. ABC is a triangle (which is not isosceles) such that the base BC is 12·2 centimetres long, and the sum of the sides AB and AC is 19 centimetres. Draw a triangle to satisfy these conditions; bisect the angle at A by AG meeting BC in G; draw GL perpendicular to AB. Measure AL.

How would you draw the triangle if the base and the sum of the sides were given as above and the angle ABC were also given in magnitude? Prove your construction.

11. A cubical block, measuring 3 inches along each edge, is divided into two equal wedges by a plane through two opposite edges. Draw full size one of the faces formed by the division, and measure the length of a diagonal of this face.

Section 19. Evaluation of Formulas

FOR GIVEN NUMERICAL VALUES OF THEIR VARIABLES.

1. To find the height of a flagstaff a man places a stick by the side of the staff and measures the lengths of the shadows cast by both. The height in feet is equal to $\frac{b \times c}{d}$ where b is the length of the shadow of the flagstaff, d that of the stick, and c the height of the stick, all in feet. Find the height of the flagstaff if c is 5 ft., d is 7 ft. 3 in., and b is 127 ft. 6 in.

At what time of day are the shadows least?

2. In finding the cubic contents of a tree, the following formula is used:

$$\text{Cubic contents in cubic feet} = L\left(\frac{G+g_1+g_2}{12}\right)^2$$

where L is the length of the tree, G is the girth at the middle, g_1 and g_2 the girths at the two ends; all expressed in feet. Find the cubic contents of a tree 20 feet long, and of girth 31 inches at the middle, 34 inches at one end, and 29 inches at the other.

3. A flywheel is a disc 1·18 metres in diameter, and 14 centimetres thick. A cubic centimetre of the iron it is made of weighs 7·2 grams, and the volume of a disc of diameter a cm. and thickness b cm. is $0{\cdot}785 \times a \times a \times b$ c. cm. Find the weight of the flywheel in kilograms to three significant figures.

4. The number of cubic feet of timber in an oak log, which tapers uniformly, may be calculated from the formula

$$0{\cdot}2618h \times (aa + ab + bb)$$

where a and b are the numbers of feet in the diameters of the ends of the log, and h is the number of feet in the length of the log. Find what is the weight of such a log of oak whose length is 15 feet, and the diameters of whose ends are 3 feet and 2 feet respectively, assuming that a cubic foot of the oak weighs 50 lb. Give your answer to the nearest cwt.

5. Put two pins into squared paper 12 cm. apart, and another pin 8 cm. from each. Tie a thread round the triangle thus formed, taking care that it is taut. Remove the last-mentioned pin, place the point of your pencil against the thread, and, keeping the thread taut, move the pencil round so as to trace out a closed curve. Remove the pins and thread, and find the area of the figure by counting squares.

Measure the greatest and least diameters d_1 d_2 of the figure, and compare the value of the product $\frac{\pi}{4}d_1d_2$ with the value you have found.

6. If $\pi = 3{\cdot}14159265\ldots\ldots$ and $e = 2{\cdot}71828182\ldots\ldots$, find (i) to 3 places of decimals, (ii) to 3 significant figures, the value of

$$\pi + e, \quad 10\pi - e, \quad \pi \times e, \quad \frac{\pi}{e}, \quad \frac{e}{25\pi}.$$

7. Find the value of $(a^2+3b)/(b^2+3a)$ correct to three significant figures when $a=8{\cdot}702$ and $b=5{\cdot}678$.

8. Having given $x=1{\cdot}73$, $y=0{\cdot}87$, $z=3{\cdot}58$, find the value of

$$\frac{(z-x)(z-y)}{x-y}.$$

9. If $3x=1$, $4y=3$, $5z=2$, find the value of

$$\frac{\frac{1}{x}}{\frac{1}{z}-\frac{1}{y}}+\frac{y^2-x^2}{y^2-z^2}.$$

10. Find the number of gallons of water which can be contained in a bowl $5\frac{1}{2}$ in. deep and $23\frac{1}{4}$ in. broad, it being given that the volume of a bowl a in. deep and b in. broad is

$$0{\cdot}5236a(a^2+3b^2)$$

cub. in. See the Tables inside the front cover.

11. A bucket is made of galvanised iron a millimetre thick. The top is a circle whose circumference is 110 cm., the bottom a circle whose circumference is 60 cm., the length of the slant side is 30 cm.; there is no lid. Find the weight of the bucket. The volume (base included) is given by the formula

$$\pi(r^2+Rh+rh)t,$$

where R r h t are the radius of the top, the radius of the bottom, the length of the slant side, and the thickness of the metal. A cubic centimetre of the galvanised iron weighs 7·8 grams.

12. If an oak beam L inches long, B inches wide, and D inches deep is fixed at one end, and a weight W cwt. is placed at the other end, it will break should W exceed $15\times B\times D\times D\div L$. Find, with as little work as possible, whether such a beam 10 feet 8 inches long, 1 foot 10 inches wide, and $4\frac{1}{2}$ inches deep will break if a weight of 2 tons is placed at the free end.

13. Where a railway goes round a curve the outer rail is raised to keep the train on the line. When the curve has a radius of R feet, the distance between the two rails is W feet, and the customary speed of the trains is V miles an hour, the outer rail should be $\dfrac{W\times V^2}{1{\cdot}25\times R}$ inches higher than the inner rail. Calculate how much higher the outer rail should be on a curve of radius

580 feet, if the rails are 4 ft. $8\frac{1}{2}$ in. apart, and the customary speed is 40 miles an hour.

14. An American company promoter plans a gigantic airship which is stated to be approximately cylindrical, 700 feet long, and 80 feet in diameter. I know that an airship of cylindrical shape of length k feet and diameter d feet has a cubic capacity of $0{\cdot}785ddk$ cubic feet and that the total weight of such an airship and its load, estimated in ounces, must not exceed the total capacity measured in cubic feet. From the formula given, find what is the greatest weight that the airship and load may have. Give the result in tons.

15. Find the value of

$$1 + mn + m(m-1)nn/2 + m(m-1)(m-2)nnn/6$$

(i) when $m=2$ and $n=3$, and also (ii) when $m=24$ and $n=1/4$.

16. The rectangle $abcd$ (see the figure) represents, on a scale of 1 centimetre to 10 feet, the plan of a plot of ground ABCD. A is the lowest point; B is 1 foot above A; C is 2 feet above B; D is 1 foot below C. Determine the number of cubic

Fig. for Qn. 16.

yards of earth that must be removed in order to reduce the plot to the level of its lowest point.

You may assume that if in levelling a triangular field of area S the three corners are lowered by amounts p q r then the volume V of the earth removed is given by the formula

$$V = S(p+q+r)/3.$$

17. A cask is 4 feet 6 inches deep, its greatest diameter is 2 feet 3 inches, and the diameter of each end is 2 feet. Calculate, to the nearest whole number, the number of gallons which it will hold. The volume of a cask of depth h feet, of which the greatest and least diameters are D and d feet, is approximately $0{\cdot}7854 \times h \times \frac{D+d}{2} \times \frac{D+d}{2}$ cubic feet, and a cubic foot is 6·23 gallons.

Also calculate the error caused by taking this formula instead of the following more accurate formula: Volume of cask of depth h feet, of which the greatest and least diameters are D and d feet, is

$0{\cdot}05236 \times h \times (8 \times D \times D + 4 \times D \times d + 3 \times d \times d)$ cubic feet.

18. A hollow sphere of iron is of uniform thickness. Its external diameter is 9 inches and its weight 72 lb. Find the volume of the internal cavity, to the nearest cubic inch. The iron weighs 480 lb. per cubic foot, and the volume of a sphere of diameter d is $\frac{1}{6}\pi d^3$.

If the diameter may be wrong by 0·1 inch, the weight of the sphere by an ounce, and the weight of a cubic foot of the iron by 10 lb., estimate the possible error in your result.

19. If v and V are the volumes of a certain quantity of gas, under constant pressure at 0° and t° Centigrade respectively, then

$$V = v \times (1 + t/273).$$

Find, to the nearest tenth of a cubic centimetre, the volume occupied at 18° Centigrade by some gas which occupied 100 c. cm. at 0° Centigrade.

20. A shot-gun, into whose barrel will exactly fit a sphere of lead weighing $\frac{1}{N}$th of a pound, is termed an 'N-bore' gun. The diameter of the barrel of a 12-bore gun is 0·729 inch. What is the diameter of the barrel of a 20-bore gun?

21. The coal used by a steamer during a voyage of 2400 sea miles costs $2{\cdot}5 \times vv$ pounds, where v is the speed in knots (*i.e.* sea miles per hour). The other expenses of the voyage amount to £200 for each day. Consider the three cases in which v has the values 10, 15, 20, and find for each case (i) the number of days the voyage lasts, (ii) the cost of coal, (iii) other expenses, (iv) the total cost of the voyage.

22. A stone is thrown horizontally from the top of a vertical cliff 50 metres high with a speed of 11 metres per second. Its horizontal distance from the top of the cliff increases steadily so that after t seconds the distance is $11t$ metres. Its vertical depth below the top of the cliff increases in such a way that after t seconds the depth is $4{\cdot}9tt$ metres. Calculate the position of the stone at intervals of half a second, and on squared paper show its path on a scale of 1 to 400, that is, using 1 cm. to represent 4 metres.

23. The interval between the ticks of a pendulum k cm. long is either $\pi\sqrt{\dfrac{k}{g}}$ seconds or $2\pi\sqrt{\dfrac{k}{g}}$ seconds, where $g=981$. Determine which; given that, roughly, a pendulum a metre long beats seconds.

24. Calculate the length of one of the two chains required for a suspension bridge of 300 feet span from the following:

Length of chain in feet is $2\sqrt{\left(\dfrac{1}{2}C\right)^2+\dfrac{4}{3}V^2}$ where C is the span in feet, and $V=C/13$.

25. Assuming that when a train is running at v miles per hour, the number w of tons of coal consumed in each hour is given by $560w=\sqrt{v^3}$, what (to the nearest cwt.) is the difference between the amounts of coal consumed on a journey of 200 miles by two trains, one travelling at the rate of 35 miles an hour, and the other at the rate of 50 miles an hour?

26. Put 10 for x in the expression $ax^4+bx^3+cx^2+dx+e$, and write down the value of your result when

$$a=1,\quad b=0,\quad c=9,\quad d=4,\quad e=2.$$

Section 20. Parallels.

1. Draw a straight line XY, to lie roughly diagonally across the page, and fix a point O in it, about the middle of the page. Then find two points A and B, on the same side of XY, to satisfy the following conditions:

A is to be $1\frac{1}{2}$ in. from B, and 1 in. from the line XY. B is to be $2\frac{1}{2}$ in. from O, and 2 in. from the line XY.

Write down the steps by which you make your construction, but do not give a proof of their correctness.

Measure the length of AO.

2. ABC is a triangular field in which AB is 120 metres long, BC 85 metres long, and AB and BC are at right angles. The field is to be divided into three equal parts by drawing lines parallel to BC. If these lines meet AB in P and Q, it is known that

$$\frac{AP}{1} = \frac{AQ}{\sqrt{2}} = \frac{AB}{\sqrt{3}}.$$

Calculate the lengths of AP and AQ to the nearest metre, and draw a plan of the field, showing the lines of division, and representing 10 metres by 1 centimetre.

3. Prove that a triangle is isosceles if the angles at the base are equal.

ABC is a triangle, straight lines are drawn through C parallel to the bisectors of the angles at A and B to meet AB produced in D and E. Prove that DE is equal to the perimeter of the triangle.

4. A strip of paper of uniform width is cut into two parts along a straight sloping line, as EF in the first diagram. One part of the strip is then turned over and the two cut edges are again fitted as in the second diagram.

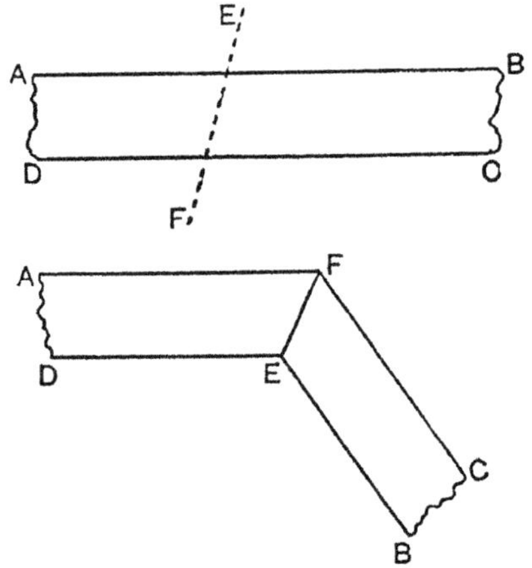

Fig. for Qn. 4.

Find what angle EF must make with the edges AB and DC of the strip, so that the angle DEB in the second diagram may be 90°. You may find out by experimenting with tracing paper, then by reasoning justify the statement you make.

In making a rectangular picture frame from a long strip of wood, the pieces cannot be turned over on account of the moulding. Give a sketch showing all the saw cuts that would be made, and shade the parts cut to waste.

5. Draw a straight line AB and take a point C not lying on it. From C draw a line making an angle of 40° with AB.

6. Draw a straight line and take two points not lying on it. Through each point, with the help of your protractor, draw a line inclined at 60° to the original line, these two lines not being parallel. By measurement compare the lengths of the sides of the triangle formed by the three lines.

7. ABCD is a croquet lawn on which the boundary AD runs due north, and AB runs due east, from A. On the lawn are four balls, P, Q, R, S. P is 6 feet from AD and 9 feet from AB; Q is 38 feet from AD and 6 feet from AB; R is 20 feet from P and 30 feet from Q; and S is 8 feet from P and 28 feet from Q.

Draw a figure to scale showing the position of each ball, and state your construction for finding the positions of R and S.

8. Draw two straight lines OX and OY at right angles, OX to the right, OY up, and find a point P 4 inches from OX and 1 inch from OY. Now imagine P to remain fixed while YOX is revolved, counter-clockwise, about the point O. Determine by drawing (i) the distance of P from OX when OX has turned through 60°; (ii) what further amount of revolution would make OX pass through P.

Section 21. Simple Graphs.

1. You have 60 hurdles, each 1 metre long, with which to make a rectangular enclosure. Copy the following table, and fill in the missing numbers:

Number of hurdles in one side of the enclosure	6	10	12	14	16	18
Area of the enclosure in square metres - -	144	...	...	...	...	...

With these data draw a graph showing the area of the enclosure, as the number of hurdles in one side increases from 6 to 18. For the number of hurdles let 5 mm. represent 1; for the area let 5 mm. represent 10 sq. metres.

What is the area of the largest enclosure you can make, and what is then the number of hurdles in one side?

2. A publisher is publishing a book and means to sell it at 6, 12, 18, 24, or 30 pence a copy, whichever will bring the greatest profit. He estimates that he could sell 25,000 copies at 6d. each, and that at higher prices the sale would be smaller, as in the table below. He knows also that to produce 25,000 copies will cost him 4d. each, and smaller numbers will be dearer as in the table.

Determine at which price he must sell the book to realise the most profit, and what his profit then is.

Selling price in pence	6	12	18	24	30
Estimated sale	25,000	20,000	12,000	6000	3000
Cost per copy in pence	4	5	6	9	12

3. The table below gives the expenses and receipts for a certain newspaper for various numbers of copies produced:

Number of copies	1000	2000	3000	4000
Expenses in £	85	110	135	160
Receipts in £	60	99	138	177

Draw a graph showing the relation between the number of copies produced and the expenses; taking 5 cm. horizontally to represent 1000 copies, and 5 mm. vertically to represent £4.

On the same diagram and with the same scales, show the relation between the receipts and the number of copies.

From the graphs estimate as nearly as you can the smallest number of copies that must be produced, if the paper is to pay.

4. A suspension bridge consists of a horizontal roadway 200 yards long suspended from a chain by 199 vertical steel rods each distant 1 yard from the next. The ends of the chain are attached to two pillars, each 85 feet high, at the ends of the bridge. Having given that the 10th, 25th, 50th, 60th, 74th, and 100th rods are 70·75, 52·19, 28·75, 22, 15·07, and 10 feet in length, draw a figure on squared paper to represent the roadway, chain, and pillars, and find from your figure the lengths of the 87th and 175th rods.

Find, in any way you please, the area represented by the figure, and show that the number of yards of steel rods used is nearly equal to the number of square yards in the area.

5. In a certain examination the *total* number of candidates who obtained less than certain specified marks is given in the following table:

Number of candidates	27	94	214	392	632	1041
Obtaining a smaller number of marks than	10	20	30	40	50	60

Express this by a curve on squared paper, being careful to make it clear what axes and units you adopt.

Candidates obtaining less than 25 marks fail, while those who obtain 45 or more are placed in the first class. Determine by means of the curve the approximate number of the remainder.

6. The following table gives the number of men out of 10,000 men in the U.S. Army who are of the given heights. Draw a graph representing these results, and from it find approximately the number of men of height 64 in., 69 in., 72 in.:

Height. in.	No.	Height. in.	No.	Height. in.	No.
61	100	66	1399	70	884
62	171	67	1584	73	118
63	368	68	1531	74	61
65	1051				

7. The distances travelled by a moving point in 1, 2, 3, 5, 7 seconds from the beginning of reckoning are 2, 8, 18, 50, 98 feet respectively. Represent this statement graphically by means of a space-time graph; and from your graph determine the distances travelled in (i) 4 seconds, (ii) $5\frac{1}{2}$ seconds.

8. A manufacturer received an order to make a great number of lidless tin boxes, each box to have a square base and to hold 48 cu. in. In order to find what dimensions would enable him to make the boxes with the least amount of material, he drew a graph showing the area of the surface for various values of the side of the base. Denoting the height of the box by h inches and the side of the base by b inches, show that $h=\frac{48}{b^2}$, and that the total area of the surface in square inches is b^2+4bh. Complete the table given below, draw the graph showing how the area of the tin required varies as b has values ranging from 3 to 8 inches, and from the graph determine the length of the side of the base of the box that would require the smallest amount of tin. With your graphical result as a guide, and assuming the side of the base to measure an exact number of tenths of an inch, calculate the area of tin for a few sizes of the base, and so determine the best dimensions for the box.

b	3	4	5	6	7	8
b^2+4bh	73	...	...	...	...	...

9. The area of the surface of an open cylindrical tank, whose volume is 220 cu. ft., is known to be $22\left(\frac{r^2}{7}+\frac{20}{r}\right)$ sq. ft., where r is the radius in ft.

Find the value of $\frac{r^2}{7}+\frac{20}{r}$ when r is 1, 2, 3, and 4 respectively. With these and other similar results plot a graph and find which value of r makes the expression in brackets smallest.

Hence find the smallest surface-area the cylinder can have.

10. A tradesman, selling off, reduces all his prices 30 per cent. Draw a graph from which can be read off the selling price of all articles up to ten shillings in original value. Read off the selling price of articles, whose original values are 1s. 3d., 3s., 5s., 8s. 6d.

11. The weight of a foot length of cast-iron pipe when the metal is half an inch thick is given by the following table:

D	1	1·5	2	2·5	3	3·5	4	4·5	5
W	7·4	9·8	12·3	14·7	17·2	19·6	22·1	24·5	27·0

D being the diameter of the bore in inches and W the weight of the pipe in lb. Draw a graph to represent this relation, and from it give the weight of a foot length when the diameter of the bore is $3\frac{1}{4}$ inches. Represent 1 inch by 4 cm. and 1 lb. by 1 cm. (or if your paper is too small use smaller scales), and show the scales clearly on the figure.

To check your result calculate the weight of the pipe for a bore of $3\frac{1}{4}$ inches by means of the formula $W=1{\cdot}56\pi(D+0{\cdot}5)$.

12. The freezing and boiling points of water are represented by 0° and 100° on the Centigrade scale, and by 32° and 212° on the Fahrenheit scale. Draw a graph connecting these scales, taking one inch vertically to represent 20° Centigrade and one inch horizontally to represent 40° Fahrenheit, or such other scales as the size and ruling of your paper allow.

By your graph convert 20° Centigrade to the Fahrenheit scale.

13. Oxford is $63\frac{1}{2}$ miles from London by rail. A train starting from Oxford arrives at London at 8.48 p.m. Another train, starting from London at 7.30 p.m., passes the first 31 miles from London, and arrives at Oxford at 8.45 p.m. Supposing both trains to run at uniform speeds without stopping, find the rate at which the first train travels, and the time at which it starts from Oxford.

14. Two trains, moving at the rates of 45 and $47\frac{1}{2}$ miles an hour, pass simultaneously, but in opposite directions, through a station at 9 a.m. Draw a graph to show their distance apart at any time up to noon. Use the graph to find the distance apart at 10.48 a.m., and the time at which they are 200 miles apart.

15. The maximum number of marks in an examination is 178. It is desired to reduce the maximum to 100. Construct a graph for the purpose. Find from your graph the revised marks of two candidates who obtained 87 and 113 marks originally.

16. A servant's wages are £15 a year, but at the end of half a year the rate is raised to £16 per year. Draw a graph from which we may read the approximate amount earned by her up to any particular time in the year.

Owing to the alteration of wages the servant will have earned £15 sooner than she expected. Find from the graph how many days sooner this will be.

17. The cost of coal at a certain place A is 15s. per ton. The cost of carriage along a railway through A is 3d. per ton per mile. Draw on squared paper a diagram from which may be found the cost of coal delivered from A at any place on this railway less than 80 miles from A.

Eighty miles from A on the railway there is a place B at which the cost of coal is 18s. 6d. per ton. Find, by means of your squared paper, the position of a town on the line between A and B at which the cost of coal is the same whether it is brought from A or B.

18. Draw a figure to a scale of 1 inch to 1 mile, to represent the position of the following places: P is 3 miles North of a point C, Q is 3 miles South of C, and R is 10 miles East of C.

Water is to be conveyed by pipes from a reservoir at R to the two towns P and Q. The pipe runs along RC to some point D, and then divides into two branches, DP and DQ. Take the point D at distances of 0, 1, 2, 3, ... 10 miles from R, and in each case find by measurement the total length of RD + DP + DQ. Draw a graph showing how the total length of the pipes depends on the position of D, and from your graph determine where the branching should be, in order that the total length of the pipes may be the least possible.

19. In the figure PQ is a ladder 25 ft. long leaning against a wall at P. OX is a line drawn perpendicular to the ladder from

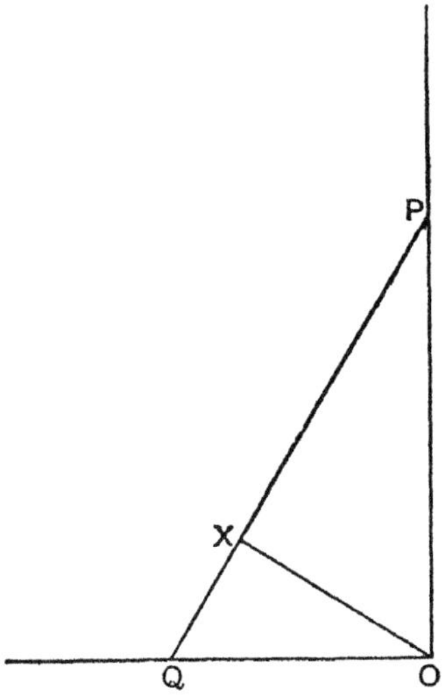

Fig. for Qn. 19.

O, the bottom of the wall. Draw the ladder in position with its end Q at various distances from O, and draw the perpendicular OX in each case. Make a graph showing the length of OX as a function of the distance OQ, and find the position of Q when OX is greatest.

Section 22. The Angle-sum Property

OF A TRIANGLE, AND CERTAIN OTHER PROPERTIES.

1. O is a point within the triangle ABC. Prove (i) that the angle BOC is greater than the angle BAC, (ii) that OB and OC are together less than AB and AC.

2. Prove that the exterior angle of a triangle is greater than either of the interior opposite angles.

ABC is a triangle and O is any point inside it. A straight line is drawn through the points A O. On this line two points D and E are taken such that BD = BO and CE = CO. Prove that one, at least, of the points D E must be outside the triangle ABC.

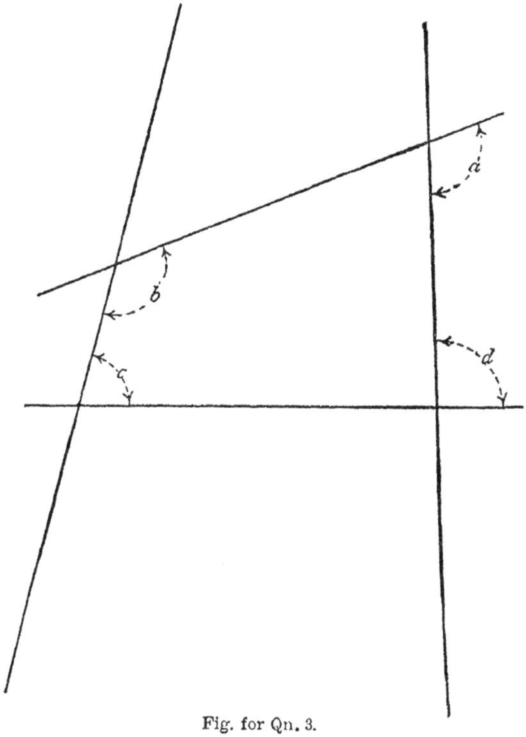

Fig. for Qn. 3.

3. Four straight straws are arranged as shown in the figure, and pinned together at the points of crossing so as to form an irregular plane quadrilateral with pivoted corners.

Prove that the difference between the angles *a* and *b* is always equal to the difference between the angles *c* and *d*.

Measure the angles *a* *b* *c* *d* of the figure in the position shown, and record their values in degrees.

4. Is it possible to make a pavement consisting of equal equilateral triangles, leaving no gaps?

Is it possible to do so with equal regular polygons of (i) four, (ii) five, or (iii) six sides? Give reasons for your answers.

5. Construct with ruler and compasses an isosceles triangle having a vertical angle of 45°, and the base equal to 7 centimetres.

Describe and justify your construction.

6. In a certain triangle each of the angles at the base is treble the vertical angle. Find the number of degrees in each angle, to the nearest degree.

7. In working a geometrical problem a boy found that two angles P and Q together were 136°, while Q and R together were 116°. What conclusion might he draw about P and R? Could he infer anything about Q?

Having afterwards found that P and R together were 108°, what could he now conclude about P and Q?

Determine from the above data whether P Q R could form the angles of a triangle.

Illustrate your answer as far as possible by sketches. Your reasoning must be clearly shown.

8. With only ruler, compasses, and set square, draw two straight lines AB AC, such that the angle BAC is 45°.

Draw also, by means of the same instruments, two lines BD BE, meeting AC in D and E, such that the angle BDA is 75°, and the angle BEA is 120°.

Prove that the angle ABE is 1/8 of the angle AEB, and 1/3 of the angle BAC.

9. An acute-angled triangle ABC has the side CB greater than the side CA, and a circle described with centre C and radius CA cuts AB again at D. If the angle CAB is x degrees and the angle ABC y degrees, find expressions in terms of x and y for the angles ACD and BCD.

Hence show how to construct a triangle in which CB = 10 cm., CA = 7 cm., and the difference between the angles CAB and ABC is 20°.

10. Draw an irregular quadrilateral ABCD; bisect the two adjacent angles, DAB and ABC, by lines AE and BE meeting in E. Measure the angle AEB and compare it with the sum of the angles BCD and CDA.

Repeat with a second quadrilateral.

What relation do you find between the angles in each case? Prove that it is true in general.

11. If a straight line fall on two parallel straight lines, then it shall make the alternate angles equal to one another, and the exterior angle equal to the interior opposite angle on the same side.

The base BC of a triangle ABC is divided at F and H. It is found that each of the three parts BF FH HC measures three inches. EHF is an equilateral triangle. Show that ABC is an equilateral triangle if EC and EB bisect the angles B and C.

12. Explain the terms: isosceles triangle; plane surface; right angle; hypotenuse; a proof by "reductio ad absurdum."

Prove that the straight line joining the middle point of the hypotenuse of a right-angled triangle to the opposite angle divides the original triangle into two isosceles triangles.

13. ABC is a triangle in which AB is produced through B to D and BC is produced through C to E. If the sum of the angles CBD and ACE is equal to 3 right angles, find, with accompanying proof, what kind of triangle ABC is.

14. Two triangles, ABC and DEF, are such that BC = EF and the angle ABC is equal to the angle DEF. Show that the angle BAC is twice the angle EDF if DE is equal to the sum of AB and AC.

15. If the lines AD BE CF bisecting the angles of a triangle intersect in a point O, and the six angles at O thus formed are all equal, show that the triangle is equilateral.

16. AB and AC are two straight lines meeting at A. Show how to find a point P in AB such that PB may be equal to the perpendicular from P on AC.

17. Two opposite angles of a quadrilateral are equal; prove that the angles between the pairs of opposite sides (produced to meet) are equal.

18. Prove that all the interior angles of any rectilineal figure, together with four right angles, are equal to twice as many right angles as the figure has sides.

If the bisectors of the angles of a quadrilateral form another quadrilateral, prove that a pair of opposite angles of the latter will be together equal to two right angles.

19. If one angle of a triangle be greater than another, then the side opposite to the greater angle shall be greater than the side opposite to the less.

BC is a diagonal of a square ABXC. D and E are points in AC and BC, such that BD bisects the angle ABC, and DE the angle BDC. Prove that CD is greater than CE, and CE than DE.

20. Give (without proof) any practical test you know of for finding, by the drawing and measurement of lengths, whether two given straight lines may be considered parallel.

Show that if all the angles of a hexagon are equal, the opposite sides must be parallel.

21. Prove that the exterior angle of a triangle is equal to the sum of the two interior opposite (*i.e.* non-adjacent) angles.

A B C D E F G are the angular points of a regular polygon. Find the number of degrees in each of the angles of the quadrilateral BDEG.

22. ABCDE is a convex equilateral pentagon, and the angle A is equal to the angle of a *regular* pentagon. If CD is parallel to BE, prove that ABCDE is also equiangular.

23. Find the sum of the interior angles of a polygon of 14 sides.

24. Assuming that the three angles of any triangle are together equal to two right angles, find the sum of the interior angles of a convex polygon of n sides.

E F are points within a quadrilateral ABCD, such that FD and AE do not intersect. Explain clearly how far the above proposition is applicable to the rectilinear figure ABCDFEA.

25. From the vertex of a triangle which is not isosceles, two straight lines are drawn, one bisecting the vertical angle, the other perpendicular to the base. Which of these lines is the nearer to the greater of the two sides containing the vertical angle? Test the point by drawing a figure, and then give a general proof of your conclusion.

Find, either by calculation or by measurement, the angle between the bisector and the perpendicular when the base angles of the triangle are 52° and 64°, and give an expression for the angle between them when the base angles are A and B degrees.

26. A triangle is formed by the bisectors of the interior angles at B and C of a triangle ABC, and the bisector of the exterior angle at A. Prove that its angles are $\frac{1}{2}B$ $\frac{1}{2}C$ $90° + \frac{1}{2}A$ where A B C are the angles of the triangle ABC.

27. State clearly with the aid of a figure the special properties of the interior angles made by a straight line falling across two other lines which (i) are not parallel, (ii) are parallel. No proofs are required.

In any triangle ABC, prove that the internal bisector of the angle B and the external bisector of the angle C cannot be parallel, and also prove that their point of intersection lies on the same side of BC as A.

28. O is the mid point of the side BC of a triangle ABC, BC measuring 16 cm. and OA 5 cm. Draw the triangle with A in a number of different positions relative to BC, always nearer to B than to C. From measurements show the angle CAO as a function of the angle BAO.

Is the angle CAO ever greater than angle BAO? Give the answer from the graph.

Section 23. Construction of Formulas from adequate Data.

1. Using any convenient length as unit, mark out a rectangle 4 units long and 3 units broad, and count the number of squares it contains. Do the same with rectangles 9 units by 5 units and 14 units by 12 units.

Deduce a method of calculating the area of a rectangle whose length and breadth are given. And give a formula for the area in square inches of a rectangle a inches long and b inches broad; and a formula for the area in acres of a rectangular field c yards long and d yards broad.

2. Express a length of a metres b decimeters c centimetres in metres.

Give, in square metres, the area of a room p metres q decimetres r centimetres long, and s metres t decimetres u centimetres wide.

3. A rectangular oil tank is 2 metres long, 1 metre 60 cm. wide, and 1 metre 30 cm. deep. How many litres of oil can it hold, and how many cylindrical oil drums, each 40 cm. in diameter and 60 cm. high, could be completely filled from the full tank? See the Tables inside the front cover.

A rectangular oil tank is a m. long, b m. c cm. wide, and d m. e cm. deep. How many litres of oil can it hold, and how many cylindrical drums each f cm. in diameter and k cm. high can be filled from it? (This answer need not be an integer.)

4. A path half a mile long is to have a kerb made on one side. The kerbstones used are b inches long, and c of them have already been supplied. How many more are wanted?

Give the numerical result when $b = 20$ and $c = 1000$.

5. A grocer buys a lb. of butter at b pence per lb., and c lb. at d pence per lb., and sells the whole for £g. Find expressions for the average buying price and selling price, in pence per lb.; also for his percentage gain on the cost price.

Give numerical results when $a = 100$, $c = 60$, $b = 11$, $d = 13$, and $g = 10$.

6. Give an expression for the number of bricks measuring 8 by 4 by 2 inches that it takes to build a wall a yards long, b feet high, and c inches thick. Give the numerical result to the nearest thousand when $a = 70$, $b = 6$, $c = 16$.

7. Draw a graph of $3x - x^2$ from $x = 0$ to $x = 3$, and from the graph find the two values of x which make $3x - x^2$ equal to 1.

A wire 6 feet long is bent so as to enclose a rectangle. If one of the sides of the rectangle is x feet long, find an expression in terms of x for the area of the rectangle. Now state how the wire must be bent if the area of the rectangle is to be one square foot.

8. A piece of wire 2 feet long is bent into the shape of a rectangle. If one side of this rectangle is x inches long, shew that its area is $x(12 - x)$ square inches.

Representing an inch of length by an inch on one axis and 2 square inches of area by an inch on the other axis (or using other large scales), plot a graph to show the area of the rectangle for all values of x between 3 and 9. From your graph state for what values of x the area of the rectangle is (i) as large as possible, (ii) equal to 33 square inches.

9. A rectangular field measures a yards by b yards and has an area of p acres. Give an expression for p in terms of a and b, and use the expression to find p to the nearest whole number when $a = 400$ and $b = 300$.

A square field measuring c yards in the side has an area of q acres. Express q in terms of c, and draw a graph of q in terms of c up to the value 500 of c, showing clearly the scales you use.

10. The base of a tank is a square of side x feet, and the total area of the bottom and sides (excluding the lid) is 36 square feet. Prove that its volume is $\frac{1}{4}x(36 - x^2)$ cubic feet.

By drawing a graph of this expression for values of x between 0 and 6, find the side of the base of the largest tank with a square base which has this superficial area.

11. A square garden of side 50 ft. is to be laid out with different plants for the border and centre, the two being divided by a path 5 ft. wide. If the border has a uniform width of x ft., find the cost in shillings of laying out the garden, the cost of the border being $2\frac{1}{2}$d. per sq. ft., of the path $4\frac{1}{2}$d. per sq. ft., and of the square centre bed $1\frac{1}{2}$d. per sq. ft.

If the minimum width of the border is 5 ft. and the maximum width 12 ft., plot a curve showing the cost of different arrangements, and from it find the greatest cost of the garden, drawing a corresponding plan to scale.

12. A room is a feet long, b feet wide. The walls are to be lined with deal to a height of d feet at the rate of 9 pence per square foot; find an expression for the cost in pounds. The floor is to be carpeted at the rate of 6 shillings a square yard; find an expression for the cost in pounds.

Give numerical results when $a = 15$, $b = 11$, $d = 4$.

13. A man has £a in one bank and £b in another. If he has £c more to deposit, how should he divide it between the two banks so that (i) the two accounts may be equal, (ii) the first account may be double the second?

14. The outer edges of a picture frame measure p and q inches respectively. If the distance between the inner and outer edges of the frame is $1\frac{1}{2}$ inches, find the total length of the inner edges, and the area enclosed by them.

What is the area when $p = 8$, $q = 12$?

15. A rectangular plate of metal is x inches long, y inches broad, and z inches thick. Find an expression for the cost of the plate if the metal costs a shillings a pound, and weighs b pounds per cubic foot.

Evaluate the expression when

$$a=0{\cdot}5, \quad b=540, \quad x=8, \quad y=6, \quad z=0{\cdot}01.$$

16. The average age of a class of n girls is 16 years 5 months. If four additional pupils of ages 17 years, 16 years 10 months, 16 years 6 months, and 16 years 4 months, join the class, find the new average age in years and months, or in years, or in months.

What is n if this new average is 16 years 6 months?

17. In a class there are k children 8 years and 2 months old, m children 8 years and x months old, and n children 7 years and 5 months old. What is the average age?

18. From ground floor to first floor of a building there are n steps. The first is a inches high, and the other $n-1$ steps rise by b inches (see the sketch). Give in inches the height above the ground floor of the 4th, 5th, and 10th steps, and also of the first floor.

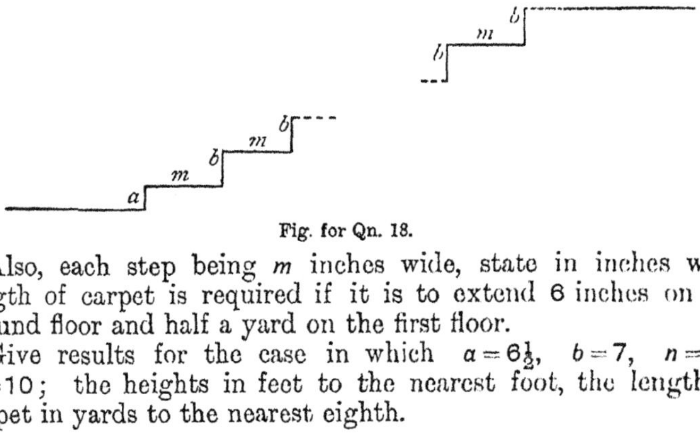

Fig. for Qn. 18.

Also, each step being m inches wide, state in inches what length of carpet is required if it is to extend 6 inches on the ground floor and half a yard on the first floor.

Give results for the case in which $a=6\frac{1}{2}$, $b=7$, $n=20$, $m=10$; the heights in feet to the nearest foot, the length of carpet in yards to the nearest eighth.

19. Each angle of a regular polygon measures n degrees. Find an expression for the number of sides the polygon has.

The number of sides must of course be a whole number. Give three values of n which will make it so.

20. A boy is instructed as follows: "Think of a number, treble it, add eight to the result, multiply the sum by 2, subtract 7, divide by 3; give me the answer."

Find a simple rule by which the number first thought of can be readily deduced from the answer given by the boy.

21. Express a farthing as a decimal of £1 correct to 5 places of decimals.

Show that if your expression to five decimal places were absolutely correct n farthings would be equal to £$(n+1)/1000$ correct to the nearest thousandth for integral values of n from 13 to 37 inclusive, but that actually it is true for values from 13 to 35 inclusive.

Express 2s. $7\frac{3}{4}$d. as the decimal of £1 correct to the nearest thousandth.

22. Three tramps meet on a road; the first has a loaves, the second b loaves, and the third has $a+b$ pence. The loaves are shared equally between the three of them, and the third tramp divides all his money between the other two in payment for his share. How much money should he give to each?

23. Two men rent a field for £50. The first turns x cattle into the field for a days, while the second turns y cattle into the field for b days. Find how much each should contribute to the rent if they pay in proportion to the use they have had of the field.

24. A certain alloy is composed of x per cent. of tin and y per cent. of copper. Find the value of z oz. of alloy if the price of tin is a shillings per lb., and of copper b shillings per lb.

What relation is there between x and y?

25. A man has x sovereigns and y pence. He dies, and leaves to each of his m sons one-tenth, and to each of his n daughters one-twelfth of his possessions, while his wife gets the remainder. How much does she get?

26. A candidate attempts to prove that *the difference of the cubes of two numbers differing by six, together with* 108, *is nine times the sum of their squares,* as follows. Why is this insufficient? Give a sufficient proof.

"Let the numbers be 5 and 11.

"The difference of the cubes $=11^3-5^3=1331-125=1206$.

"The sum of the squares $=11^2+5^2=146$.

"But $1206+108=1314=9\times146$.

"Similarly of the numbers 7 and 13 or 11 and 17, the theorem can be proved true.

"Thus it is always true."

27. The average salary of 13,300 headmasters of elementary schools is £156. 11s. 6d., the average salary of 13,100 assistant masters is £112. 4s. 4d. What is the average salary of all those teachers, omitting fractions of a penny.

If the number of headmasters was x, and the number of assistants y, what would be the average salary?

28. In a long account the error of writing £3. 3s. instead of writing 3s. 3d. occurred 960 times. But the account added up to £33 too little because another error, that of writing 4s. 5d. for £4. 5s., occurred a sufficient number of times. How many times did this second error occur?

Work the same sum with a substituted for 960, £b £c £d and £e substituted respectively for £3. 3s., 3s. 3d., 4s. 5d., and £4. 5s.

29. A dealer buys a sheep at p shillings each, and b sheep at q shillings each; find (i) the average price which he pays for a sheep, and (ii) at what price per sheep he must sell the whole number, in order that his profit may be $5(a+b)$ shillings on the whole transaction.

Verify your results when $a=7$, $b=8$, $p=45$, $q=55$.

30. A has a pounds b shillings and c pence, B has c pounds b shillings and a pence. If c be less than a, a be less than 12, and b be less than 20, find how much money A has more than B; expressing it in pounds shillings and pence.

31. a men and b women can reap a field in x days, while c men and b women can do the same piece of work in y days. Find how long one man working alone would take.

32. A and B working together can complete a piece of work in z days, B and C can together complete the same piece of work in x days, and C and A in y days. How long would A, B, and C take to complete the work if they were all working at once?

33. From a pounds b shillings c pence is taken away a less sum, namely c pounds b shillings a pence. The remaining money is spent in buying loaves at $a-c$ pence apiece. How many loaves are bought?

34. On the Northern Railway of France, passengers are allowed 30 kilograms of luggage free, and excess luggage is charged at the rate of 1 centime per 10 kilograms per kilometre. A tax of 10 centimes is also charged for registration. Give a formula for the amount in shillings charged for carrying p lb. weight of luggage m miles (if this weight exceeds the free limit); given 1 shilling = 125 centimes, 1 kilogram = 2·2 lb., 5 miles = 8 kilometres. From this formula calculate to the nearest penny the charge for conveying 100 lb. of luggage from Paris to Calais, the distance being 185 miles.

35. An article which cost £a is sold at a profit, for £b; find the gain per cent.

If the cost price and the selling price were both increased by £x, determine whether the gain per cent. would be increased or diminished, and by how much.

Evaluate each expression to the nearest integer when

$$a = 36 \quad b = 42 \quad x = 4.$$

36. On measuring a rectangular sheet of paper its length is found to be a cm., and its breadth b cm. From these measurements its area is calculated. If each of the above measurements is x cm. less than it should be, by what percentage of itself is the calculated area in error?

Give a numerical result to the nearest integer for the case in which $a = 33$ $b = 20$ $x = 0{\cdot}5$.

37. The sides of a rectangle are measured as a inches and b inches respectively, and its area calculated. If the true lengths are within x per cent. of these measurements, express in sq. inches the greatest possible error in the calculated area.

Give this error to the nearest 100th of a sq. inch when $a = 9{\cdot}7$ $b = 8{\cdot}6$ $x = 1$.

38. If 12 eggs cost x pence, and if at the same rate y eggs are bought for one shilling, find the relation between x and y.

Express this relation by means of a curve on squared paper, between the limits $x = 4$ and $x = 36$, and read from the curve the price per dozen when 7 are sold for a shilling.

39. A man walks at x miles an hour, and the length of his step is y inches. Find an expression for the number of steps which he takes in a minute, and evaluate the expression when $x = 4$ and $y = 31$.

40. A walks a mile in x minutes, and B takes y minutes for the same distance. In z hours A has walked one mile further than B. Find z in terms of x and y.

Evaluate z when $x=16$ $y=19$; giving the value of z hours to the nearest minute.

41. A motor-car was timed over a measured distance of a yards; the time was given as x seconds, the error made in taking the time may have been as much as y seconds. What were the limits of the speed of the car, in miles an hour?

Find, to one decimal place, the limits of speed when

$$a=440 \quad x=35 \quad y=\tfrac{1}{2}.$$

42. Two trains start at the same time from places a miles apart and approach each other, one going at p miles an hour and the other at q miles an hour. Show that they will meet after $a/(p+q)$ hours.

Hence state how long after their start two trains will meet, which start from places 216 miles apart and approach each other at speeds of 36 and 45 miles an hour.

43. A and B run a race which A wins by z minutes. What (in yards) was the distance run if A's rate was x miles an hour and B's rate was y miles an hour?

44. Two men are riding round a racing track in opposite directions; one completes a laps in p minutes and the other completes b laps in q minutes. Find the time between two successive meetings, and the distance travelled by each between two successive meeting places.

45. Prove that the sum of any number of two digits and the number with the same digits reversed is a multiple of 11, and that their difference is a multiple of 9.

Find the number of two digits which is equal to six times the sum of its digits.

46. A manufacturer buys crude alkali at p pence per lb. After purifying it, he sells it at p shillings per lb., making a clear profit of q per cent. on his total outlay. What is the cost in pence per lb. of purifying the alkali?

47. Persons enter a college on January 1st in each year for a one-year course. The numbers entering in three consecutive years were a b c respectively. Express in symbols (i) the percentage increase in the entries in the second, and in the third year, on those in the preceding year; (ii) the average number of students during the period.

48. The total receipts of a railway for one year are £p, of which £a is received for 1st class, £b for 2nd class, £c for 3rd class passengers, and the remainder for goods. In the following year the receipts for 1st class passengers rise x per cent., 2nd class y per cent., 3rd class z per cent., and goods w per cent. What is the increase per cent. of the total receipts?

49. Brine is formed by dissolving p pounds of salt in q pounds of water; how many pounds of salt are there in q pounds of brine?

50. A can beat B by x yards in a race of 100 yards. B can beat C by y yards in the same distance. By how much could A beat C?

Find the value of your answer when $x = y = 5$.

51. A town A is on the bank of a river whose length is r kilometres. Its length from the source to A is a kilometres. What is its length from A to the mouth?

If in the foregoing we know that the river rises at a height of x metres above the sea and that the town is at a height of y metres, find expressions to denote (α) in metres per kilometre, and (β) in simple fractional form, the average fall of the river:

(i) from the source to A,
(ii) from A to the mouth,
(iii) throughout its whole course.

52. A man can row at the rate of a miles an hour on a lake where there is no current. How long will it take him to row a mile downstream and back on a river which flows at the rate of b miles an hour? Is your result greater or less than the time which he would take to row two miles on the lake?

Section 24. Congruence of Triangles.

1. Describe a triangle of sides 2·3, 3·5, and 4 inches.

Bisect the greatest angle, and measure the length of the segment of the bisector intercepted by the opposite side.

2. Take three straight lines, any two of which are greater than the third, and make a triangle having its sides equal to them.

Show how your construction will fail if any two of the given lines were together equal to the third.

3. From how many of the five sets of data below can triangles be made? When the triangle cannot be made, explain why not.

(i) $a=10$ cm., $b=8$ cm., $C=63°$.
(ii) $a=8$ cm., $A=50°$, $B=60°$, $C=70°$.
(iii) $A=42°$, $B=62°$, $C=70°$.
(iv) $A=42°$, $B=62°$, $C=76°$.
(v) $A=42°$, $a=5$ cm., $b=10$ cm.

4. In a triangle $a=10$ cm., $b=7$ cm., one angle is 95°. There being no restriction as to which angle of the triangle has the value 95°, discuss how many distinct triangles can be made.

5. State as many different sets of conditions as you can, each set of which is sufficient to prove two triangles ABC and DEF equal in all respects.

Examine the case in which $AB=DE$, $AC=DF$, and angle $B=$ angle E.

If $AB=DE=4$ in., $AC=DF=5$ in., angle $B=$ angle $E=120°$, are the triangles necessarily equal in all respects?

6. X, Y, and Z are the middle points of the sides of an equilateral triangle PQR. Prove that the triangle XYZ is equilateral.

7. If two sides of a triangle be equal, prove that the angles which are opposite the equal sides are also equal.

Draw an isosceles triangle with sides 15·6 8·4 8·4 cm. in length. Bisect the sides and form another triangle by joining the points of bisection; see by measurement whether this triangle is isosceles.

8. A triangular set square ABC, right-angled at A, is laid flat on a table. The set square revolves through a right angle about the point A, which is fixed, and takes up a new position AB′C′. Prove that BB′ is perpendicular to CC′.

9. Show how to bisect a given rectilineal angle without the use of a protractor; and show why your construction is correct.

Show that the bisectors of two adjacent supplementary angles are perpendicular to each other.

10. If two triangles have two sides of the one equal respectively to two sides of the other but the contained angles not equal, prove that that triangle in which the contained angle is greater has the greater third side.

In the triangle ABC it is given that $AB = 2 \cdot 5$ in., $AC = 1 \cdot 9$ in., and the angle BAC is known to be less than 125° but greater than 70°. Find graphically the limits between which the third side BC must lie.

11. If two triangles, ABC and PQR, have the sides AB and AC equal to the sides PQ and PR respectively, and the angles at C and R right angles, prove that the triangles are equal in all respects.

If, however, the angles at C and R are unequal and supplementary, prove that the angles at B and Q are equal and that the sum of the angles at A and P is twice the complement of either B or Q.

12. Draw a straight line ABCD, and mark the points A B C D, so that $AB = 2$ inches and $BC = CD = 1$ inch.

Draw AP perpendicular to any line BP drawn through B.

Construct an isosceles triangle having P for the middle of the base, A one extremity of the base, and the side opposite to A passing through C. Justify your construction of this triangle.

Measure the distance of the vertex from D.

13. ABCDE is an equilateral pentagon; the angles at B and C are respectively 100° and 110°. The triangle AED is folded about AD so as to form a new pentagon with a re-entrant angle. Construct these pentagons. Measure the angles at A and D in the new pentagon.

14. Draw a triangle ABC in which $AB = AC = 6$ cm. and $\angle BAC = 45°$, without using the protractor. From B draw BD at right angles to AC, and take a point E in AD such that $ED = DC$. Join BE.

Prove that the angle AEB is equal to five times the angle ABE.

15. ABC is a triangle. AD bisects the angle A and meets BC in D. Prove that the area of the triangle ABD is greater than the area of the triangle ACD if AB is greater than AC.

16. Draw an isosceles triangle ABC having $AB = AC = 7$ cm. and $BC = 4$ cm. Find a point D in AB and a point E in AC such that $BD = DE = EC$.

17. Draw any triangle ABC. Bisect BC in E, and AB in G. Join AE and produce it to F so that $EF = EA$. Join CG and produce it to H so that $GH = GC$.

Lay your straight edge against H B F, and see if they are collinear, *i.e.* lie in a straight line. Then by general reasoning show that they are so.

18. One of the rules of a certain game played in a square courtyard is that all the players start from a certain point in one of the walls, run to each of the other walls in turn, and so home again.

Show that if the players, in running from one wall to the next, always run in straight lines parallel to one or other of the two diagonals of the courtyard, they will not only reach the point from which they started, but that from whatever point in the first wall the start is made they will always run the same total distance.

19. A triangular piece of paper ABC is folded along a line AD, passing through the vertex A and a point D in BC; and the folding brings C to C′ (see the sketch).

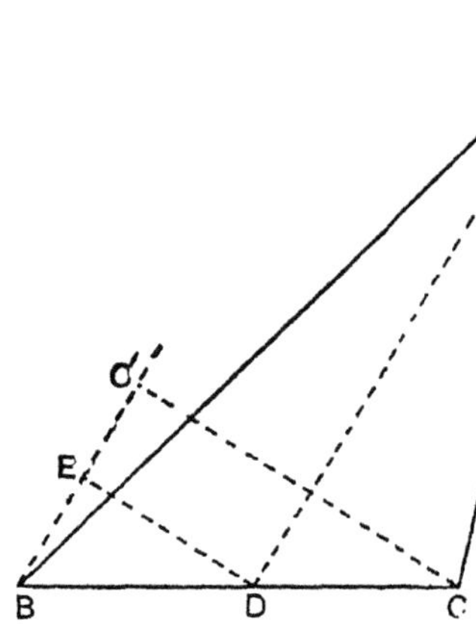

Fig. for Qn. 19.

Take a triangle having $AB = 6{\cdot}6$ in. $BC = 3{\cdot}7$ in. $CA = 4{\cdot}6$ in., and by experiment (*i.e.* by folding or preferably by drawing) find how AD must lie in order that BC′ may be parallel to AD.

20. Prove that when two triangles have the three sides of one respectively equal to the three sides of the other, they are equal in all respects.

In the quadrilateral ABCD the opposite sides AB and CD are equal, and the diagonals AC and BD are equal. Prove that the angles ABC and BCD are equal.

21. A sheet of paper folded once is pricked through at two points with a pin and is then unfolded and laid flat on a table. A and B are the holes on one side of the crease, and X and Y are the corresponding holes on the other side of the crease, so that the lines AX and BY are bisected at right angles by the crease. Prove that AY and BX meet on the crease.

22. A quadrilateral ABCD (not a parallelogram) is such that the sides AB and CD are equal and make equal angles with the side AD. Prove that the line joining the middle points of AD and BC is perpendicular to BC and AD.

Find the length of this line by drawing a figure to scale, when the lengths of the sides AB BC CD DA are 1 foot 1 inch, 2 feet, 1 foot 1 inch, and 2 feet 10 inches respectively.

23. A triangle ABC is given. Find which point of the perimeter of the triangle is nearest to any given point O, discussing the different cases, and especially marking the region in which O must lie if A is the nearest point.

24. Draw a straight line 4 inches long and divide it by geometrical constructions into five and six equal parts. Measure and state the distance between every pair of consecutive points of division.

25. Draw any triangle ABC with unequal sides. Suppose the triangle to revolve in the plane of the paper with the point A fixed. Show its position when it has completed a third of a revolution. *N.B.*—A protractor is not to be used in the solution of this question.

26. If two triangles have two sides and the included angle of the one respectively equal to two sides and the included angle of the other, prove that they are equal in all respects.

AB and A′B′ are two equal straight lines. The straight lines bisecting AA′ and BB′ at right angles meet at O. Show that the angle AOB is equal to the angle A′OB′.

27. Trace the figure on transparent paper. Prick the points A B C through on to a sheet of paper and draw the triangle. Replace the transparent paper on your figure and pin it down at P. Now rotate the transparent paper through a right angle in the direction of the arrow, and prick through the tracing of the triangle. Mark with X Y Z the new positions of A B C respectively.

State how you know when you have rotated the transparent paper through 90°.

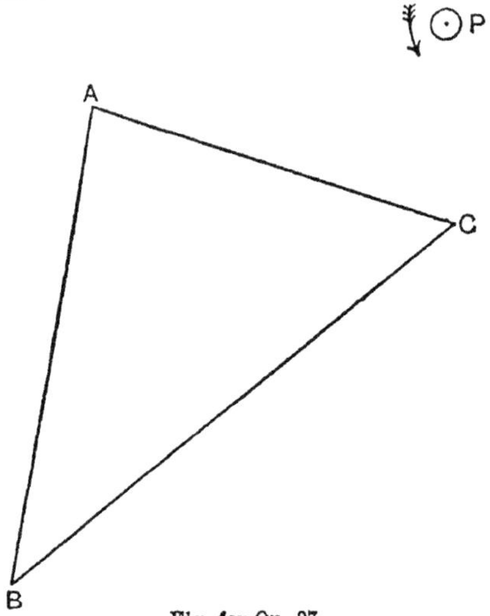

Fig. for Qn. 27.

To what points is it possible to bring the tracing of A by rotation about P? and the tracing of B? If you have no transparent paper, but are given the spot to which the tracing of A is to come by rotation about P, how will you find the spot to which B will come?

Put the tracing of the triangle back in its original position over ABC. Find the point about which it must be rotated to bring the tracing of B on to C and the tracing of C on to B.

28. Draw two congruent triangles ABC and XYZ, such that the points A B C and the points X Y Z are in the same direction round. Find a point S such that SB = SY and SC = SZ. Prove that the angle SCB is equal to angle SZY and SA to SX.

Suppose a piece of transparent paper is laid on the figure and fastened by one drawing pin so that it can revolve round this drawing pin while remaining in contact with the paper. Suppose also that when the transparent paper is in its first position a tracing is made of the triangle ABC. Where must the drawing pin be placed if it is found that as the transparent paper revolves, the tracing of ABC presently coincides with XYZ?

29. ABC and A′B′C′ are two different positions of a paper triangle in the same plane and with the same side of the paper upwards. Find a point O which is equidistant from A and A′ and also from B and B′. Join O to A A′ B B′ C C′, and then prove that the angles AOA′ BOB′ COC′ are all equal, and that O is equidistant from C and C′.

30. The sides AB and BC of a triangle ABC are bisected at D and E. Draw DO and EO perpendicular to AB and BC respectively, and let them intersect at O. Show that the angles OAC and OCA are equal.

31. Equilateral triangles CEA and AFB are described outwardly on the sides CA and AB of any triangle ABC. Show that BE and CF are equal.

32. O is the middle point of a straight line AB, and C is any point in AB produced. Show that OC is half the sum of AC and BC.

Draw the line, making AB 11 cm. long and BC 5 cm. Through C draw three straight lines making angles of 20° 35° 50° with CA, and drop perpendiculars upon them from A B O. From measurements check the statement that "the perpendicular from O on any line through C is half the sum of the perpendiculars from A and B on the same line."

Show how the property stated in the first paragraph may be looked upon as a particular case of that stated in the second.

33. ABCD and ECGF are two squares external to each other. Show that the perpendiculars from E and G respectively on CB and CD, produced if necessary, are equal.

34. ABC is an isosceles triangle having AB = AC. Through A a straight line is drawn parallel to BC and any point P is taken on it; prove that BP + PC > BA + AC.

[If you treat this as a corollary to another proposition, the proposition as well as the corollary is to be proved.]

35. If P be any point within a triangle ABC, prove that the sum of BP and PC is less than the sum of BA and AC.

Also prove that if Q be any point on the external bisector of the angle A, the sum of BQ and QC is greater than the sum of BA and AC.

36. Prove that the perpendicular bisectors of the sides of a triangle are concurrent.

37. Prove that the three bisectors of the angles of a triangle are concurrent.

Section 25. Transformation of Formulas;

i.e. THE CHANGING OF THE SUBJECT OF A FORMULA.

1. The volume V in cubic inches of a piece of wire of length k inches and diameter d inches is given by

$$V = 0 \cdot 785 \times d \times d \times k.$$

From this find an expression for the length k in terms of the volume V and the diameter d.

Certain wire has a diameter of 1/32 of an inch. Give in cubic inches (in decimal form) the volume of 1000 inches length of the wire. Also find in inches and in miles the length which has a volume of 1000 cubic inches.

2. x the depth of water in a certain trough, in inches, and p the number of gallons of water in the trough, are connected by the relation $p = 11 \cdot 4 \times x$. Find (to the nearest whole number) the number of gallons when the water is 8·2 inches deep.

Also from the given relation express the depth of the water in terms of the number of gallons, and calculate (to the nearest inch) the depth for which the quantity of water is 100 gallons.

3. A glass h inches high, a inches in diameter at the top, and narrowing to b inches at the bottom holds $h(a^2 + ab + b^2)/132$ pints. What will a glass hold (to the nearest tenth of a pint) for which $h = 4 \cdot 2$ $a = 3 \cdot 6$ $b = 1 \cdot 8$?

Denoting the number of pints a glass will hold by p, and the other dimensions as before, express the depth of the glass in terms of p, a, b.

If $a = 3 \cdot 6$ and $b = 1 \cdot 8$, what depth will give the glass a capacity of one pint?

4. A small object a mm. long, when examined with a magnifying glass held b mm. from it, appears to have a length of c mm. given by

$$c = \frac{af}{f - b}$$

f being a number that depends on the kind of magnifying glass that is used. The value of f being 20, find the value of c when $a = 1$ and $b = 17$.

Express a in terms of the other quantities, and find the real length of an object which appears to be 5 mm. long, f being 20 and b 17 as before.

5. (i) Calculate the weight of a cast-iron pipe from the following dimensions: Inside diameter 15 inches, thickness 0·8 inch, and length 8 ft. Density of metal = 465 lb. per cu. ft. It is assumed that there are no flanges or other enlargements at the ends. The volume of a length k inches of a pipe of internal diameter d inches and thickness t inches is $3{\cdot}14(d+t)tk$ cubic inches.

(ii) What is the external diameter of a pipe of internal diameter d inches and thickness t inches?

(iii) The area of the surface of a cylinder being 3·14 times the product of the diameter by the length, write down expressions for the areas of the internal and external surfaces of a pipe of length k inches, internal diameter d inches and thickness t inches.

(iv) It is required to coat a mile of the pipe specified above inside and outside with some protective material at the rate of 20 sq. ft. for 7d. Calculate the cost.

(v) From the formula given above, show that, w lb. being the weight of a cubic foot of iron, the weight in pounds of a foot length of iron pipe of internal diameter d inches and thickness t inches is

$$\frac{3{\cdot}14}{144}wt(d+t).$$

6. A square reservoir measures a yards each way, the water is b inches deep, and the quantity of water is c cubic feet. Express the side of the reservoir and the depth of the water in feet, and find an expression for c in terms of a and b.

Also express the depth of the water (in inches) in terms of the quantity of water (in cu. ft.) and the side of the reservoir (in yards).

What depth of water gives 1000 cu. ft. of water in such a reservoir whose side measures 40 yards?

7. Posts are placed at the side of a railroad at equal distances apart. If the distance between two consecutive posts is p miles and the time taken by a train in passing over this distance is q seconds, what is the speed of the train in miles per hour?

Find the speed in the particular case when

$$p=\tfrac{1}{4} \quad \text{and} \quad q=30\tfrac{1}{2}.$$

If a train travels r miles in s seconds, find how many yards it travels in one minute.

8. (i) It is known than when a certain tackle is used to raise heavy weights, the weight W lb. that can be raised by a pull of P lb. is given by the formula

$$W=4P-70.$$

Find what weight a pull of 140 lb. will raise.

Also (ii) express P in terms of W, and (iii) find what pull is needed to raise a weight of 400 lb.

9. The diameter d of a pipe in inches, its length L in yards, G the number of gallons of water flowing through per minute and H the head of water (in feet) which will cause this flow are connected by the relation $(3d)^5 \times H = L \times G^2$. Express H in terms of the other quantities, and find the head required to force 100 gallons per minute through a 4-inch pipe 1000 yards long.

Section 26. The Parallelogram.

1. Describe a parallelogram, two adjacent sides measuring 4·2 and 5·7 centimetres, and the included angle 103°.

2. Construct a parallelogram, having an angle of 49°, a side of 75 mm., and a diagonal of 58 mm.

Measure the other side and the other diagonal as accurately as possible.

3. If a straight line, cutting two other straight lines, make an exterior angle equal to the interior and opposite angle on the same side of the line, prove that the two straight lines are parallel.

AB is at right angles to CD, and EF is at right angles to GH. If AB be parallel to EF, prove that CD is parallel to GH.

4. Show how to trisect a line by geometrical construction, and give a proof of the validity of your construction.

5. AB is a straight line. An arc is described with centre A, and another of different radius with centre B, to meet the first at C and D. Prove that CD is perpendicular to AB.

Having CD drawn, state and prove a simple and accurate construction for making a perpendicular to AB at A with ruler and compasses only.

6. Part of a window with diamond-shaped panes is represented by the rough sketch. Draw it complete on a scale of 1 cm. to 4 inches. Give the number of whole panes and the number of half panes of glass required for it, showing clearly how you make the calculation. Two of the whole panes are shown shaded in the figure.

Draw one of the panes full size, and give the length of its sides and the size of its angles.

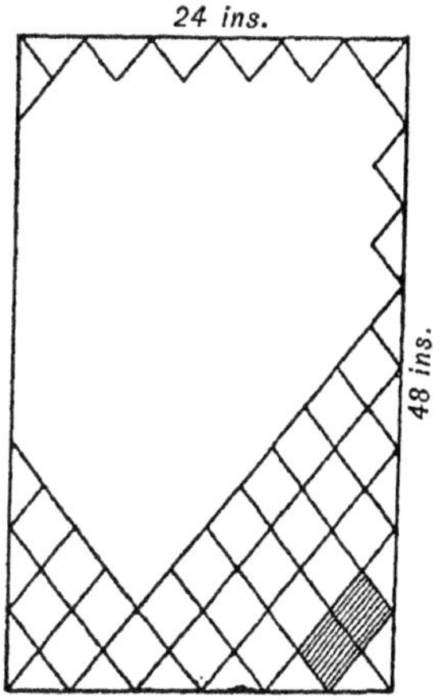

Fig. for Qn. 6.

7. Prove that equal triangles on the same base and on the same side of it are between the same parallels.

ABC is an equilateral triangle. Any point D is taken on AC produced. BF and BE are cut off from BC and BA, produced if necessary, equal to CD. Show that CEFD is a parallelogram.

8. Define a rhombus, and prove from your definition that the diagonals of a rhombus bisect one another at right angles.

Is it ever possible to describe a circle about a rhombus?

9. Find the locus of a point which is equally distant from two given straight lines.

ABC is a triangle. Find points D E F in the sides BC CA AB respectively such that CDFE is a rhombus.

10. Construct a circle with a diameter equal to two inches. Let O be the centre of this circle, and P Q R any points on its circumference. Draw OO′ PP′ QQ′ RR′ all parallel to each other and in the same direction and 2·5 inches in length. Measure O′P′ O′Q′ O′R′. What conclusion as to the position of the points P′ Q′ R′ can you make? Prove it.

11. ABCD is a parallelogram, E the middle point of AB, and F the middle point of CD. Let DE and BF cut the diagonal AC at O and Q. Show that AC is trisected at O and Q. First satisfy yourself of the truth of the statement by drawing a few figures and measuring. Then give a general proof. (Hint: AO and OQ are equal, because AE = EB and DE ∥ FB.)

12. If a straight line cut two parallel straight lines, the alternate angles are equal.

The side CA of a triangle ABC is produced to a point P such that PA is less than AC. Through P draw a straight line to meet BC and be bisected by AB.

13. ABCD is a parallelogram, and PQ is a straight line through A, which does not meet the parallelogram. Prove that the perpendicular drawn from C to PQ is equal to the sum of the perpendiculars drawn from B and D to PQ.

14. Prove that the straight line which bisects the two sides of a triangle is parallel to the base.

AOB and COD are two straight lines meeting in O, and P is a point which does not lie on either of the lines. Show how to draw a straight line from P to AB which shall be bisected by CD.

15. In the figure (p. 137) AD and BF are perpendicular to AB, and AC = CD = BE = EF. Prove that CF and DE are parallel and that they divide AB into three equal parts.

State clearly how a similar method may be used to divide a given line into *n* equal parts.

16. Construct a diagonal scale to measure sixteenths of an inch.

17. Prove that the opposite sides and angles of a parallelogram are equal to one another and each diagonal bisects the parallelogram.

AX and AY are two given straight lines. Find a point P in AX and a point Q in AY such that AP and AQ are equal, and PQ is of given length.

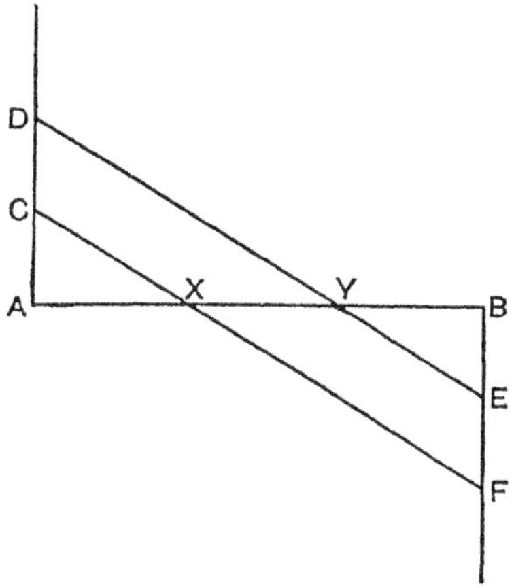

Fig. for Qn. 15.

18. If two triangles have two sides of one equal to two sides of the other, but the contained angle in the first greater than the contained angle in the second, show that the base of the first is greater than the base of the second.

Prove that the greater diagonal of a parallelogram passes through the smaller angles.

19. ABCD is a parallelogram. Points P Q R S are taken on AB BC CD DA respectively such that AP = BQ = CR = DS. Prove that PQRS is also a parallelogram.

20. Prove that in any quadrilateral the lines joining the mid-points of opposite sides bisect each other.

If they bisect each other at right angles, prove that the original quadrilateral has its diagonals equal.

21. ABCD is a parallelogram. If the straight line bisecting the angle A meets BC, prove that the straight line bisecting the angle B will meet CD produced, but that if the former line meets BC produced, the latter line will meet CD.

22. Prove that the diagonals of a parallelogram bisect each other.

OX and OY are two straight lines and E is a point between them. Show that it is impossible for more than one straight line whose extremities lie on OX and OY to be bisected at E.

23. How many diagonals has a rectangular block, and how many diagonals have the plane faces altogether?

Prove that any two diagonals of the block bisect each other.

24. On a loose sheet of paper make a parallelogram ABCD, having AB 6 in. long, BC 3 in. long, and $\angle A$ of 60°. Cut out this figure; your ruler will serve as a paper cutter.

Fold BC on BA and call the point where C falls E. Fold DA on DC and call the point where A falls F. What is the new figure thus formed? Prove your statement.

Fold this new figure about EF. How do B and D fall?

Open the paper out and make in your book an accurate drawing of it, showing creases by dotted lines.

Now keep BEF flat on your desk, turn BCF about BF and DEF about EF till they meet along CF and DF. Then turn ADE about DE till AD falls along BC. What other pair of lines fall together?

Measure the height of the pyramid you have made. Find the volume of the pyramid, the volume being $\frac{1}{3} \times h \times A$ cubic inches, where h is the height in inches and A is the area of the base in square inches.

25. K E F G are points on the sides AB BC CD DA of a parallelogram. If KE and GF are parallel to the diagonal AC and equal to each other, show that the lines through F and E parallel to AD and AB respectively cut at a point lying on AC.

26. Three straight lines are parallel, and another straight line meets them at H K L, so that HK = KL. Prove that if any other straight line meets these lines at P Q R respectively, then PQ = QR.

Show also that if, conversely, it is given that HK = KL and PQ = QR, then it does *not* follow that the three straight lines are parallel, but that if two of them are parallel then the third is parallel to them.

PART IV.

SECTIONS 27–33. NUMERICAL EQUATIONS, AREAS, NEGATIVE QUANTITIES AND PYTHAGORAS.

Section 27. Linear Equations.

Subsection (i). LINEAR EQUATIONS INVOLVING ONLY ONE UNKNOWN QUANTITY.

1. 750 persons were present at a lecture, for which p of them paid a shilling each and the remainder half-a-crown each. Give the total takings in pounds.

If the p persons had paid half-a-crown each, and the others a shilling each, how much greater would the receipts have been?

Evaluate the two results if $p = 450$.

Suppose, again, that the total cost of the lecture was £45. How many half-crown, and how many shilling, tickets would have had to be sold just to cover expenses, the total number sold being 750?

2. I have 30 miles to go to the railway station, and $4\frac{1}{2}$ hours to get there in. On my bicycle I go at the rate of 15 miles an hour, but my front tyre is likely to burst, in which case I shall have to walk with the machine at the rate of 3 miles an hour. How many miles must I ride before I am free from anxiety as to whether I shall catch the train?

3. A retail bookseller is supplied by a wholesale firm, who allows discount off the published price of certain books (class A) at 40 per cent.; of all others (class B) at $33\frac{1}{3}$ per cent. At the end of a year he finds that he would have paid $2\frac{1}{2}$ per cent. less than he actually did, had he accepted the tender of another wholesale firm, who offered a uniform discount of $37\frac{1}{2}$ per cent. What percentage of his purchases (reckoned at published price) belonged to each class? Give your answer to the nearest integer.

4. A man shouts and hears an echo after an interval of $1\frac{1}{5}$ seconds. He then walks 70 yards in the direction from which the echo came, and finds he now hears an echo after an interval of $\frac{4}{5}$ of a second. Calculate the velocity of sound.

5. Ice is made by cooling water in a "freezer." Water in freezing expands 10 per cent. of its volume. How much water when frozen will just fill a 5-gallon freezer?

6. If s denotes the sum of the angles of a triangle, and is known to have the same value for all triangles, obtain an equation for s by taking a triangle, dividing it into two parts by a line through the vertex, and considering the sums of the angles of the whole triangle and of its two parts. Hence prove that s is equal to 2 right angles.

Apply the same method to a quadrilateral, dividing it into two quadrilateral parts by a line meeting a pair of opposite sides.

7. A man purchases ice at 30 cents per 100 lb. At what rate must he sell it, after it has lost 10 per cent. of its weight by melting, to gain 20 per cent.?

8. When a photograph is to be taken, the plate on which the picture is made must be placed at a distance x from the lens, this distance being connected with the distance y of the object from the lens by the relation $1/x+1/y=1/f$, where f is a length depending on the lens used. For $f=4$ inches, find the distance of the plate from the lens when the distance of the object is (i) 6 feet, (ii) 20 feet, (iii) 100 feet, and show that the distance of the plate for 100 feet is nearly the same as for any greater distance.

9. A small object a mm. long, when examined with a magnifying glass held b mm. from it, appears to have a length of c mm. given by

$$c=\frac{af}{f-b}$$

f being a number that depends on the kind of magnifying glass that is used. The value of f being 22, find the value of c when $a=1$ and $b=18$; and find at what distance from the glass the object must be placed to look twice as long as it is.

10. If a beam of oak L feet long is submitted to a pull of T tons per square inch it is found that it increases in length by m feet; L T $\cdot$ m being connected by the equation $m=k\times L\times T$, where k is a constant. A beam of oak 6 feet long increases one quarter of an inch in length under a pull of $2\frac{1}{2}$ tons per square inch; what is the value of k, and under what pull per square inch will a beam 9 feet long increase $\frac{5}{8}$ inch in length?

11. In the accompanying figure A B C D are four cog wheels so arranged that A drives B, and C drives D. B and C are fastened rigidly together. Find out by measurement, using

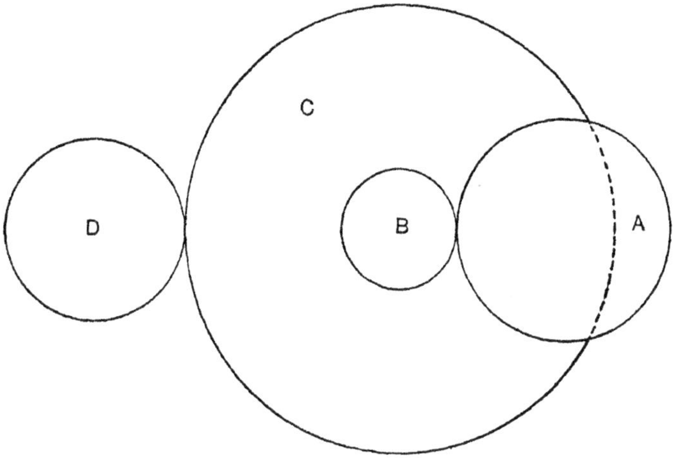

Fig. for Qn. 11.

the metric scale, how many revolutions D will make to every revolution of A.

If A revolves x times per second, show that the number of revolutions of D per sec. is $x \times \frac{a}{b} \times \frac{c}{d}$ where a b c d are the diameters of the wheels A B C D.

What must be the diameter of D in order that it may revolve 20 times as fast as A, the diameters of the wheels A B C being those shown in the figure?

12. Trams run at 10-minute intervals at 10 miles an hour. A man who walks at the rate of 4 miles an hour wishes to go from the starting point A to a stopping point B on the route. What must be the distance AB if he just misses a tram, walks on, and arrives at B at the same moment as the next tram?

13. At what advance must a shopkeeper mark goods costing 90 cents that he may allow a 20 per cent. discount and yet gain 25 per cent.?

14. A man starts daily on a journey of three miles, walking at the rate of 4 miles an hour, but entering an omnibus whose speed is 10 miles an hour as soon as one overtakes him, and completing the journey in it. Obtain a formula connecting T the number of minutes occupied by the whole journey, with t the number of minutes occupied in walking.

Draw a graph of this relation, figuring the scales neatly on the axes. If the whole journey takes half an hour, how far does he walk?

15. A reservoir with a rectangular horizontal section, 50 feet by 20 feet, is discharged by a pipe 8 in. in diameter and 1000 feet long, and the surface of the water is lowered five feet in twenty minutes. Find the mean head of water from the formula

$$v = 50\sqrt{\frac{dH}{L+54d}}$$

where v is the average velocity of discharge in cubic feet per sec. during the 20 minutes, H is the mean head of water in feet, L the length of the pipe in feet, and d the diameter of the pipe in feet.

SUBSECTION (ii). LINEAR EQUATIONS INVOLVING MORE THAN ONE UNKNOWN QUANTITY.

16. A mixture is composed of 17 parts (by weight) of a substance A and 24 of a substance B. Given that the volumes of equal weights of A and B are in the ratio 3 : 7, find the proportions of A and B in the mixture when estimated by volume.

17. The three sides of a triangle are x y z centimetres long respectively. Having given

$$12x = 5y \qquad 2x + 3 = z \qquad x + y - z = 4,$$

find the values of x y z, and draw the triangle.

18. The sum of 17 shillings is to be paid with x florins and y shillings. (i) Write down an equation connecting x and y. (ii) When $x = 7$, what is the value of y? (iii) Write down a second equation connecting x and y when 13 coins are to be used, and find the values of x and y that satisfy both equations.

19. A man driving to a railway station at a rate of 8 miles an hour discovers when 12 miles from home that he has left one of his bags behind. He sends word back by a motor-bicycle that happens to be passing, which travels at the rate of 20 miles an hour; and, five minutes after the news is received, the bag is despatched by motor car at the rate of 40 miles an hour. At what distance from home does the bag overtake the owner?

20. A man finds that he consumes a gallon of spirits and $3\frac{3}{4}$ lb. of tobacco in the same time, and that last year he spent £3. 2s. 10d. on tobacco and spirits together. Spirits are now dearer by 3s. 9d. a gallon and tobacco by 8d. a pound, and he calculates that his expenditure for this year will be £3. 10s. 11d. if he gives up both tobacco and spirits for six weeks during the year, but that he will save £1. 1s. 8d. if he gives up spirits entirely, but maintains his usual consumption of tobacco throughout the year. Find his consumption of spirits and of tobacco during last year, and their prices per gallon and per lb. Assume that a year consists of 52 weeks.

21. Two trains are travelling in opposite directions at uniform speeds. They each pass a telegraph post in the same time, and they pass each other in five seconds. If the lengths of the trains added together be 264 yards, and the faster train's speed exceed that of the slower by 25 per cent., find the length and speed of each train, in feet and feet per second.

22. If the sides BC CA AB of a triangle ABC are produced in these directions, prove that the sum of the three exterior angles so formed is equal to four right angles. (No other proposition on the angles of a triangle may be assumed.)

The triangle ABC has $\angle ABC = 40°$ and $\angle ACB = 80°$; points P Q R are taken on BC CA AB respectively, so that $\angle APC = x°$, $\angle BQA = y°$, $\angle CRB = z°$. Calculate the angles of the triangle formed by AP BQ CR, and hence find the values of x y z if this triangle is equilateral and BQ is perpendicular to AC. Finally, draw a large figure in which x y z have these values, and so verify your work

23. In the formula $z = ax + by$, a and b are numbers, and x and y are prices. Having given

(1) when $x =$ 3s. 5d. and $y =$ 2s. 2d., $z =$ £21. 3s. 4d.;
(2) when $x =$ 3s. 2d. and $y =$ 2s. 1d., $z =$ £19. 18s. 9d.;

determine the values of a and b; and the value of z when $x = 3$s. 0d. and $y = 2$s. 0d.

24. A man takes a boat and rows up stream. When he has gone $1\frac{1}{2}$ miles up stream he passes a bottle floating down with the stream. He goes on up the river for another 18 minutes and then turns. When he gets back to the starting-place he finds that the bottle has just arrived there. Assuming that his speed through the water is uniform, find the speed of the stream in miles per hour.

25. Assuming that the formula $\sqrt{(ax - bx^2)}$ represents for all values of x greater than unity the price in pounds of a season ticket for three months for a distance of x miles on a railway; determine a (to three significant figures) and b (to one figure) from the data:

Distance - - -	55 miles.	26 miles.
Price - - -	£6	£4. 4s.;

and find the price of a season ticket (to the nearest shilling) for a distance of 30 miles.

26. In still water a wave travels at the rate of $\sqrt{32d}$ feet per second, d being the depth of the water in feet. It is observed that the waves formed by a large stone dropped into a stream travel at the rate of 15 feet per second down stream and $6\frac{1}{3}$ feet per second up stream. Find the depth of the stream and its rate of flow.

27. From the relation $v = u + at$ express t in terms of v u a; then substitute this value in $d = ut + \frac{1}{2}at^2$ and so obtain the relation $v^2 = u^2 + 2ad$. Find the values of t and d when $v = 100$, $u = 84$, $a = 32$.

28. Solve $x = 5(x - y + 7)$ $y = 7(y - x - 5)$.

29. Solve the equations:

(i) $\frac{x-1}{6} + \frac{3x}{4} = \frac{2x+7}{3}$;

(ii) $2x + 5 = 3y$ $4y - 6 = 3x$.

30. Find values of x and y from

$$\frac{x+3}{y+3} = \frac{x-7}{y-5} \qquad \frac{x}{7} - \frac{y}{6} = \frac{1}{2}.$$

31. Find x y z if

$$x + y + z = 5 \qquad x + 3y + 7z = 23 \qquad 2x + z = 3y.$$

32. Find a b c if the expression $\dfrac{x+a}{bx-c}$ has the values 2 1, 0·8, when $x=1, 2, 3$ respectively.

33. The equations $3x=2y$ $\dfrac{x}{2}+\dfrac{y}{3}=a$ $4x-y=2$ can be satisfied by the same values of x and y. What is the value of a?

34. In an athletic meet, team A won the championship with a total of 33 points, made up as shown in the following table. Team B won second place with a total of 25 points, and team C third place with a total of 23 points. How many points does each place count?

Team.	1sts.	2nds.	3rds.	Total Score
A - -	5	2	2	33
B - -	3	3	1	25
C - -	1	4	6	23

35. Portland cement is built on the following formula: Per cent. of silica × 2·8 + per cent. of alumina × 1·1 = per cent. of lime. It is desired to make Portland cement of a clay having 12·5 % lime, 44·5 % silica, and 10·1 % of alumina; and a marl having 51·2 % lime, 4·1 % silica, and 0·66 % alumina. Find how many parts of marl should be mixed with 100 parts of clay.

36. Archimedes suspected a crown of being made of a mixture of copper and gold, and proved his contention. Suppose that he found the weight of the crown to be 1140 grammes and of the water displaced by it 63 grammes; gold being 19·3 and copper 8·9 times as heavy as water. Find how many cubic centimetres of each metal there were in the crown, and its weight if it had been pure gold. A cubic centimetre of water weighs a gramme.

SUBSECTION (iii). RIDDLE QUESTIONS.

Many of the questions in this subsection are inverted arithmetical questions, natural data being suppressed and the student asked to find them. The subsection may be omitted without loss.

37. A school met morning and afternoon from Monday to Friday. On Thursday evening, the mistress found that the average attendance for the four days was exactly 46. On Friday every child on the books attended both morning and afternoon. If the average number present at each meeting of the school during the five days was exactly 47, what was the number of children on the books?

38. A man bought 2000 apples. He sold 500 at a gain of 40 %, 500 others at a gain of 30 %, 500 others at a gain of 20 %, and the remainder at a loss of 10 %, and realised a profit of 100d. At what price per apple did he buy the apples?

Also find at what price per apple the remainder should have been sold to give him a profit of 25 % on the whole.

39. There are 24 men too many to form a battalion as a solid square, or 29 too few if one more man be allowed to each side. How many men are there in the battalion?

40. In 1906 a man travelled 6400 miles at an average cost of 4½d. a mile and an average pace of 10 miles an hour. He then hired a motor car, and in 1907 travelled the same number of miles at 6d. a mile, but he saved £30 in hotel expenses, and he saved time by travelling at 15 miles per hour. He calculates that he has gained £16. 13s. 4d. during the year by his purchase. What is the value of his time per hour?

41. A grocer blends teas which cost respectively 1s. 8d. and 1s. 9d. per pound. If he sells 450 lb. of the blended tea at 1s. 10d. he gains £2. 18s. 4d. In what proportion did he mix the teas?

42. Passengers are allowed a certain amount of luggage free on a train. Two passengers paid 6s. 6d. each for excess luggage for the same journey. If one of them travels alone and takes with him the luggage of both, he has to pay 15s. for excess luggage. If the combined luggage weighs 952 lb., how much is allowed free?

43. A passenger with two trunks weighing in the aggregate 360 lb. finds he will be charged £1. 0s. 8d. for excess luggage, whereas if the trunks had belonged to two passengers they would only have been charged 4s. and 7s. 4d. respectively. What weight of luggage is allowed per passenger free of charge?

44. A man had a certain income. Of it £120 was free of income tax, and on the remainder he had to pay 9d. in the £. He saved one quarter of what was left after paying the tax. His income has risen by £100, and he is now allowed only £70 free of tax. After paying the income tax he now saves one-third of what is left. His yearly saving is now £73. 10s. more than it formerly was. Find his original income to the nearest pound.

45. A man's total income is made up partly of his salary and partly of income derived from investments, his salary being four times as great as the rest of his income. He has to pay income tax at the rate of 9d. in the pound on his salary and at the rate of 1s. 2d. in the pound on the other part of his income. He is allowed an abatement of £120 from his salary (*i.e.* he does not have to pay income tax on this portion of his salary), and the total amount of income tax that he pays is £18. 8s. 4d. Find his total income.

46. Express algebraically a number whose unit digit is $2x+5$, and whose 10-digit is $x+2$. What remainder will there be if the number is divided by 6?

Of two numbers one exceeds the other by 2. Show that the difference of their squares is always divisible by 4, and find the numbers if this difference is also divisible by 7 and the numbers are between 30 and 40.

47. A man has five hours to spare, and he spends the time by going out in a coach, and back in a motor-car, travelling respectively at the rate of 10 and 15 miles an hour. How far did he go out?

48. A party of boys wished to divide some apples, and found that if each took 3, there were 4 apples over, but that if each took 4, only one apple was left for the last boy. How many boys and how many apples were there?

49. A man buys two horses, and pays altogether £75. He sells one at a loss of 20 per cent. and the other at a profit of 50 per cent., and receives altogether £81: find how much he gave for each.

50. The cost of 11 cows, 23 sheep, and 9 pigs is £132. 3s.; that of 4 cows, 8 sheep, and 5 pigs is £50; while 7 cows, 12 sheep, and 10 pigs cost £85. Find the cost of each kind of animal.

51. A number of sovereigns and shillings are together worth £27. 10s. and together weigh 7425 grains. How many are there of each? A sovereign weighs 123 grains; a shilling weighs 87 grains.

52. A brother and sister were asked their ages. The boy said, 'Half my age is one year more than half hers.' The girl said, 'In 6 more years, 10 times my age then will be equal to 9 times his age then.' What were their ages?

53. Two persons A and B go out shopping A spends 7s. 6d. and B spends 5s.; it is then found that A has $2\frac{1}{2}$ times as much as B. A then lends B 2s. 6d. and B spends 7s., after which A has 20 times as much as B. Find how much money each had at first.

54. The headmistress of a private boarding-school has 45 and 50 pupils respectively in two consecutive years. She finds that this addition of 5 pupils increases her yearly expenditure by £131. 5s., but diminishes the average cost per head by £1. 10s. What was the total expenditure in each of the two years?

55. A number is expressed by two digits, and another number by the same two digits interchanged. The difference between the numbers is 18, and the sum of the four digits is 16. Find the numbers.

56. There are two numbers whose sum is 125. Also 3/4 of the one number exceeds 2/3 of the other by 13. Find the numbers.

57. If the breadth of a room were increased by 5 feet and the length diminished by 5 feet, the area would be increased by 10 square feet. But if the length and breadth had each been increased by 5 feet, the area would have been increased by 230 square feet. Find the length and breadth of the room.

58. A mixture P is formed of 23 parts of a liquid A and 37 parts of a liquid B, and another mixture Q is formed of 15 parts of A and 29 parts of B. If 7 parts of P are mixed with 18 parts of Q, find the percentages of A and B in the resulting mixture, giving them to two significant figures.

59. The ages of four brothers are respectively a b c d years. The age a of the eldest is equal to the combined ages of the others. In how many years will his age be only half their combined ages?

60. A father leaves his estate of £25,000 among his four sons, directing that the third is to have two-thirds as much as the second, the fourth three-quarters as much as the third, and the eldest £5000 more than the amount divided among the others. Find, to the nearest shilling, how much each gets.

61. In leaving his fortune to his four sons, a man directed that the eldest should have one-third more than the second son, while the fourth should have one-third less than the third son; in addition, he desired that the eldest and youngest sons should have between them half of the whole property. The estate being valued at £17,403. 13s. 9d., find the share of each son to the nearest shilling.

62. The distance by road from a village to a railway station is d miles; a field path is 2 miles shorter than the road. Give expressions for the time taken (i) to drive by road at 7 miles an hour; (ii) to walk by the footpath at $4\frac{1}{2}$ miles an hour. If at these rates the same time is taken for walking as for driving, find the length of each route, and then verify your results.

63. A passenger finds that his train actually arrives at four o'clock. If he had travelled at 60 miles an hour he would have arrived at half-past three, while if at 40 miles an hour he would have arrived at a quarter past four; at what time did he start?

64. A rides from P to Q at the rate of 12 miles an hour and returns at the rate of 8 miles an hour. B rides both ways at the rate of 10 miles an hour, and takes 10 minutes less than A to do the double journey. Find the distance from P to Q.

65. A rides from P to Q on his bicycle in 2 hours 40 minutes; B goes by train in 2 hours 48 minutes. If the average rate of the train be 5 miles an hour quicker than that of the bicycle, but the length of the journey by train 16 miles greater than that by bicycle, find both rates and distances.

66. Running a certain course, starting level, A finishes 176 yards in front of B. With a minute's start, B would have finished 160 yards in front of A; and with 160 yards' start he would have finished three seconds later than A. What is the length of the course, and what is the time of each (in seconds) for the full course? [It must be assumed that the speed of each is constant over the course.]

Section 28. Loci.

1. By means of your protractor draw two straight lines AOB and COD, intersecting at O at an angle of 50°. On each side of AOB draw a straight line parallel to AOB, and at a distance of 1 cm.; on each side of COD draw a parallel at a distance of 1 cm.; state how you draw these parallels.

In the same way draw parallels to AOB and COD at distances of 2, 3, 4 cm.

Mark all the points given by your construction which are equidistant from AOB and COD; and draw freehand the locus of all points equidistant from AOB and COD.

Discuss by general reasoning the nature of the locus and the magnitude of the angle at which the two parts of the locus intersect.

2. Draw and state in words, without proof, the complete locus of a point, which moves so as to be always equally distant from two given intersecting straight lines.

Prove that any point which is not on your locus is not equidistant from the two straight lines.

3. Show how, given a triangle ABC, to find points in BC and BC produced which are equidistant from AB and AC.

In what special case can only one such point be found?

4. Show that there are just four points which are equidistant from three straight lines which do not pass through a point and are not parallel.

5. Two points A and B are $1\frac{1}{2}$ inches apart. Construct half a dozen positions of a point which moves so that its distance from A is always double its distance from B; and draw the locus freehand.

6. Draw a straight line XQ, 9 inches long, and mark the position of a point Z, $1\frac{1}{2}$ inches distant from the line and near the middle of the line. Draw a straight line AB on tracing-paper; mark a point N on this line $2\frac{1}{2}$ inches from A. Move the tracing-paper about so that AB always passes through Z, and A is always on the line XQ. Find in this way the locus of N; that is, for each position of A determine the position of N by pricking through the tracing-paper, and then draw a smooth curve through the points thus found.

Give reasons for thinking that the curve would never meet XQ.

7. Near the lower edge of a page draw a horizontal line MN right across the page, and from about its middle point X erect a perpendicular XF 2 cm. long. On the tracing-paper draw 10 circles having the same centre, the radii being 1, 2, 3, 4, etc., cm. respectively; then place the tracing-paper so that the circumference of the smallest circle passes through F and touches MN, and in this position prick through the centre of the circle. Deal with the second and the remaining circles in a similar way, determining the positions of the centre points on both sides of FX. Draw a freehand curve through the points thus obtained.

8. A dog at a point A catches sight of his master at B, 400 metres away. The master is walking at the rate of 2 metres a second to a point C, in a direction at right angles to AB, and 150 metres from B. The dog sets off towards B at the rate of 5 metres a second. After 10 seconds the dog changes his course towards the point where his master is at that moment, and runs towards this point for the next 10 seconds. He continues running at the same speed, and changing his course in this way every 10 seconds. Show, on as large a scale as your paper allows, the path of the dog while his master walks from B to C.

9. A ladder 20 metres long is set against a house, the bottom of the ladder being 2 metres from the wall, and a man goes 14 metres up it. Then the ladder slips. Draw, on as large a scale as your paper allows, the path of the man as the ladder slips.

10. A tube AB, five inches long, rotates at a uniform speed about its middle point. A marble moves through the tube at uniform speed, traversing the tube in the time which the tube takes to perform a complete revolution. Draw the path of the marble.

11. On transparent paper draw lines OX and OY, meeting at an angle of 30°, making OX 10 cm. and OY 7·5 cm. long; joining these lines draw others X_1Y_1 X_2Y_2 X_3Y_3 .. parallel to XY and cutting OY at intervals of 0·5 cm. Draw a semi-circle bounded by a diameter, and mark T the centre and S any point within the boundary. Place the transparent paper on this semi-circle with O above the point S, and pin it at this point. Now rotate the transparent paper, and as any point on OX (say X_1)

comes over the boundary of the semi-circle, prick through Y_1 the corresponding point on OY; when X_2 is on the boundary prick through Y_2; and so on. Draw the complete locus of Y as X traces out the given figure.

12. Draw a circle of radius 2 inches. Take a diameter AB, and divide it into ten equal parts. Call the points of division C_1 C_2 .. C_9. Draw the set of chords $D_1C_1E_1$ $D_2C_2E_2$ .. at right angles to AB, meeting the circumference in D_1 E_1 D_2 E_2 etc. F_1 and G_1 are the feet of the perpendiculars let fall from C_1 on AD_1 and AE_1 respectively. Similarly let fall perpendiculars C_2F_2 and C_2G_2 on AD_2 and AE_2, and so on.

Draw a fair curve through all the F and G points thus obtained. Measure the greatest breadth of the curve, and write down your result. How would you obtain more points on the curve if required? Where do you think the curve cuts the line AB?

13. In the sketch AB is a horizontal rail 15 ft. above the ground. A bar DE, 12 ft. long, is suspended from the rail, and while its upper end D moves along AB at the rate of one inch per second, the bar rotates about D at the uniform rate of one degree per second in the plane CAB and in the direction shown by the arrow.

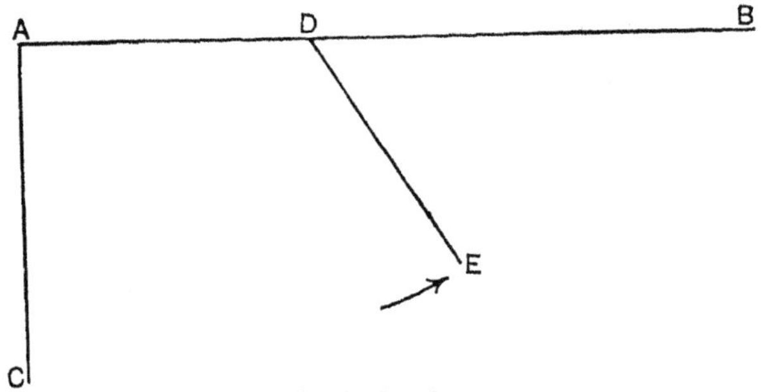

Fig. for Qn. 13.

Supposing that initially D coincides with A, and DE is vertical, trace the path of E until DE is horizontal. Use a scale of 1 cm. to the foot.

From your diagram find the distance of E above the ground, and its horizontal distance from AC, after one minute.

14. In the sketch AB is 1 inch, AC is $1\frac{1}{2}$ inches, and BP is 5 inches. AB rotates and BP always passes through the fixed point C. Draw the path of P for one complete revolution of AB.

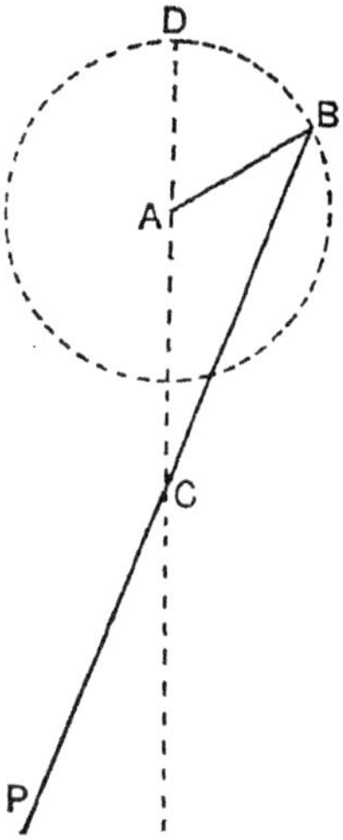

Fig. for Qn. 14.

15. In the sketch, OADB is a square and OAPB a quadrant. Think of two lines each moving uniformly and both starting together from the position OA; one, MN, moving parallel to OA, and reaching its final position BD, in the same time that the other, OP, takes to revolve about O to its final position OB. Their point of intersection C traces out a curve. Obtain this curve for a quadrant of radius 9 cm. without your protractor. Use the curve to draw the angle AOC when MN has moved (i) 3·2 cm., (ii) 6·4 cm. Trace the angle XYZ on transparent paper, use the curve to divide it into two parts in the ratio of 2 to 3, and state your method.

Now use your protractor to measure the two angles AOC which you drew, and the two parts into which you divided XYZ.

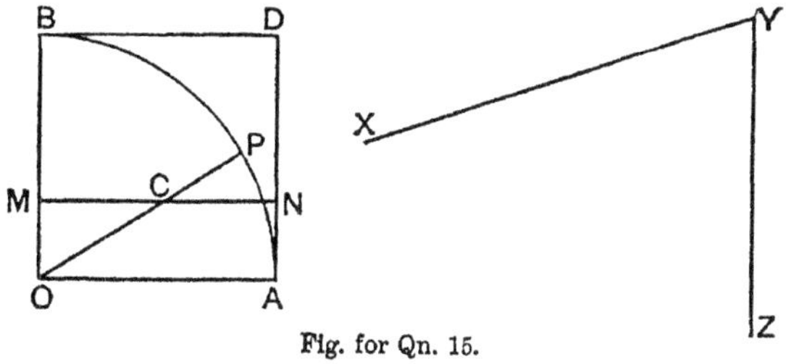

Fig. for Qn. 15.

16. The sketch shows a piece of mechanism, consisting of

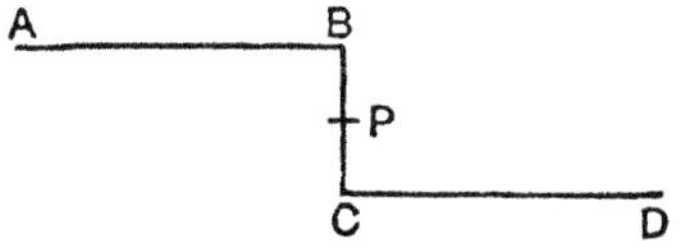

Fig. for Qn. 16.

3 straight rods freely jointed at B and C. The rods AB and CD can move round the fixed points A and D, carrying BC with

them. Draw the path of P, and find the greatest angle through which AB can turn. The dimensions are AB = CD = 2 ft., BP = PC = 6 in., and in the given position BC is at right angles to AB and CD.

17. One end A of a rod AB 4 feet long describes a circle of radius 6 inches, while the other end B moves along the fixed line CD as shown in the sketch. The furthest position of B from the circle is on the dotted line shown. Draw the curve described by the middle point of AB on a scale of 1 inch to a foot.

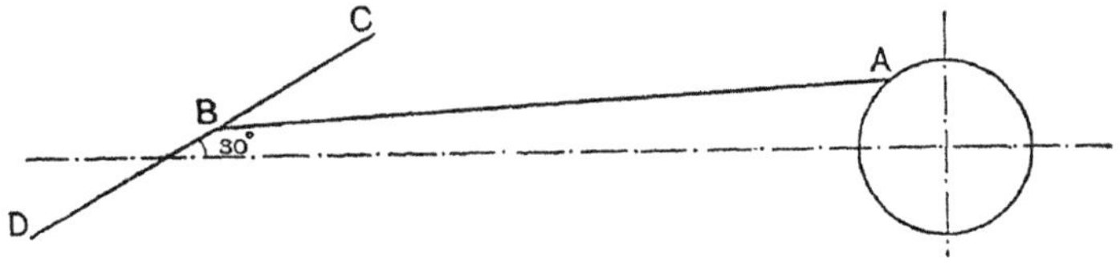

Fig. for Qu. 17.

18. A board has cut in it two grooves crossing at right angles. A rod AB has on it two knobs at A and B which slide along the grooves. Investigate the kind of path which any point P of the rod describes. For this purpose draw two straight lines at right angles to represent the grooves, and draw on tracing-paper a line with the knobs A and B marked to represent the rod.

Draw the path described by P for a number of different positions of P on the rod, some between A and B and some outside. For what positions of P on the rod are the ovals flatter, and for what positions are they rounder? All these ovals are called ellipses.

Section 29. Areas.

(i.) RECTANGULAR AREAS AND VOLUMES.

1. A closed box is made of wood x inches thick. The outside length breadth and depth are a b c inches respectively. Write down expressions for the inside length breadth and depth.

Give an expression for the whole volume (wood and inside volume together), an expression for the inside volume, and an expression for the volume of wood in the box.

Arrange the last expression according to powers of x.

2. In the annexed figure, ABCD is a cardboard mount, EFGH being a picture in the centre. The sides of the rectangle EFGH are produced as shown, and it is found that $NF = HL = a$, $KG = EM = b$, $NE = GL = c$, and $KF = HM = d$, all these lengths in inches.

Express the areas ABCD and EFGH measured in square inches, in terms of a b c d.

Verify by algebra that the difference of these areas is equal to the sum of the areas of the four rectangles AF HC BG DE.

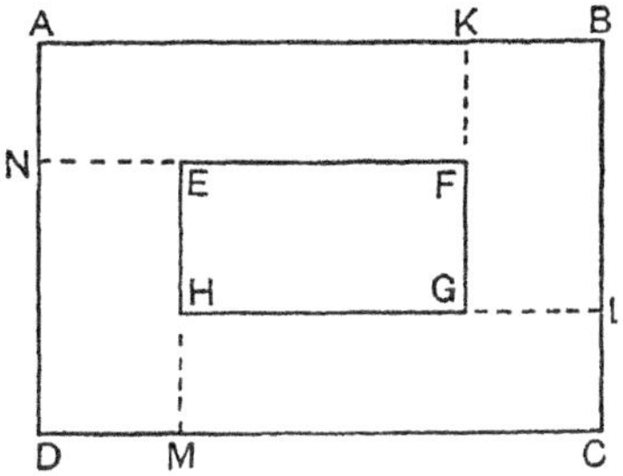

Fig. for Qn. 2.

3. The dimensions in inches of five rectangular pieces of card are 4 by 4, 2 by 2, 2 by 2, 4 by 2, 4 by 2. With these (omitting none) can be formed either (i) two separate squares, or (ii) one square and two equal rectangles. Draw the figure for each of these two combinations, and state, in Euclid's words if you like, the geometrical property which is thus illustrated.

4. A square pond measuring a yards each way receives the drainage of b acres. It is found that a thunderstorm raised the level of water in the pond by c inches. What depth of water fell, if it may be supposed that it fell uniformly over the whole area, and that all the water which fell found its way into the pond?

Give a numerical result for the case $a = 40$, $b = 20$, $c = 14$.

5. Prove by means of a figure that the difference of the squares on two straight lines is equal to the rectangle contained by the sum and difference of the lines.

6. AB is a straight line divided at C; DE is another straight line divided at F. Show that the area of the rectangle contained by AB and DE is the sum of the areas of the four rectangles contained by AC and DF, by AC and FE, by CB and DF, and by CB and FE.

7. If a straight line be divided into three parts, prove that the square on the whole line exceeds the sum of the squares on the three separate parts by twice the sum of the three rectangles contained by each pair of the parts.

8. Show that if a straight line be divided into two parts the rectangles contained by the whole and each of the parts are together equal to the square on the whole line.

A B C D are four points in a straight line such that AB = BC = CD, and E is the middle point of CD. Prove that the square on AB and the rectangle AB.BE are together equal to twice the rectangle AE.ED.

9. Give a geometrical illustration of the identity

$$(a+b)(a-b)+b^2=a^2.$$

State clearly which are the lines corresponding to a, b, $a+b$, $a-b$, and the rectangles corresponding to a^2, b^2, $(a+b)(a-b)$.

10. State, and prove by reference to a figure, the geometrical theorem corresponding to the algebraic formula

$$(a+b)(c-d) = ac-bd+bc-ad.$$

11. If a and b are lengths, give a geometrical interpretation of the theorem $(a+b)(a-b)=a^2-b^2$, and prove the theorem so enunciated.

12. Enunciate the geometrical theorem corresponding to the following algebraical identity:

$$(a+b)^2=a^2+2ab+b^2$$

and draw an illustrative diagram.

13. CD is a straight line bisected at the point A and divided unequally at the point B. Prove that the rectangle CB.BD is equal to the difference of the squares on AD and AB.

Hence show how to divide a given straight line into two parts such that the rectangle contained by them shall be the greatest possible.

14. Prove geometrically that if A B C are three collinear points in this order,

$$BC^2 = AB^2 + AC^2 - 2AB.AC.$$

Also state the form which the result takes if A lies between B and C, and draw a corresponding diagram.

15. If A P B be three collinear points in this order, prove that $AP^2 + PB^2 + 2AP.PB$ is equal to AB^2.

If the three points are not collinear, prove that the former quantity is greater than the latter.

16. Show how to illustrate by a geometrical figure the identity

$$xy - ab = (x-a)(y-b) + b(x-a) + a(y-b).$$

A teacher illustrating this identity chose a large sheet of postage stamps to represent the product xy, and commenced his illustration by tearing off one stamp to represent the subtraction of ab. In what respects will his demonstration fail as a general illustration?

(ii.) AREA OF TRIANGLE.

17. A plate of metal has the dimensions shown on the rough sketch. Make an accurate full-size drawing of the plate, draw and measure the length of the perpendicular from A to the

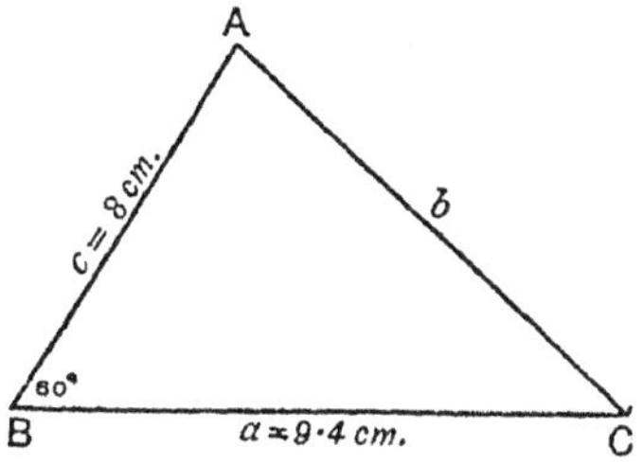

Fig. for Qn. 17.

opposite side, and find the area of the surface ABC; also measure the length of the side b.

Check your two results by substituting in the following:

(i) Area of $ABC = \frac{ca}{2} \times 0{\cdot}866$; (ii) $b^2 = c^2 + a^2 - ca$.

18. Show how to construct a triangle, two of whose sides are 3 and 5 inches long, and whose area is 6 square inches. Perform the construction, and show experimentally that two such triangles may be constructed, one of which is right-angled and the other obtuse-angled.

19. Draw a triangle having two sides of lengths 3·4 and 5·2 inches, and an area of 6 square inches. How many such triangles may be drawn, not identically equal?

20. Construct a triangle having given that the area is 2·7 square inches, the base is 2·36 inches in length, and one of the angles at the base is 30°. Measure the two sides of the triangle. Explain and justify your construction.

21. The area of a triangle ABC is 12 square in., and the base BC is 6 in., this being the longest side, while AC is the shortest side. Prove that the vertex must lie on a segment of a straight line; constructing an accurate figure and measuring the length of the segment.

22. Construct a rectangle with one side 7 cm., equal in area to an equilateral triangle of side 10 cm. Measure the other side of the rectangle.

23. Prove that a straight line drawn through the middle point of one side of a triangle parallel to a second side bisects the third side.

24. Prove that the straight line joining the middle points of two sides of a triangle is parallel to the third side and equal to half of it.

25. Show clearly in a diagram, giving your reasons, whereabouts a point P must lie inside a triangle ABC in order that each of the triangles PBC PCA PAB may be less in area than half the triangle ABC.

26. Draw a triangle ABC with sides 3 cm., 4 cm., and 5 cm. Construct a second triangle PQR, so that A B C are the middle points of its sides, and prove the correctness of your method.

What is the area of the triangle PQR?

27. *a b c d* are equidistant parallel lines; from a point O on *a* straight lines OBCD and OPQR are drawn cutting *b* at B and P, c at C and Q, d at D and R. How many times is the area OPB contained in the area ORD? Justify your statement.

28. ABC is any triangle with the angles A and B unequal. The angle BAD is made equal to the angle B, and AD equal to BC. Prove that the triangles BCD and ADC are equal, and that DC is parallel to AB.

29. Construct a triangle whose sides BC CA AB are respectively 14, 13, 15 cm. long. Measure the perpendicular from A on BC, and find a point D in BC such that the difference of the areas of the triangles ABD ACD is 12 square cm.

30. If the diagonal AC of a quadrilateral ABCD divides it into two parts of equal area, prove that the diagonal BD is bisected by AC.

If, in addition, AC bisects the angle BAD, prove that the triangles ACB and ACD are equal in all respects.

31. If each of the diagonals of a quadrilateral divides it into two triangles equal in area, prove that it is a parallelogram.

32. Show that a parallelogram can be formed by joining the middle points of the sides of any quadrilateral.

33. Define a parallelogram, and from your definition prove that a diagonal of the figure divides it into two parts of equal area.

Also prove that if one diagonal of a quadrilateral divides it into two parts of equal area it need not be a parallelogram, but that it must be a parallelogram if in addition one pair of opposite sides are parallel.

(iii.) AREA OF PARALLELOGRAM.

34. Prove that parallelograms on the same base and between the same parallels are equal in area, and show that a similar theorem is true for triangles.

A surveyor wishes to find the area of a field which is in the shape of a parallelogram ABCD. He finds AB and AD to be each 35 chains and the angle BAD to be 80°. Make a correct drawing of the field to the scale of 1 inch to 10 chains.

Find the number of acres in the field to the nearest acre.

35. Construct a parallelogram whose diagonals include an angle of 76° and are 8·4 cm. and 5·6 cm. long. Measure its acute angle and find its area.

36. Draw a parallelogram having two sides of lengths 6 and 8 cm., and one diagonal of length 9 cm. Upon a line of length 2 in. describe a rectangle equal in area to the parallelogram.

37. In solving a problem in which a rectangle with a side of 3 cm. had to be drawn equal to a square of area 64 sq. cm., a student gave the following construction: "Draw the square OXZY with side 8 cm. Along one side OY, set off OA of length 3 cm., and join AX. Draw YP parallel to AX and meeting OX produced in P. Then OP is the other side of the required rectangle."

Carry out this construction, and find roughly by measuring OP whether the construction is correct. State whether you consider the method valid, and justify by a general proof whatever statement you make.

38. Prove that the complements of the parallelograms about the diagonal of any parallelogram are equal to one another.

Show how to find from this proposition the approximate value of $3{\cdot}7 \times 4{\cdot}2 \div 2{\cdot}9$ by means of a geometrical construction.

39. Construct a parallelogram ABCD of area 4 sq. in., the side AB being 2·2 inches long and the angle ABD being 33°.

40. Show how to describe a rectangle equal to a given triangle.

Show how to construct a triangle having two sides equal to two given lines and its area equal to that of a given triangle.

41. Draw a square on a line 3 inches long. Construct a parallelogram, whose sides are 3·7 inches and 2·8 inches in length, equal to the square.

Justify the method of your construction. Measure the longer diagonal.

42. The sides of a triangle are 5, 8 and 11 centimetres, and one diagonal of a rhombus equal to the triangle in area is required to be 9 centimetres. Construct the rhombus, explaining carefully all the steps.

43. Show how to construct a parallelogram that shall be equal to a given triangle, and shall have one side of given length and one angle of given magnitude.

44. Define a parallelogram, and prove that if the opposite sides of a quadrilateral are equal the figure must be a parallelogram.

Construct the greatest possible parallelogram whose diagonals are 4·7 cm., 6·8 cm., and measure its sides.

45. Prove that the area of any parallelogram is equal to the area of the rectangle contained by one side of the parallelogram and the perpendicular distance between that side and the opposite side.

If ABCD be a parallelogram in which the distance between AB and CD is greater than AB, prove that the distance between BC and DA is less than BC.

(iv.) VARIOUS AREAS.

46. Draw a rectangle ABCD, in which $AB = 4 \cdot 16''$, $BC = 4 \cdot 3''$. Take points P Q R S in AB BC CD DA, such that $AP = 1 \cdot 76''$, $BQ = 0 \cdot 7''$, $CR = 1 \cdot 76''$, $DS = 1''$.

Calculate the area of PQRS in square inches.

47. ABCD is a square. O is the middle point of AB. A square PQRS is inscribed in ABCD having the angular point P in AO, and Q in BC. Prove that $AP = BQ$, and also that the nearer P is to O the smaller is the square PQRS.

48. In the figure ABCD represents a square of side 9 inches. $AP = AS = CQ = CR = x$ inches. Prove that PQRS is a rectangle, and by calculating the areas of the four triangles, or otherwise, show that its area is $18x - 2x^2$ square inches.

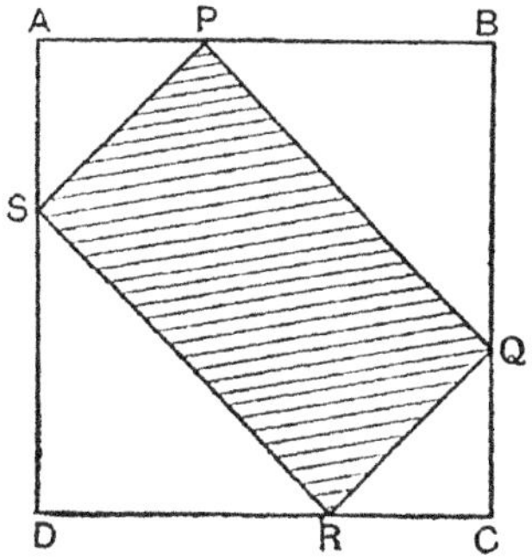

Fig. for Qn. 48.

Give x the values 0, 1, 2 . . . 9, and calculate the corresponding areas of PQRS. Put the results in tabular form, and show by a graph the relation between the area of PQRS and the length of AP.

Find from your graph what value of AP gives the greatest area for PQRS, and what are then the lengths of its sides.

49. Show that if a triangle and a rectangle be upon the same base and have equal altitudes, the rectangle will be double the triangle.

Prove that the area of a quadrilateral which has two sides parallel is equal to the rectangle under half the sum of these sides and the distance between them.

50. The parallel sides of a trapezium are 6·7 cm. and 9·8 cm. in length; the two remaining sides are 5·5 cm. and 4·2 cm.

Draw the trapezium, and from your figure calculate approximately its area.

Prove that the angles of this trapezium will be unaltered if each of the parallel sides is increased by a length x cm., but the area will be increased by a quantity kx sq. cm., where k is a constant. Find the value of k in any way you please.

51. Show that triangles of the same or equal altitude on equal bases are equal to one another.

A four-sided figure has two of its sides parallel; show how through the middle point of one of those sides to draw a line bisecting the quadrilateral.

52. Make a quadrilateral ABCD having AB = 9 cm., BC = 11 cm., the angle ABC = 90°, the angle BCD = 75°, and the angle CDA = 90°. Measure AD and DC and then calculate the area of the figure.

Also find the area of the figure by counting squares.

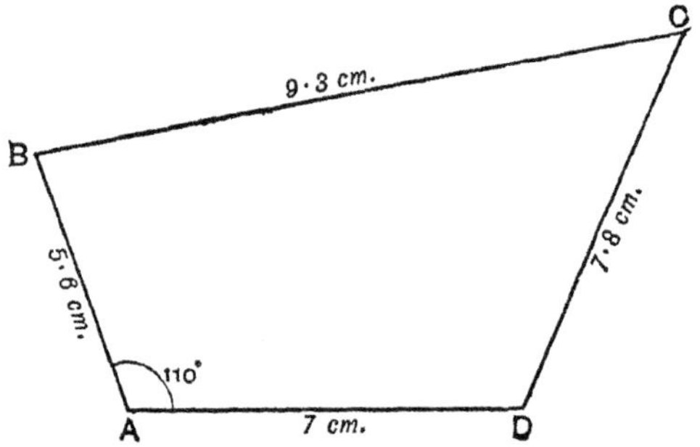

Fig. for Qn. 53.

53. Draw full-size in your book the quadrilateral shown in the rough sketch, explaining your method.

Measure each of the angles of the quadrilateral and give their sum.

Join AC; and find the area of each of the triangles ABC and ADC and of the quadrilateral.

54. ABCD is a lawn in the shape of a quadrilateral, such that AB = 38 ft., BC = 32·5 ft., CD = 24·8 ft., DA = 35 ft., and AC = 45 ft.

Draw a plan of the lawn to the scale of 10 ft. to the inch, and from your figure calculate the area of the lawn.

55. Construct a quadrilateral ABCD of which two sides AB and BC—lengths 1·1 and 2·4 in.—contain an angle 130°, and the other two sides CD and DA are of lengths 2·5 and 3 in. Describe a triangle of equal area to it on a base of 4 in., and state the height of the triangle.

56. Prove that a parallelogram is divided into two equal triangles by either diagonal.

AOC and BOD are the diagonals of a quadrilateral ABCD. The lengths AO, BO, CO, DO are 2·7, 3·3, 2·7, 3·1 inches respectively. Prove that the figure ABCD cannot be a parallelogram but is bisected by the line BD.

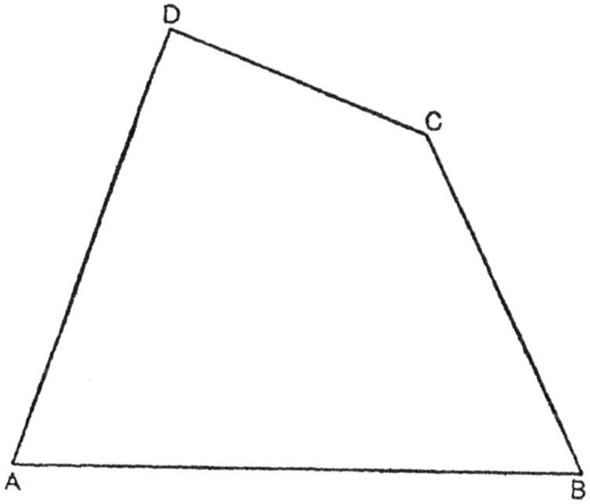

Fig. for Qn. 57.

57. In the figure ABCD represents a quadrilateral field. Transfer the figure to your book and find a point P in the side

AB such that the area of the triangle DAP is one-third of the area of the quadrilateral ABCD.

If 1 inch in the figure represents 200 feet, find the area of the field in acres.

58. ABCDE, shown in the figure, is the plan of a field. The scale is 5 centimetres to 100 yards. Calculate, to the nearest acre, the area of the field.

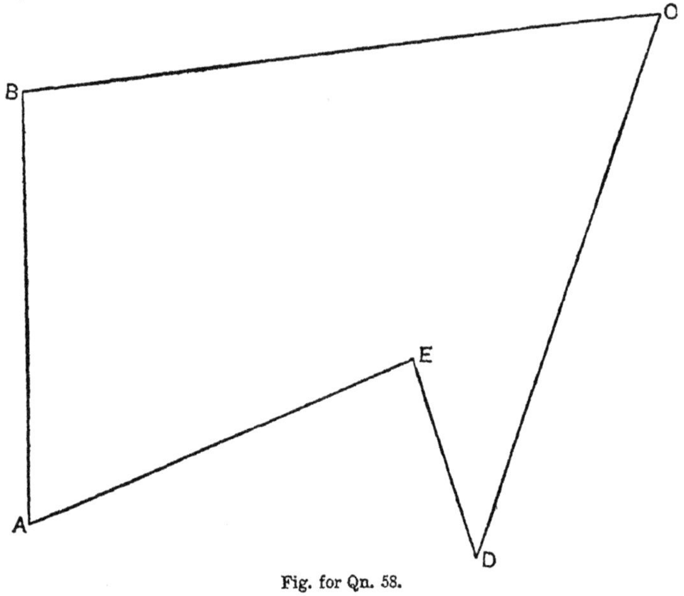

Fig. for Qn. 58.

Take sufficient measurements from the figure to make a copy thereof in your book. Make such a copy, stating exactly what measurements you took.

Determine on your copy, by geometrical construction, the position of a point P in BC such that the areas of the triangles ABP and ADE may be equal. Explain and justify your construction.

59. A rectangular piece of cardboard (of uniform thickness), sides 33 cm. and 22 cm., weighs 50 grams. Out of it is cut an irregular figure which weighs 18·3 grams. What is the area of the irregular figure?

60. Take two lines OA and OB inclined at an angle of 60°. Make OA = 8 cm., OB = 12 cm. Produce OA to C making AC = 8 cm., and OB to D making BD = 4 cm. At any point K of AC erect a perpendicular KN (lying between OA and OB), and mark off on it KL KM KN equal respectively to 1 2 3 cm. At any point S of BD erect a perpendicular SR (lying between OA and OB), and mark off on it SP SQ SR equal respectively to 6 4 2 cm. Draw a parallel to OA through L, and a parallel to OB through P; let them meet in X. Proceed similarly with the points M and Q; N and R; calling the points of intersection of the parallels Y and Z. Join XY and YZ.

What does your result suggest about the locus of a point V such that the sum of the areas of two triangles with bases AC and BD and a common vertex V is equal to 16 square centimetres? Prove your result.

Section 30. Negative Quantities.

1. In the figure the thick line AB represents a level railway which is carried partly through cuttings and partly on an embankment. The irregular line represents the surface of the ground. If the scale of horizontal distances is half an inch to a mile, and the scale of vertical distances half an inch to 100 feet,

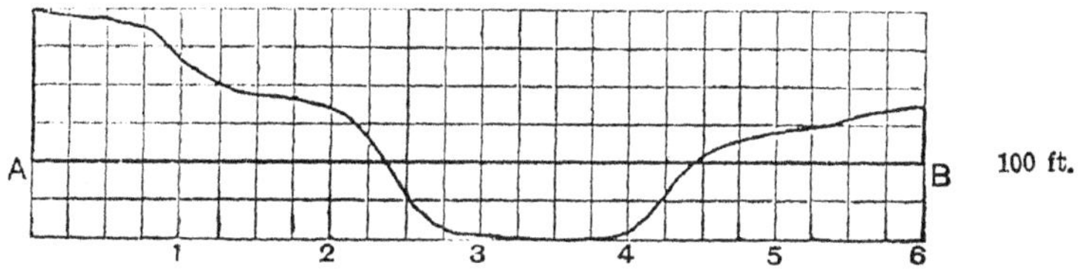

Fig. for Qn. 1.

draw up a table showing the height of the ground above the level of the railway at intervals of half a mile starting from A, and indicating points where the ground is below the level of the railway by prefixing the sign −.

2. Two thermometers A and B are both graduated, but on A the freezing point is marked 32° and the boiling point 212°, while on B neither of these points is indicated. The following pairs of corresponding readings of the two thermometers are observed:

A	59°	77°
B	15°	25°

Show, by means of a graph, the readings of B corresponding to those of A between $-50°$ and $212°$, and from your graph ascertain

(i) the freezing point and the boiling point on B,

(ii) the reading of each thermometer corresponding to zero on the other,

(iii) the temperature for which the readings of A and B are the same.

3. The sides of a rectangle measure p and q inches respectively. From a point within the rectangle four straight lines are drawn perpendicular to the sides; find the sum of these lines in terms of p and q. Prove that the sum is the same for all positions of the point within the rectangle.

If the point is outside the rectangle, and perpendiculars are dropped from it on the sides, produced if necessary, show that the sum of the perpendiculars can still be expressed in terms of p and q by regarding some of the perpendiculars as positive and others as negative.

4. Perpendiculars are drawn from the four vertices of a parallelogram upon a straight line which does not cut the parallelogram. Prove that the sum of the lengths of the perpendiculars drawn from one pair of opposite vertices is equal to the sum of the lengths of those drawn from the other pair of vertices.

How must the statement be modified in the case when the straight line upon which the perpendiculars are drawn cuts the parallelogram?

5. Find the two products obtained by multiplying $1+2x\pm5x^2$ by $6+7x\pm5x^2$, taking both upper signs, or both lower signs, in the third terms.

Find the value of the difference of these two products when $x=0{\cdot}1$.

Hence, if the dimensions of a rectangle are known to the nearest tenth of an inch as 6·7 inches by 1·2 inches, find the range of error possible in estimating the area.

6. Trace on the same diagram the graphs for values of x between $x=-5$ and $x=5$ of

$$y=1/x \quad \text{and} \quad y=3(1+x/4).$$

From your diagram determine the values of x for which

$$1/x=3(1+x/4).$$

7. Taking 2 cm. as your unit of length, trace the graph of $y=\dfrac{10}{x^2+2}$ from $x=-4$ to $x=+4$.

What is the greatest value of y? To what does the value of y approach when x becomes extremely large?

8. Plot the graph of $y=1{\cdot}5-2{\cdot}4x-2x^2$, taking an inch as the unit of measurement for x and for y, and x ranging in value from -2 to 1.

Find where on your graph y is equal to x, and verify by substitution in the equation.

9. Give any reasons you can which support the statements

$$a-(-b)=a+b, \qquad (-a)(-b)=ab.$$

For what values of x is the expression $(x-5)(x-1)$ positive?

10. The expression $(ax^2+b)/x$ has the value $1/2$ when $x=1$, and the value 7 when $x=2$. Use these facts to write down two equations, and by solving the equations find the values of a and b.

Using these values of a and b, find the value of the expression $x/(ax^2+b)$ when $x=-1$ and when $x=4$.

11. Two variable quantities x and y are connected by the equation $y=\dfrac{x+a}{bx+c}$.

When $x=1$, $y=0$; when $x=2$, $y=5$; and when $x=3$, $y=-2$. Find the values of a b c; and the value of y when $x=-2$.

12. From the statement $ax^2+bx=cx-1/d$ find

(i) d, when $a=1$, $b=2$, $c=-1$, $x=-2$;

(ii) a and c, when $x=0{\cdot}4$, $b=3$, $d=-1$, $a+c=2{\cdot}25$.

13. If $y=\frac{a(x+b)}{x-a}$ find $x+y$ in terms of x.

Express your result in its simplest form when $a=\frac{550}{17}$, $b=-\frac{250}{11}$.

Section 31. Specifications.

QUESTIONS INVOLVING THE CONSTRUCTION OF FIGURES FROM ADEQUATE SPECIFICATION.

1. Copy accurately on your paper the given quadrilateral ABCD, taking whatever measurements you wish.

Give in appropriate units the measurements you took, and say how you drew the figure.

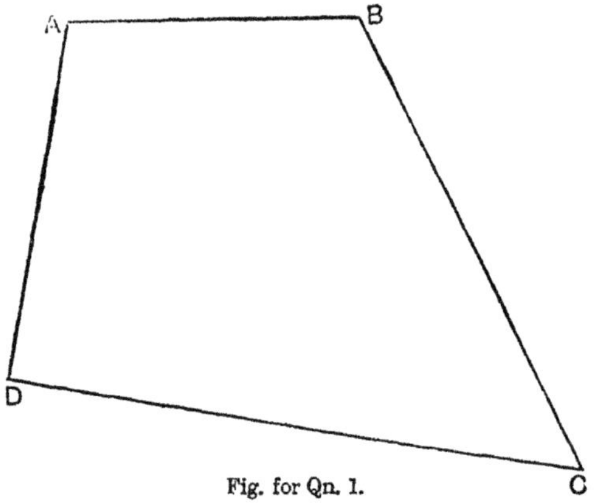

Fig. for Qn. 1.

Mention two other sets of measurements that could have been used instead. How many measurements must there be in each set?

2. Take, and record in your book, sufficient measurements to enable you to copy the given figure (p. 169).

From your measurements draw the figure in your book.

To check the accuracy of your work take two measurements of the original figure, which you did not use in copying the figure; take the corresponding measurements of your copy; and record all four.

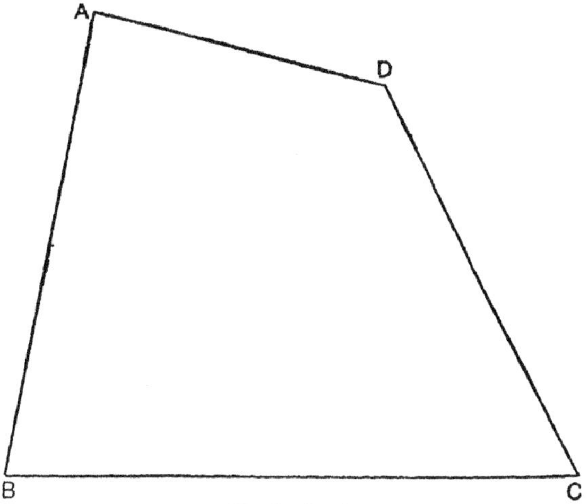

Fig. for Qn. 2.

3. It is required to make a plan of the rectangular field ABCD,

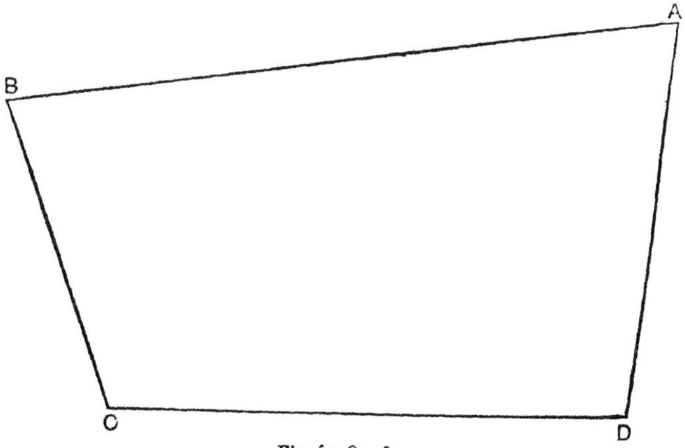

Fig. for Qn. 3.

shown in the figure, of which it is known that CD runs east and west and is 120 yards long.

Supposing that you have appliances for measuring lengths only, what is the smallest number of additional measurements that you must take?

If, on the other hand, you wish to measure angles only, what is the smallest number of additional measurements you must take?

Give reasons for your statements, and determine from the figure the length of each line which you would measure in the first case, and the size of each angle which you would measure in the second case.

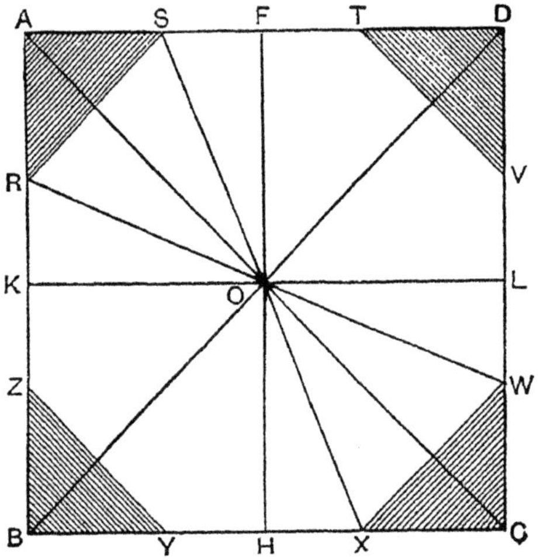

Fig. for Qn. 4.

4. A boy wishes to plane down the edges of a square pole, so that the pole may have eight equal faces. He proceeds as follows:—Let ABCD (see sketch) represent the end of the pole. Draw the diagonals, AC and BD, intersecting at O, and through O draw FH and KL parallel to the sides of the square. Bisect the angles AOF and AOK by lines OS and OR, meeting the sides of the square in R and S. Join RS. In a similar manner obtain the lines TV WX YZ. Then RS ST etc. are the eight sides required.

Carry out the construction, making the side of the square

3 inches long. Prove that RS = ST = TV = etc.; assuming that lines passing through O are bisected at that point.

Also find by measurement the area of the triangle RAS and so estimate roughly what fraction of the pole must be planed away.

5. A village B is 3 miles east of another village A. A school has to be built which shall not be more than 2 miles from either village. Mark on a plan the area in which the school may be built. If the school is built at a point which is 2 miles from each of the villages, find by measurement from your plan how far it is from a straight road from A to B.

6. ABCD is a quadrilateral figure such that the angles at A and C are right angles. Draw a plan of the figure having given that BD measures 4·15 inches and that the angles ABD and ABC are 30° and 75° respectively. Prove that two of the sides of the quadrilateral are equal.

Measure the diagonal AC, and find the area of the quadrilateral in any way you please.

7. AB is a vertical pole, BCD is a straight horizontal line, the angles ACB and ADB measure 60° and 45°, while CD measures 5 metres. Construct on a scale one-hundredth of full size the triangle ACD, and hence determine the height of the pole AB.

8. Construct a quadrilateral ABCD in which AB = 7·2 cm., AD = 4 cm., CD = 9·6 cm., ∠A = 126°, and the diagonals AC and BD are perpendicular to each other. Find the area of the quadrilateral.

9. A field is in the form of a quadrilateral ABCD. Draw a map of it upon the scale of 1 in. to 100 yd., given that: AD = 175 yd., DB = 300 yd., CD = 225 yd., and that the angle at A is 60°, that at C 90°. Find the area in acres and roods.

10. The area of a parallelogram is 10·5 sq. cm.; one side is 4·2 cm.; one diagonal is 3·3 cm. From these data two parallelograms of different shapes can be constructed; construct them.

11. The area of a triangle is given as 3·5 sq. in., and two of its sides are given 2·2 in. and 4 in. Show that there are two triangles within these data, and find by construction their third sides.

12. A B C are 3 buoys marking the corners of a triangular yacht race-course round an island. The angles A B C of the triangle ABC are found to be 75°, 63°, and 42° respectively. P is a flagstaff on the island, from which A and B can be seen, and the distances of P from A and B are observed by a range-finder to be 650 yards and 535 yards respectively, and the angle APB to be 137°.

Find the length of one lap of the course from measurement of a careful drawing.

13. Two quadrilaterals are supposed to be identical in size and shape. State the least number of measurements you would have to make on each in order to prove the supposition true; giving a set of measurements, and indicating how the proof would run.

14. It is related in a certain novel that a miser buried a box of treasure in a cellar and wrote a memorandum of the position of the box in the words "Twenty-three and a half from N.W., sixteen and three-quarters from N.E." He afterwards explained that these were the distances in feet of the box from two corners of the cellar.

Illustrate the above statement, as you understand it, by a rough sketch, taking the cellar as rectangular. Give complete answers to the following questions *with reasons for any statement you make*:

(i) Is there sufficient information here to enable any one knowing the cellar to find the box?

(ii) Is there enough information for you to represent the position of the box by a drawing to scale?

(iii) What can you infer from the above statement as to the dimensions of the cellar and the distances of the box from the four walls?

Finally, take 5 mm. to represent a foot and the distance between the specified corners as 30 ft., and give a drawing to scale locating the box. From your drawing state another set of measurements that would fix its position.

15. If you were required to make an accurate plan of a five-sided field, explain how you would proceed, giving a numerical example.

16. The base BC of a triangle is 7·5 cm., the side AB exceeds the side AC by 3 cm., and the angle ABC is 41°. Construct the triangle and measure its other sides and angles.

17. In the symmetrical framework shown in the sketch, PQ and PR are 12 feet long; PS ST TP are 6 feet long; and the distance QR is 20 feet. Draw the framework on as large a scale as your paper allows, stating how you do so; and measure the length SQ.

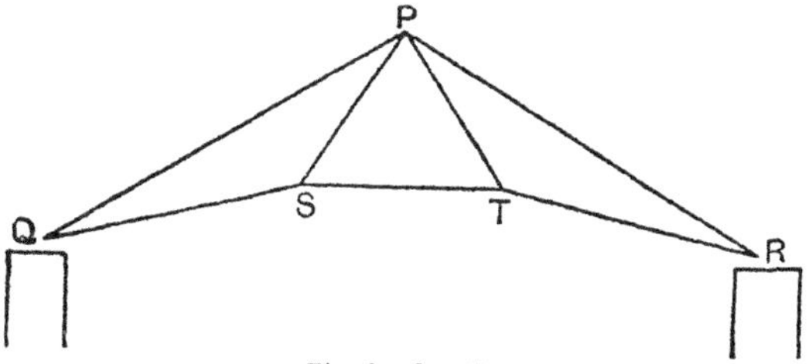

Fig. for Qn. 17.

18. Some books which are each 9 inches tall and 2 inches thick are too tall for the shelf intended for them. It is found, however, that three books can just be placed in the shelf if they are inclined at an angle of 30° to the vertical. Draw a plan showing the three books in position on a scale of 3 inches to a foot. Find the length and height of the shelf in any way you please.

Find the length of a shelf which would contain *n* books of the above dimensions, arranged as described in the first part of the question.

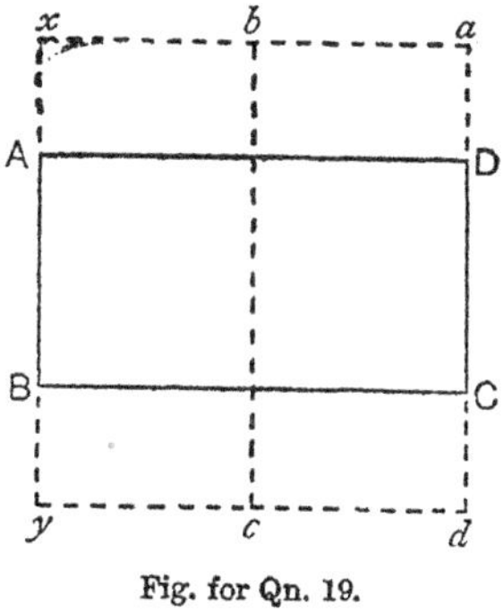

Fig. for Qn. 19.

19. The top *axyd* (see the sketch) of a card table when opened out is 34 inches square. It is hinged along *bc*, and *xy* can be folded back to fit along *ad*. The part *abcd* can now be turned about a pivot and brought into the position ABCD so that the table takes up only half the room it does when open.

Draw the figure on a scale of one-tenth, and with transparent paper find where the pivot must be. Finally show how the position of the pivot may be determined by a geometrical construction, and justify the construction.

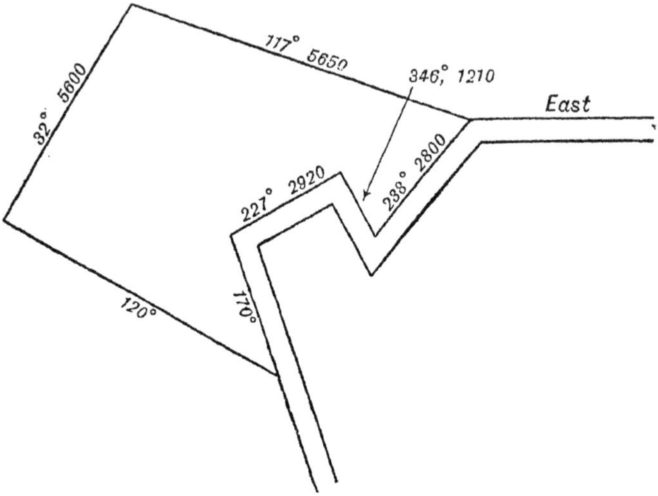

Fig. for Qn. 20.

20. From the information given in the sketch, draw the enclosure and road shown, on as large a scale as your paper allows. The road is 5 chains broad. The lengths are shown in links. The angle written against a line is the angle it makes with the north, measured towards the east.

Section 32. Further Graphs.

1. A rod AB, 8 cm. long, rotates about the point A. A second rod BC, 10 cm. long, has one end B hinged to AB, while the other end C slides along the fixed line AO. Draw the triangle ABC, taking 30°, 60°, 90°, 120°, 150° in succession as the value of A, and draw a graph showing how the length AC varies as the angle A increases from 0° to 360°.

Find when the angle C is greatest, giving the value of C and the corresponding value of A.

Draw the triangle ABC when C = 28°, explaining your construction. How many different triangles are there that have AB = 8 cm., BC = 10 cm., C = 28°? Give the values of the angle A.

2. From the table given below plot curves showing for the given period (i) the hour of sunrise, (ii) the hour of sunset, (iii) the duration of daylight. Deduce the date of (iv) the latest sunrise, (v) the earliest sunset, (vi) the shortest day.

	Sunrise. h. m.	Sunset. h. m.
Nov. 26	7 37 a.m.	3 57 p.m.
Dec. 3	7 48 ,,	3 51 ,,
Dec. 10	7 58 ,,	3 47 ,,
Dec. 17	8 4 ,,	3 47 ,,
Dec. 24	8 8 ,,	3 50 ,,
Dec. 31	8 9 ,,	3 56 ,,
Jan. 7	8 7 ,,	4 4 ,,
Jan. 14	8 4 ,,	4 14 ,,
Jan. 21	7 56 ,,	4 26 ,,

3. The base BC of a triangle ABC is 8 centimetres, and its perimeter (the sum of its three sides) is 24 centimetres. What are the greatest and least possible values of the side AB? Draw six different triangles on the same base BC, each having a perimeter of 24 centimetres; measure their altitudes and calculate their areas. Now draw a curve showing the relation between the area of the triangle and the length of the side AB, and from your curve determine how the triangle should be drawn so as to have the greatest area.

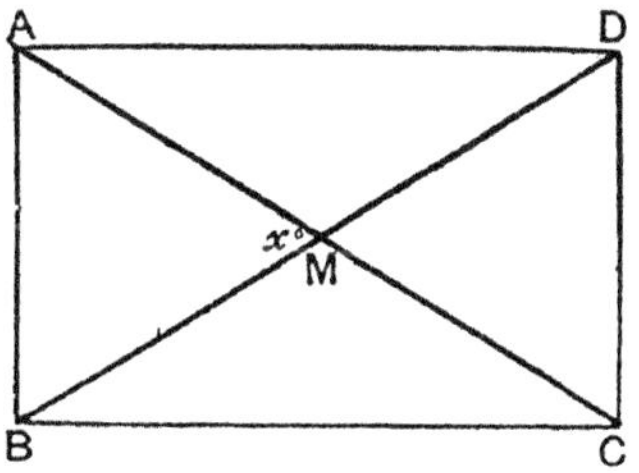

Fig. for Qn. 4.

4. Two rods each 20 cm. long are hinged together at their middle points, and their ends are joined by elastic threads (see the figure). Sketch the arrangement, and on your figure write the value of each angle in terms of the angle x degrees between the rods.

Give from the figure as many instances as you can of (i) an isosceles triangle, (ii) a right-angled triangle, (iii) a pair of congruent triangles.

What kind of figure do the four elastic threads form? Find the area of this figure for a few values of x, and give a graph of the area as a function of x from $x=0$ to $x=180$. Take 1 cm. to represent 10 degrees for the angle-scale, and 1 cm. to represent an area of 10 sq. cm. for the area-scale.

5. In the figure the angle between the fixed straight lines OA and OB is 45°, and P is a fixed point such that A, its projection on OA, is distant a from O. Through any point Q in OA, distant x from O, a straight line QR is drawn perpendicular to OA, meeting OB in R. Find an expression for the area of the triangle PQR in terms of a and x.

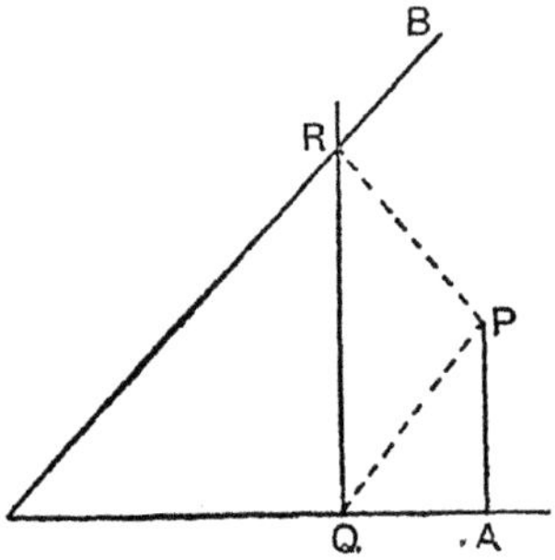

Fig. for Qn. 5.

Taking OA as equal to 4 inches, plot a graph showing how the value of this area varies as Q moves from O to A, and from the graph deduce (i) the position of Q for which the area PQR has a maximum value, (ii) the position of Q for which the area PQR is equal to 1 square inch.

6. The following table gives the square roots S of certain numbers N:

S	1	2	3	4	5	6	7	8
N	1	4	9	16	25	36	49	64

Plot the graph on squared paper, showing the relation between N and S, and hence find approximately the square root of 34.

7. Plot on squared paper the curve $y=x^3$, and thence find approximately the cube root of 100.

8. A man calculates that if the tax on a certain commodity were x pence per lb., the consumption would be $100-10x$ per cent. of what it is without any tax. Trace a graph showing the total amount of duty received for different rates of taxation, supposing that 100 lb. of the commodity would be consumed if no tax were imposed. For what rate is the total duty a maximum?

9. The coal used by a steamer during a voyage of 2400 nautical miles costs $2{\cdot}5\times v^2$ pounds, where v is the speed in knots (*i.e.* nautical miles per hour). The other expenses of the voyage amount to £200 for each day. Consider the three cases in which v has the values 10, 15, 20, and find for each case (i) the number of days the voyage lasts, (ii) the cost of coal, (iii) other expenses, (iv) the total cost of the voyage.

Make a rough graph connecting the speed and the total cost, using 5 cm. to represent £400, and 1 cm. to represent 1 knot. Now select several other numbers for v near that portion of the graph where the total cost seems likely to be least, and draw this part of the graph more carefully on the same diagram.

Hence find the least possible cost for the journey, and the speed at which the vessel must run for this cost.

10. On a certain steamer the cost of coal is reckoned at £$x^2/8$ per hour, where x is the rate of steaming in miles per hour. Wages and other expenses amount to £5. 5s. per hour at all speeds.

(1) Express in pounds the cost of a journey of 100 miles at x miles per hour.

(2) Draw a graph showing the cost of this journey at any speed from 4 to 12 miles per hour.

(3) From your graph determine the most economical speed and the cost of the journey at that speed.

11. The cost of the material for making a cylindrical tank of given height can be represented by a formula $ax+bx^2$, where x is the radius of the base in feet. For a tank radius 5 ft. the cost is £6. 5s., for one of radius 11 ft. it is £17. 1s.

Find the values of a and b, and draw a graph showing the relation between the radius and the cost.

12. The cost of a certain steamer voyage is the sum of two quantities, one of which is proportional to the time taken and the other to the average speed during that time. When the voyage takes 10 days, the cost is £700; when it takes 14 days the cost is £884. Find the cost when it takes 8 days.

Also find the number of days which gives the smallest possible cost, and the corresponding cost.

13. Make a table giving the value of x^2, $-3x$, x^2-3x+2 when x varies from -1 to $+4$, taking $1/2$ as the interval between each pair of values.

Use this table to draw a curve, which will show how x^2-3x+2 varies, while x increases from -1 to $+4$, taking one inch as unit.

Find from your figure the values of the expression x^2-3x+2 when (i) $x=\frac{1}{4}$, and (ii) $x=2\frac{1}{4}$, and verify your results by substituting these values of x in the given expression.

14. Draw in one figure and with the same scale and axes the graphs $y=3-x/2$ and $y=5/x$ from $x=1$ to $x=6$, taking one inch as unit.

Either directly from these graphs, or by means of their tables of values, form the graph of $y=5/x-3+x/2$ and find its minimum or turning point as nearly as you can from your figure.

15. Draw the graph of the expression $2+12x-x^3$ and hence find, as nearly as you can, for what values of x the value of the expression lies between ± 10.

16. Draw the graph of $y=x^2(6-4x)$, and find the value of y which is greater than all those immediately preceding and following, and also the value which is less than all those immediately preceding and following.

17. Draw the graph of $3x^2+2x-5$, and hence find for what values of x the value of the expression lies between ± 1.

18. By drawing and measuring complete the table indicated below, and hence draw a graph showing the relation between the area of an equilateral triangle and the length of its side.

Length of side in inches -	0	1	2	3	4	5
Area of triangle in sq. in. -	...	...	1·7	...	...	10·8

Read off from the graph

(1) the area of an equilateral triangle of side 2·5 in.,

(2) the side of an equilateral triangle of area 5 sq. in.

19. Draw a straight line CD 30 cm. long; at C erect a perpendicular CA 12 cm. long, and at D a perpendicular DB 20 cm. long (see the figure). Consider the length of a path from A to some point P on CD, and thence to B. In order to consider how the length of the path APB varies for different positions of P, tabulate (from measurement) the values of AP + PB when CP = 0, 3, 6, 9 ... 30 cm. Draw a graph to scale, showing how the value of AP + PB varies with the value of CP; hence find the position of P which makes AP + PB a minimum. When you have found this position for P as accurately as you can, measure the angles APC and BPD and record them.

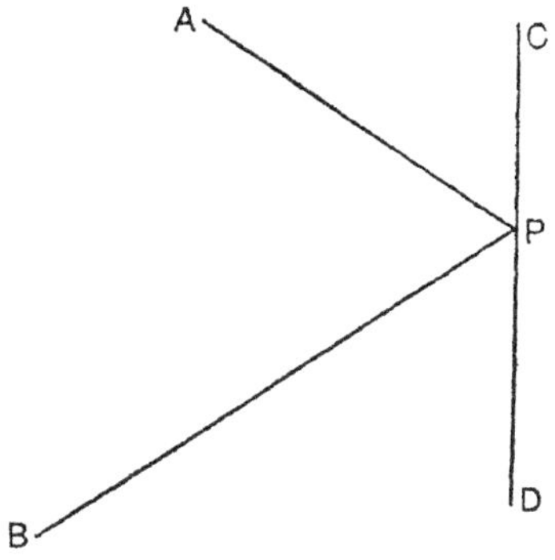

Fig. for Qu. 19.

Give a general proof that for any other position of P (in CD or CD produced) than that found above, AP + PB has a greater value.

If CD is a mirror and rays of light from a lamp A fall on it, the ray which is reflected to B travels in such a direction that AP + PB is a minimum. Give a geometrical construction for the ray from A which after reflection goes to B.

20. A person who wants to get some idea of the proportion of wet weather at a certain place tries the following graphical method: He finds that for the seven days of a week the number of hours of rain were 13 15 12 9 0 19 23 respectively. On a sheet of squared paper he draws two lines meeting at right angles; he represents the first day of the week by a point 13 units from one of these lines and 11 units from the other; and similarly he represents each other day of the week by a point whose distance from the first line gives the number of hours of rain and from the second line the number of hours without rain. Draw the diagram for the week, and examine whether the shape of the graph joining up the points gives the information required.

Draw the graph, and, denoting the distance of a point from the first line by x units and from the second line by y units, give the relation that holds between x and y for every point on the graph.

21. Solve *graphically*, and *also* otherwise, the equations

$$1{\cdot}4 \times x + 3{\cdot}8 \times y = 2{\cdot}1 \qquad x + y = 1$$

Take 6 inches as your unit in the graphic solution.

22. Use your squared paper to solve the equations

$$3x - 4y = 7 \qquad 4x + 3y = 8$$

23. Draw in the same diagram graphs of $y = x^3$ and $y = 3x + 1$; use the graphs to find solutions of the equation $x^3 - 3x - 1 = 0$ correct to one decimal place. State what units you use and do not take them too small.

24. Draw in one figure the graphs of $5x + 2$ and $x^3 + 1$. Hence find for what values of x the expressions are both positive.

25. Draw the graphs of the equations

$$y = 6/x \qquad y = 2x - 2$$

from $x = -3$ to $x = +3$; and find from your figure the values of x and y which satisfy both equations.

26. Draw the graph of $y = (4x^2 - 40x + 225)/20$ between the values $x = 0$ and $x = 10$, taking half an inch as unit for both x and y, and choosing as origin the lower left-hand corner of the paper.

On the same diagram, with the same origin and the same scale, draw the graph of $y = x$.

What relation exists between the straight line and the curve?

27. Trace the graphs of

$$y = \frac{x + 5}{2(x - 1)} \qquad \text{and} \qquad y = \frac{(x + 2)(x - 1)}{2} - \frac{7}{8}$$

from $x = -3$ to $x = +3$, using the same axes for both graphs.

In how many points do they meet?

28. Plot the graph of $x^3 - 2x$, taking 1 inch for scale unit. By drawing another graph on the same sheet, solve the equation

$$4(x^3 - 2x) = x^2 + 2.$$

29. Trace, to a scale "unity = 1 inch," the graph of $y = x^2/2 - 3$. With the same axes and in the same figure trace a graph suitable for giving, when considered with the above graph, the values of x which satisfy the equation $x(x^2/2 - 3) = 3/2$, and give these values from your graph.

30. The small tank shown in the figure has a square base ABCD measuring 30 cm. each way, and vertical sides AE, BF, etc., 18 cm. high, and contains water to a depth of 16 cm. How many c. cm. of water are there in the tank?

Fig. for Qn. 30.

The tank is gradually tilted about the edge BC and the water allowed to run out. What volume of water remains when the corner A is in the surface of the water? Draw the tank half-size, and measure with your protractor the angle which the edge AB then makes with the horizontal.

Give an expression for the volume of the water remaining, as the tank is gradually tilted, (i) in terms of the length of AE under water, for cases in which the corner A is covered, and (ii) in terms of the length of BA under water, for cases in which A is above the surface.

Find, by drawing, the inclination of the edge AB to the horizontal when a quarter of the water has run away, and when half the water has run away.

31. A closed tank with a sloping bottom is 56 inches long and 44 inches broad. The ends are rectangular, one of them 60 inches deep, and the other 34 inches deep. The tank is made of sheet-iron. Obtain the length of the sloping bottom by drawing; then find the area of sheet-iron used, and the capacity in gallons of the tank.

Show that when the water stands x inches high at the deep end of the tank, and x is greater than 26, the quantity of water in the tank is approximately (in gallons) $8 \cdot 9x - 116$.

Assuming that when the water stands x inches high at the deep end, x being less than 26, the number of gallons is $0 \cdot 171x^2$, find from these expressions the quantity of water in the tank when the water stands 0, 10, 20, 30, 40, 50, 60 inches high at the deep end; and show the results in a graph.

Find the height of the water at the deep end when the tank contains 40 gallons, and when it contains 200 gallons.

32. When $100y = x^3 - 100x$, calculate the values of y for $x = 8 \cdot 5$, 9, 10, 11, and draw the graph of y from $x = 8 \cdot 5$ to $x = 11$, showing the scales clearly and using 5 cm. as unit.

When $20 + z = 2x$ for what value of x is z equal to 2, and for what value is it equal to -2? And what is the value of z when $x = 8 \cdot 5$, and when $x = 11$? Show the graph of z from $x = 8 \cdot 5$ to $x = 11$ on the same diagram as the graph of y.

Show that for values of x between $9 \cdot 5$ and $10 \cdot 5$ the error made in using z in place of y is less than $0 \cdot 1$.

33. If $x = t - 1$ and $y = t + t^2$, find y in terms of x.

Plot the graph showing the relations between corresponding values of y and x as t goes through the values -3, -2, -1, 0, 1, 2, 3.

Show from your graph that y is always algebraically greater than x.

34. Plot the curves $y = x^2 - 4x + 6$ and $x = y^2 - 5y - 2$, and thence determine their points of intersection.

35. If $y = ax^2 + bx + c$, find the values of a b c, given that $y = -8$, when $x = 0$; $y = -10$, when $x = 2$; and $y = 2$, when $x = -2$. Draw a graph of the expression, and from it find the value of x corresponding to the least value of y.

36. The corresponding values of two quantities x and y are given in the following table:

x	0·30	0·40	0·48	0·54	0·60
y	0·54	0·58	0·61	0·63	0·65

By plotting these values on squared paper (taking a decimetre as unit), find the relation between x and y in the form $y = px + q$.

Find from your graph the value of x when $y = 0 \cdot 5$.

37. A North Pole expedition is undertaken by 3 sleds. Each sled carries 40 days' provisions for its complement of men and dogs. They set off together from their supply-base. Sled 1 goes to such a point that it can just fill up sleds 2 and 3 and keep enough to return to the base. Sled 2 goes on with sled 3 to such a point that it can fill up sled 3 and keep enough to return to the base. Each of the sleds 1 and 2 on reaching the base loads up again and sets off to meet the expedition on its return. How many days' journey from the base is the point sled 3 can reach?

Show how sleds 1 and 2, by making appropriate halts at the base before setting out on their second journeys, can enable sled 3 to get 40 days out from the base.

38. Taking $a=1$, $b=-4$, $c=2\cdot5$ draw the graph of $y=ax^2+bx+c$ for values of x ranging from 0 to 4; using 2·5 cm. as the unit for each scale. Discuss generally the effect on the graph of changes in the value of c.

Section 33. Pythagoras' Theorem.

1. Construct a triangle whose sides are 3, 4, and 5 inches respectively, and measure its angles.

2. Show how to construct a square having an area double that of a given square.

3. With your instruments describe a square on a base 3 inches long. Then describe another square 3 times as large as the first square, and measure its side.

4. Prove that the square on the hypotenuse of a right-angled triangle is equal to the sum of the squares on the sides.

Show how, by measurement and calculation, the converse of this theorem may be used to test whether two walls of a room are at right angles to each other.

5. A B C are three villages. The village A is 3 miles east of B and 4 miles south-west of C. Find, in any way you please, the distance between B and C correct to the nearest furlong.

6. Construct a right-angled triangle with the sides containing the right angle 4 and 7·5 cm. long. Bisect the hypotenuse and join its middle point to the opposite vertex. Measure the join and also calculate its length. State the difference, if any, between the measured and calculated lengths.

7. A room is 18 feet wide and 28 feet long. Two fireplaces 8 feet wide are built across two of the corners equally inclined to the walls. Find the area of the room.

8. Show how to describe a square on a given straight line.

ABCD is a square whose side is 21 centimetres in length; P, Q, R, S are points in AB, BC, CD, DA such that

$$AP = BQ = CR = DS = 9 \text{ centimetres.}$$

Prove that PQRS is also a square, and find the length of its side.

9. A shoe has six pairs of eyelets. The two eyelets which form a pair are $\frac{3}{8}$ inch apart when the shoe is laced, while each pair of eyelets is half an inch from the next pair. A lace measuring 20 inches is threaded through the bottom holes and carried over in the form of a letter X to the next pair of holes, and so on to the top. Find the length of lace left over at the top pair of holes to be tied into a bow.

10. Draw accurately two triangles having sides of the following lengths:

(i) $2''$, $1\frac{1}{4}''$, $1\frac{1}{4}''$;

(ii) $1\frac{1}{2}''$, $1\frac{1}{4}''$, $1\frac{1}{4}''$.

Show that these two triangles can be so placed as to make up one right-angled triangle whose sides are $1\frac{1}{2}''$, $2''$, $2\frac{1}{2}''$.

11. ABC is a triangle, right-angled at C. On AB, towards the same side as C, the square ABDE is described. A straight line is drawn through E parallel to AC; from A and D AH and DG are drawn perpendicular to this line; and from D DF is drawn perpendicular to BC, produced if necessary. Prove that the square ABDE is equal to the figure ACFDGH, and that this figure can be divided into two squares whose sides are equal to AC and BC.

12. A straight line AG is divided unequally at D, and squares ABCD and DEFG are constructed on the same side of it. Show how to find a point P in AG such that BP is the side of a square equal in area to the two original squares.

Give reasons for the construction, and show how the problem may be practically demonstrated by cutting along BP and FP and pasting the triangles BPA and FGP in fresh positions.

13. A and B are two points 18·6 centimetres apart. Two equal circles, whose radii are 10·4 centimetres in length, are described with centres A and B, cutting AB in C and D and each other in E, and EN is drawn perpendicular to CD. Calculate the lengths of the straight lines EC ED EN and the area of the triangle ECD.

14. Construct and measure a line, the square upon which equals in area the sum of the squares upon three lines of lengths 3, 4, and 5 cm.

15. ABC is a right-angled triangle having a right angle at A, and P is any point in AB. CA is produced to Q and G so that AQ and AG are equal to CP and CB respectively. Prove that PG is equal to QB.

16. Three cylinders are of the same height and of the same material. Prove that a triangle whose sides are equal to the diameters of the cylinders will be right angled if one cylinder weighs as much as the other two together. The volume of a cylinder of height h and diameter d is $0{\cdot}7854hd^2$.

17. The diagonals of the faces of a rectangular solid are 8 inches, $6\frac{1}{2}$ inches, and 5 inches. Find the lengths of the edges of the solid to the nearest tenth of an inch.

18. A tourist walks south-east for p miles and south-west for p miles and west for q miles. Express in terms of p and q the distance measured direct between the start and finish of his walk.

Calculate this distance to the nearest tenth of a mile when $p = 8{\cdot}5$ miles and $q = 10{\cdot}25$ miles.

19. PQR is an equilateral triangle divided into two equal parts by the straight line PS, and ST is perpendicular to and SU bisects PQ. Prove that the square on one of the sides of the triangle PSQ is three times the square on another side, and that the triangle PST is three times the triangle STU.

20. A parallelogram and a triangle are on the same base and between the same parallels. Prove that the area of the parallelogram is twice the area of the triangle.

ABCD is a rectangle whose sides AB and BC are 8 inches and 4 inches in length. Show that an equilateral triangle on AB is less than the rectangle ABCD.

21. AOB is an angle of 60° bisected by a straight line OC, and P is a point in OA such that OP equals ·7 centimetres. A circle with centre P and radius PO cuts OB in Q and OC in R. Find the areas of the squares described on QR and OR in any way you please.

22. AB is a straight line bisected at C. On CB an equilateral triangle BCD is described, and on AD another equilateral triangle AED is described on the side remote from C. Prove (i) that the angle EAB is a right angle; (ii) that $AD^2 = 3AC^2$.

23. If the square on one side of a triangle be equal to the sum of the squares on the other sides, prove that the triangle has a right angle.

State, without proof, the converse theorem.

A balloon ascends from a point A on the seashore and is carried by the breeze until it stops absolutely motionless, vertically above a lighthouse C, when the balloon is found to be exactly 3 miles above the sea level. The flash of a gun fired at A is noticed in the balloon, and the report is heard exactly 25 seconds later. If the speed of sound be taken as 1056 feet per second, determine how far the lighthouse is from A. Light travels so quickly that we may say the flash is seen the very instant it occurs.

24. ABCD is a quadrilateral in which the diagonals AC and BD are at right angles to each other. Prove that

$$AB^2 + CD^2 = BC^2 + DA^2.$$

If AB BC CD be 6, 7, 8 cm. respectively, use the above relation to find the length of DA to two decimal places.

Draw a quadrilateral with these measurements, having also AC = 10 cm.; measure BD.

25. AB BC CD DA are the rectangular edges of the floor of a room. Starting from a point O in the edge AB, an insect crawls straight to B and thence along the edge BC, and then to a point F in CD. It then ascends the wall vertically to a point G. By drawing or calculation find the distance of G from O. It is given that the distances OB BC CF FG are respectively 17 feet, 16 feet, 5 feet, and 15 feet.

26. ABCD is a rectangle in which $AB = 6$ inches and $AD = 3$ inches. A point E is taken in AB distant 2 inches from A, and a point F in CD distant 1 inch from C. Calculate the length of EF, and the area of the quadrilateral EBCF.

27. In the quadrilateral ABCD it is given that the diagonals are perpendicular, and also that $AB^2 + BC^2 = AD^2 + DC^2$. Prove that AC bisects BD.

28. The points A and B are fixed and P is any point on the line AB (not produced); if PQ be drawn perpendicular to and on one side of AB so that $PQ = PB$, prove that the locus of Q is a segment of a straight line passing through B.

If $AB = 3{\cdot}5$ in., use this locus to construct the position of P when $AP^2 + PB^2 = 9$ sq. in., and measure the length of AP. Each step of the construction must be explained, but no proof is required.

29. A straight line KL is 12 cm. long. At the ends of this straight line and on the same side of it are erected perpendiculars KA 6 cm. long and LB 3 cm. long. The position of a point P in KL is specified by its distance x cm. from K. Express the values of AP^2 and BP^2 in terms of x and plot graphs of these values, representing the values of x full size and the values of AP^2 and BP^2 on a scale of 1 cm. to 10 cm^2. From your graph find the value of x for which AP is equal to BP.

Show also how this position of P may be determined geometrically, and carry out the construction on a full-sized figure.

Plot also the graph of $AP^2 + BP^2$ and from it determine the positions of P for which $AP^3 + BP^2 = AB^2$.

Show also how these positions may be determined by a geometrical construction, and find them.

30. Prove that a triangle, the lengths of whose sides are $x^2 - 1$, $2x$, $x^2 + 1$, has a right angle whatever x may be. Deduce a geometrical construction for finding the square root of any number, *e.g.* 5. A drawing to scale is not required.

31. Show that, if a straight line be bisected and produced to any point, the rectangle contained by the whole line thus produced, and the part of it produced, together with the square on half the line bisected, is equal to the square on the straight line which is made up of the half and the part produced.

A straight line AB is bisected at C; show how to find a point

D in AB produced such that the rectangle AD.DB may be equal to the square on BC.

32. A and B are two fixed points and P is a variable point, so that the difference of the squares on PA and PB is constant. Show that the locus of P is a straight line perpendicular to AB.

33. Prove that the area of the rectangle contained by the sum and difference of two straight lines is equal to the difference between the areas of the squares on the two lines.

Show how to find a point E in the base BC of a triangle ABC, produced if necessary, such that $BC.BE = AB^2 - AC^2$.

Construct a figure for the case when AB = 6·5 cm., AC = 5·5 cm., BC = 7·5 cm., and measure BE.

Also find BE by calculation from the given relation.

34. AP and AQ are two straight lines of length 13 and 14 centimetres respectively. PN the perpendicular from P on AQ is of length 12 centimetres. Find the length of PQ by drawing and by calculation.

35. A straight line AB is bisected at C and produced to D. Prove that the sum of the squares on AD and DB is twice the sum of the squares on AC and CD.

AOB and COD are two diameters of a circle at right angles. OD is produced to E, and EF is drawn parallel to AB so that EF is equal to ED. Prove that the square on BE is half the square on CF.

36. A man and a boy pace two sides of a rectangular piece of ground ABCD. Along AB the man takes 60 steps and the boy 80. Along BC the man takes 50 steps; how many would the boy take?

Find by drawing how many paces each would take along the longest straight path they could take in the field, and use the relation between the three sides of a right-angled triangle to check your result.

37. ABCD is a quadrilateral such that AB = 5·2 cm., BC = 3·9 cm., CD = 2·5 cm., DA = 6·0 cm., and angle ABC = 90°. Draw the quadrilateral on squared paper, and measure the length AC and the angle CDA. Calculate the length AC, and prove that in a perfectly drawn figure the angle CDA would be exactly a right angle.

Calculate the area of the figure, and check your result by counting the number of the squares of the ruled paper which are included in the figure.

38. A figured rough sketch of a short staircase seen from the side is given. Draw it accurately to a scale of 1/10 of an inch to 1 inch, and find by measurement and also by calculation the length of the banister B resting on four supports, with its ends projecting four inches beyond the tops of the end supports.

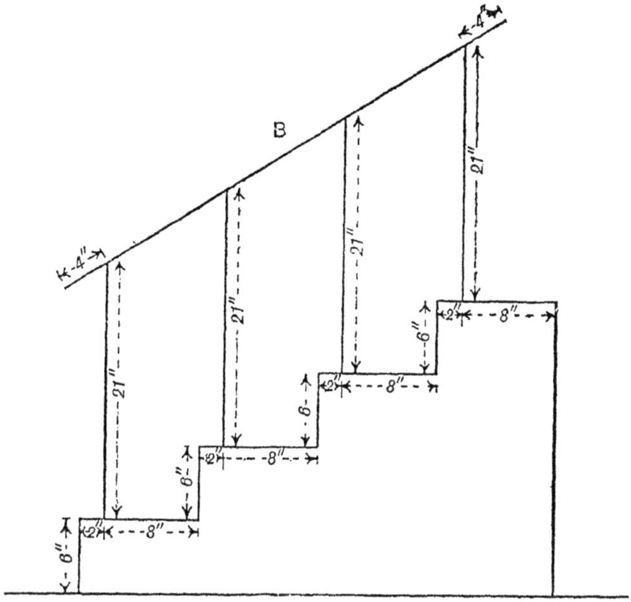

Fig. for Qn. 38.

39. State and prove geometrically a theorem relating to segments of a line corresponding to the formula

$$a^2 - b^2 = (a + b)(a - b).$$

The base AB of an isosceles triangle is 4 in. long and the equal sides BC and CA are 5 in. long. Find a point D in AB produced such that the difference of the squares on CD and CB is 21 sq. in.

40. Draw on squared paper two lines at right angles to one another, and through their intersection draw another line making an angle of 30° with one of them.

Show that this line can be treated as a graph for the multiplication or division of a quantity by $\sqrt{3}$ without calculation. From the graph find approximately the values of $8{\cdot}8 \times \sqrt{3}$ and $7{\cdot}8 \div \sqrt{3}$.

41. AP and BQ are two parallel lines of unequal length such that AP makes an angle of 45° with AB, and BQ is drawn in the same direction as AP. Prove a rule for finding the area of APQB when the lengths AP BQ AB are given.

Prove that, R being the middle point of PQ, the triangle ARB is half the quadrilateral APQB; and that if S is the middle point of AB, and AR and PS meet in U and BR and QS in T, then the sum of the triangles AUB and ATB is equal to the sum of the triangles PUQ and PTQ.

42. In the figure ABCDE is a plot of ground, the dimensions of which are given in yards. Draw the figure to scale, representing 4 yards by 1 cm.

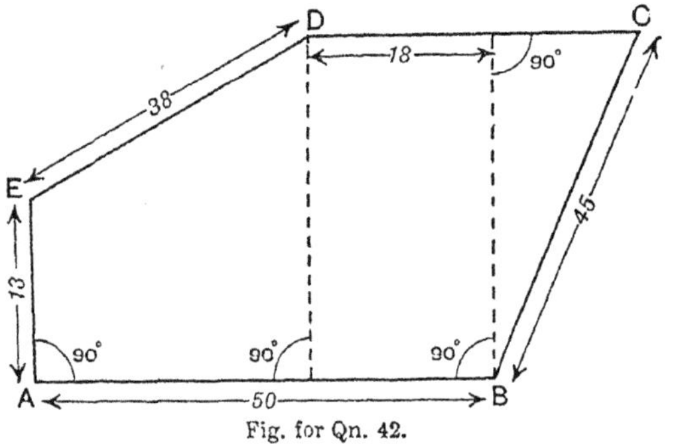

Fig. for Qn. 42.

From measurements of your figure calculate the area of the plot in square yards.

Also, from the dimensions given, and without the assistance of drawing, calculate any other necessary dimensions, and hence the area; each to three significant figures.

43. By means of a graph solve the following problem from an old mathematical book: "In the figure AB is the line of separation between two plains, one of which ACDB consists of soft sand, in which a vigorous horse can scarcely advance at the rate of 3 miles per hour; the other ABEF is covered with fine turf, where the same horse without much fatigue can proceed at the rate of 6 miles per hour. If AB is 8 miles, CA 4 miles, and BE 3 miles, find what route a horseman who has to go from C to E must pursue so as to employ the least time possible on his journey."

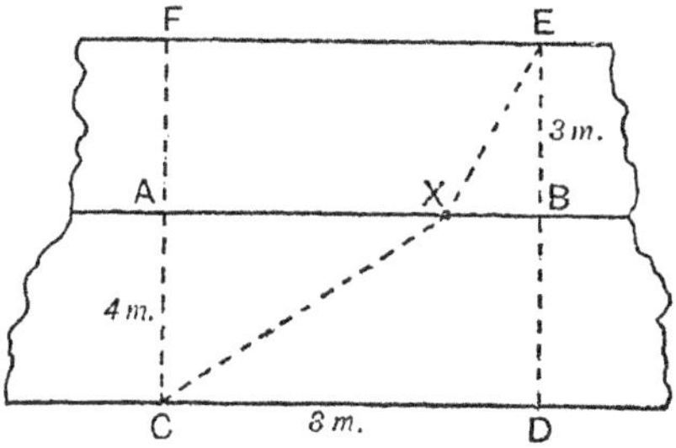

Fig. for Qn. 43.

First estimate roughly where X is likely to be, and then obtain data for the graph by calculation, expressing the time in hours and decimal parts of an hour, and tabulating the results. Scale for AX, 25 mm. for 1 mile; take values of AX from 0 to 4 miles. Scale for time, 50 cm. for 1 hour; number the scale up from 2·5 hours.

44. A room measures 25 feet long, 15 feet broad and 9 feet high. A spider in one of the bottom corners wishes to reach a fly in the opposite top corner by crawling across two adjacent walls of the room starting along the longer one. Draw a figure showing the appearance which the two walls would present if they were opened out and placed side by side in the same vertical plane; show the scale of your figure. Indicate on the figure the shortest course the spider could follow; find by calculation the distance it would have to go, and find by measurement the height above the floor at which it would have to cross the edge at which the two walls meet.

45. The figure represents a cubical box whose edge is one foot. An elastic thread inside the box is fastened at A and G. The lid EFGH opens about a hinge EF. Find, by drawing or calculation, how long the thread is (i) when the box is closed, (ii) when the lid is vertical; giving the lengths in inches.

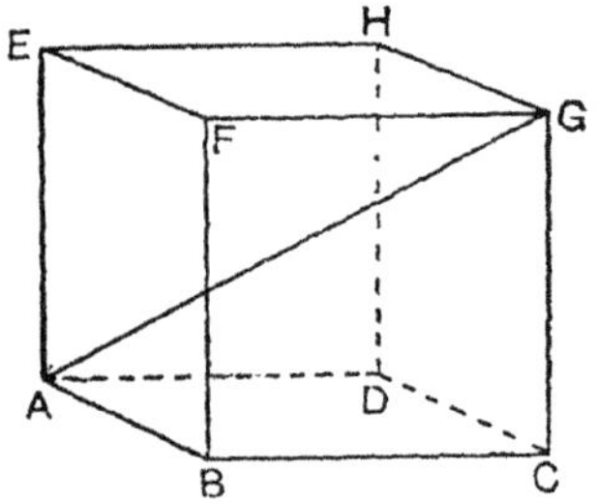

Fig. for Qn. 45.

46. A cardboard tray, with a square base and sloping sides, is to be made by cutting from a sheet of cardboard a piece shaped somewhat as shown in the sketch, and bending along the dotted lines to form the sides of the tray.

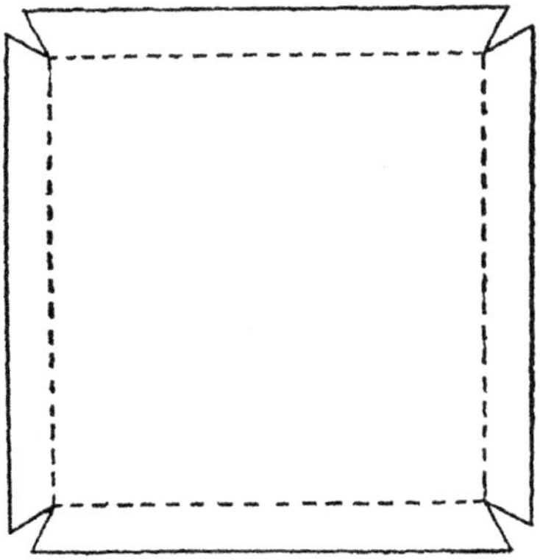

Fig. for Qn. 46.

The dimensions are to be: edge of base 4 inches, edge of top of tray $5\frac{1}{4}$ inches, vertical depth 1 inch. Find, by drawing or calculation, the breadth of the strip that forms a side of the tray.

Draw an accurate full-size figure of the piece of cardboard cut out ready for folding, and explain your method.

47. The first diagram of the figure shows the interior of a room, length 20 feet, breadth 10 feet, height 10 feet. A fly at the point P wishes to get to the point Q by crawling. The figure shows the positions of P and Q. By drawing, compare the lengths of the following three routes: (i) the route which becomes a straight line when the end walls are rotated into the

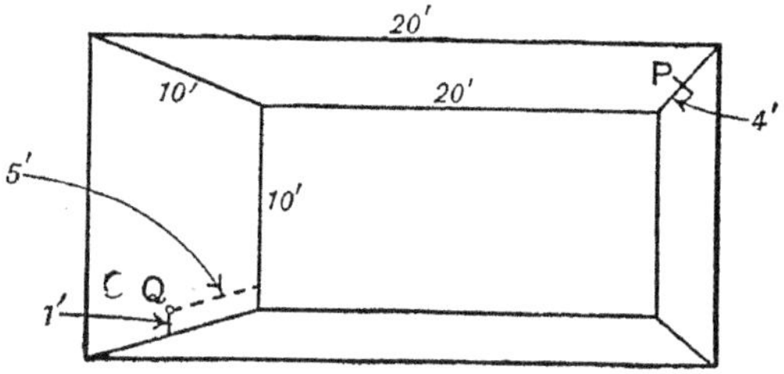

Fig. for Qn. 47, first diagram.

plane of the side wall which is nearer to P, (ii) the route which becomes a straight line when the end walls are rotated into the plane of the floor, (iii) the route which becomes a straight line when the walls are spread into the plane of the floor in the way shown in the second diagram. Represent 1 foot by 1 small

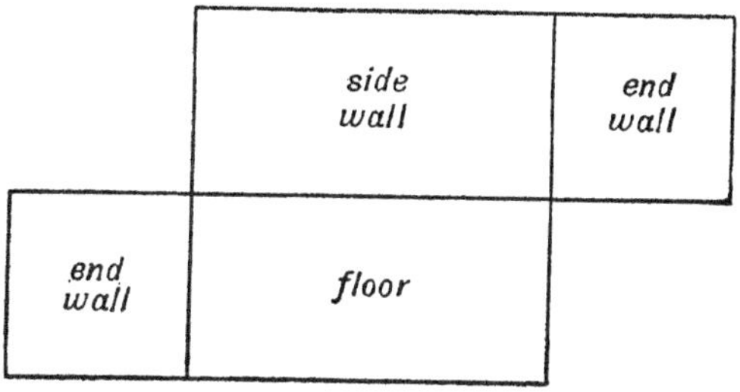

Fig. for Qn. 47, second diagram.

division of your book, and indicate the dimensions on your drawings.

Find also the distance between the two points P and Q across the room in a straight line.

Section 34. Numerical Quadratic Equations.

1. In t seconds a body falls $16 \times t \times t$ feet. How many feet does it fall in 5 seconds, and how many seconds does it take to fall 100 feet?

2. A solid sphere of lead whose diameter is $4\frac{1}{2}$ inches is converted into a lead pipe nine feet in length, whose external diameter is $1\frac{1}{4}$ inches. Find the internal diameter of the pipe.

3. At sea level water boils at a temperature of 212° Fahrenheit. At a height H feet above sea level the boiling point is lowered B degrees below 212°, where B and H are connected by the approximate formula $H = 520B + BB$. At what temperature will water boil at the top of a mountain 3500 feet high?

4. When a body is thrown up in the air with a velocity of 96 feet per second, its height above the ground t seconds after it was thrown is $96t - 16tt$ feet. Find its height after 1, and after 2 seconds. At what times is the body again at these heights?

5. The area of a rectangular field is $1\frac{1}{4}$ acres, and a longer side exceeds a shorter by 18 yards. Find the lengths of the sides.

6. Two variable quantities are so related that the sum of their squares remains constant. When the first is 39 the second is 52. Find the value of the first when the second is 25.

7. The sum of the two sides of a right-angled triangle is 17 in. and the area 30 sq. in. What is the length of the hypotenuse?

8. The length of a rectangular field exceeds its breadth by 72 feet, and the fence enclosing it exceeds in length that of a square field of equal area by 16 feet. Find the dimensions of the field.

9. A man was x years old in the year x^2 of the Christian Era. He died 22 years later in the year $1914 - x$. Find the year of his birth.

10. A certain weight is lifted from the ground through a vertical height of 9 inches in a quarter of a second by means of a contrivance projecting from the rim of a wheel which revolves in a vertical plane. At this height the weight is released and falls back to the ground in the time given by the formula $d=16t^2$, where d is the length of fall *in feet* and t is the time of fall in seconds. How many times must the wheel revolve per minute, if the time of one complete revolution of the wheel is equal to the time during which the weight is lifted and falls to the ground?

11. The formula for bodies projected vertically upwards is $s=ut-\frac{1}{2}gt^2$, where t = time in seconds, s = height in feet above starting point at time t, u = initial velocity in feet per second, and $g=32$.

Find the time a body takes to rise 192 feet vertically when the initial velocity is 128 feet per second.

12. A glass h inches high, a inches in diameter at the top, and narrowing to a diameter of b inches at the bottom holds $h(a^2+ab+b^2)/132$ pints. What will a glass hold for which $h=4$, $a=3{\cdot}7$, $b=1{\cdot}7$?

A vessel is to be made to hold a pint, to be 4·5 inches high, and to have sides sloping so that the bottom diameter is 2·3 inches less than the top diameter. Find these diameters to the nearest tenth of an inch.

13. A straight line 10 centimetres long is divided so that the square on one part is double the square on the other part. Form an equation for determining the length of the smaller part, and find its length approximately. The equation has two roots. Find a meaning for the other root.

14. Find, to the nearest hundredth, the values of x which will make $\dfrac{x+10}{x+5}$ equal to x.

15. Find to two significant figures the values of x for which $\dfrac{x-5}{2x+4}$ and $\dfrac{3x-3}{4x+2}$ are equal.

16. Solve the equation $x^2=7(x+3)$, giving the roots to the first place of decimals.

17. Solve the equations:

$$\text{(i) } x^2+12=7x; \quad \text{(ii) } x^2+11x=7,$$

giving the solutions of the second equation to three decimal places.

18. The dimensions of a rectangular block are $x-1$, x, and $x+1$ inches respectively, and the total surface is 13 sq. ft. 4 sq. in. Find x and calculate the volume in cubic feet, giving each result to two significant figures.

19. Solve the equations:

(i) $\frac{1}{4}(5x+1)-\frac{1}{2}(2+3x)=\frac{3}{2}(8x-4)$;

(ii) $2x^2-6x+3=0$,

evaluating the roots to two places of decimals.

20. Find to two decimal places the greater root of the equation

$$11{\cdot}74x^2-20{\cdot}21x+3{\cdot}45=0$$

and verify your result by substituting it in the equation and calculating the value of the left-hand member to three decimal places.

21. Solve the equation

$$5{\cdot}1x^2-16{\cdot}3x+8=3{\cdot}2x^2-10x+20{\cdot}6$$

giving each root to two significant figures, and calculate to four significant figures the value of each side of the equation when the positive root (taken to two significant figures) is substituted for x.

22. (i) Find the common solution of the equations:

$$8/x+6/y=17, \qquad 6/x-8/y=5;$$

(ii) Solve the equation: $17x^2-19x=1000$.

Give all the answers to two significant figures.

23. From each of the following statements deduce a definite value for x if you can, or explain clearly why you cannot do so:

(i) $x(2x-1)-3(4-x)=2x(x+2)+3$;

(ii) $(6x+5)(x-1)+3=(2x+1)(3x-2)$;

(iii) $(2x+1)(3x-5)=(x-1)(6x-1)+10$;

(iv) $(x-2)^2+(2x+1)^2=1$.

24. Solve the equations:

(i) $\dfrac{8}{x}-\dfrac{9}{y}=7 \qquad \dfrac{16}{x}+\dfrac{12}{y}=4$

(ii) $\dfrac{2+x}{x-3}+\dfrac{x-3}{2+x}=\dfrac{5}{2}$

25. (i) Solve the equation $x^2/0\cdot3 + x/0\cdot15 = 0\cdot272$ verifying both solutions.

(ii) Solve the following simultaneous equations, giving the values as decimals to three places:

$$x/2 - y/6 = x/3 + y/4 = 1/12.$$

26. Solve (i) $\dfrac{3}{x} - \dfrac{4}{y} = 20, \quad \dfrac{2}{x} + \dfrac{2}{y} = 25$;

(ii) $\dfrac{2x - \frac{1}{3}}{8x - \frac{1}{4}} = \dfrac{x - \frac{1}{6}}{4x - \frac{1}{5}}$;

(iii) $(x+2)(x+3) = 20$.

27. Solve the equations:

(i) $\dfrac{x+y}{8} + \dfrac{x-y}{6} = 5, \quad \dfrac{x+y}{4} - \dfrac{x-y}{3} = 10$;

(ii) $\dfrac{x^3 - 27}{x - 3} - \dfrac{x^3 + 27}{x + 3} = 6$;

(iii) $4x^2 - 4x = 624$.

28. A man borrows £100, for which he agrees to pay back £40 at the end of one year, and £70 at the end of two years. Show that if this is equivalent to paying compound interest at x per cent.,

$$40(1 + x/100) + 70 = 100(1 + x/100)^2;$$

and calculate the value of x within one-tenth.

29. The velocity of water in a rectangular mill stream whose breadth is x feet and depth y feet, is $\dfrac{500 \times R}{0\cdot5 + \sqrt{R}}$ feet per minute, where $R = \dfrac{xy}{x + 2y}$. Determine the value of R in order that the velocity may be 5 miles per hour, and then determine the depth of the stream when the breadth is 4 feet.

30. Draw in the same diagram the graphs

$$y = x^2 + x + 1 \qquad y = 2\tfrac{1}{4} - x,$$

between the limits $x = -3$ and $x = 1$, taking an inch as the unit.

Find, from your diagram, the coordinates of the points common to both curves, and verify your result by solving the equation

$$x^2 + x + 1 = 2\tfrac{1}{4} - x.$$

31. The first diagram of the figure represents a cubical room measuring 10 feet each way. CE and DF are vertical lines bisecting two of the walls. DB = 4 ft. and CA = x ft. The second diagram shows the three walls supposed to be spread out into the same plane.

The question has arisen which is the shorter route from A to B, (i) that across the ceiling shown in the first diagram, or (ii) that round the walls which becomes a straight line in the second

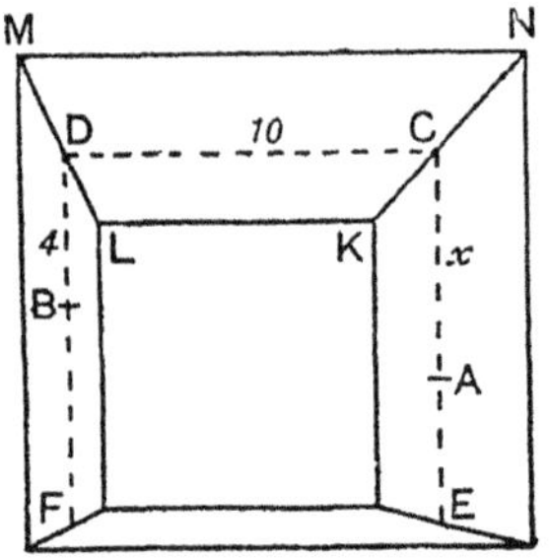

Fig. for Qn. 31, first diagram.

diagram. Take half a dozen different values of x; find, by drawing, the length of the second route for each value, and so decide the question. Sketch a graph in illustration.

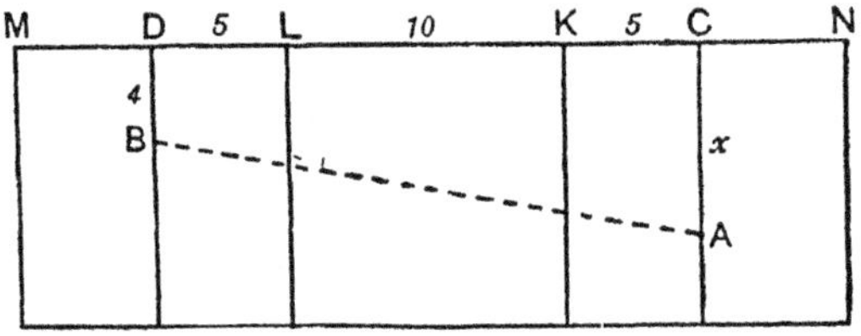

Fig. for Qn. 31, second diagram.

Then give an expression for the second route in terms of x, and by algebraic calculation determine for what values of x this route is the shorter.

32. Draw the graphs $y = 4/x$ and $y = x + 1$, from $x = -4$ to $x = +4$; and find the coordinates of their points of intersection correct to two decimal places, (1) by measurement from your figure, (2) by calculation.

33. Draw on squared paper the graph $x^2+y^2=9$. State, with your reasons, the nature of the graph.

Solve the equations: $x^2+y^2=9 \qquad 7x-8y=13{\cdot}07,$
retaining two places of decimals in your answer.

34. Solve the equations, giving the values of the roots as decimals to two places: $x^2+2y^2-x=4 \qquad y=4-2x.$

35. Solve the equations: $y=2x-1 \qquad 2x^2+5x-6=0,$
giving the value of each root correct to two significant figures.

36. Solve the equations:

(i) $\dfrac{2}{x-3}+\dfrac{3}{x-4}-\dfrac{5}{x-5}=0;$

(ii) $13(x+y)=7(1+xy), \quad 3(x+y)=10(x-y)+11;$

giving all values to two decimal places.

37. Solve the equations:

$$2{\cdot}7x+3{\cdot}4/y=4{\cdot}2x-5{\cdot}9/y \qquad x=1{\cdot}6y+7{\cdot}3$$

giving the results to two significant figures.

38. Solve the following equations, finding the roots correct to two decimal places:

(i) $\dfrac{2x-7}{3x-8}=\dfrac{2(x-9)}{3(x-11)};$

(ii) $x=y+6, \qquad x^2=y+48;$

(iii) $3x^2-7x=100.$

39. Solve the equations:

(i) $x^2-43x+432=0;$

(ii) $\dfrac{x-3}{y+2}=\dfrac{x+1}{y+8}, \qquad 2x-3y=4.$

40. Solve the equations:

(i) $\dfrac{1}{x-1}-\dfrac{2}{x+1}=\dfrac{3x^2}{x^2-1};$

(ii) $x(y-x)=9, \qquad y^2-x^2=27.$

41. Solve the equations:

(i) $(x-1)(3+1/y)=0, \quad 3x+4y+5=0$;

(ii) $27x^2+35x-41=0$,

giving all the results as decimals to three significant figures.

42. (i) Solve the equation $x+1/x=3$, finding both values of x correct to two decimal places, and verifying your solutions by calculating to four significant figures the values of the left-hand member with them substituted for x.

(ii) Solve the simultaneous equations:

$$\frac{y+z-x}{2}=\frac{z+x-y}{3}=\frac{x+y-z}{5}$$

$$x^2+y^2+z^2=138.$$

PART V.

SECTIONS 35-46. CHIEFLY THE CIRCLE AND LITERAL EQUATIONS.

Section 35. The Circle: Simple Properties.

1. Prove that the diameter is the greatest chord in a circle; and that of the others, the one which is nearer to the centre is greater than one more remote.

Draw a circle radius 3 inches, place chords AB and BC in the circle respectively equal to 4 and 5 inches, in such a way that the centre of the circle lies in the angle ABC. Measure AC, and state which of the lines AB BC CA is nearest to the centre of the circle.

2. Draw a circle of 3 inches diameter, and also two parallel chords 2 inches and $1\frac{1}{2}$ inches in length, such that the centre lies between them. Measure the distance between the chords.

3. Show how to complete a circle of which an arc is given.

4. Show how to describe a circle about a given triangle.

Show, by diagrams, the position of the centre of the circle according as the triangle is acute angled, right angled, or obtuse angled.

5. With your instruments inscribe a regular hexagon in a circle of 4 in. diameter.

6. Take any straight line and mark a point A at a distance of 4 cm. from it. Now find another point B, which is 3 cm. from A and 3 cm. from the given line. Find also a point C, 3 cm. from A and as far as possible from B. Find by measurement how far C is from the given line, and check your result by general reasoning.

7. Draw accurately a triangle ABC with its sides AB BC CA respectively 8 cm., 10 cm., and 12 cm., in length. Find a point inside the triangle at equal distances from B and C, and 7 cm. distant from A. State the steps of your construction, but do not give a proof of their correctness.

Measure the distances of the point so found from B and C.

8. Draw a straight line AB of length 8·2 centimetres, and find by construction points C and D dividing it into three equal parts. Draw CE at right angles to AB and of length equal to 5·1 centimetres. Describe a circle to pass through B D E, and measure its radius.

9. (In answering this question all construction lines are to be shown, but a written statement of the construction and a proof are not required.)

Construct a quadrilateral ABCD, having given AB = 5 centimetres, BC = CD = DA = 4 centimetres, AC = 6 centimetres.

Measure the length BD, and the angles BAC CAD.

Describe a circle about ABC, and measure its radius.

10. Show that equal chords of a circle are equally distant from the centre.

If two equal chords of a circle, AB and CD, intersect at O, show that AO is equal either to OD or else to OC.

11. ACB is a chord of a circle, C is its middle point, PCQ is another chord through C. Prove that PC is not equal to CQ, unless AB is a diameter.

12. Without using a protractor describe a regular polygon of 24 sides, and prove that your method is correct.

13. Prove that if two triangles have two angles of the one equal to two angles of the other, each to each, and a side of the one equal to a side of the other, these sides being opposite to equal angles in each, then shall the triangles be equal in all respects.

A is a point on the circumference of a given circle having its centre at O. OA is produced to B so that AB = OA. With centre B and radius BA a circle is described. C is any point on the circumference of this circle. CA is joined and produced to meet the given circle in D. Prove that CA = AD.

14. A circle passes through the four extremities of two parallel lines. What do you know about the relative position of the lines? Give reasons for your answer.

15. If from a point within a circle two equal straight lines are drawn to the circumference, prove that the line bisecting the angle between them passes through the centre, and that the chords of which these lines are part are also equal.

16. AB BC CD DE etc. represent equal pins so placed in a plane that the head of one pin coincides with the point of the next and the angles ABC BCD CDE etc. are all equal. Pins are added, as described above, until it is found that the last pin meets the first AB. What condition is necessary that the head of the last pin may coincide with the point A of the pin AB?

17. X and Y are two fixed points 10 cm. apart. A circle is described with centre X and radius 8 cm., and another circle with centre Y and a variable radius r. Between what limits must r lie that the circles may intersect?

Show that within these limits the common chord CD is always perpendicular to XY.

When CD has its maximum value, find the magnitude of the angle CYD and the value of r.

Determine r also when $CD = 5$ cm., and explain the double result.

18. Prove that any chord of a circle is bisected by the perpendicular drawn to it from the centre, and deduce the locus of the middle points of a series of parallel chords of a circle.

A straight line is drawn meeting each of three given circles. How must the circles be situated if the middle points of the three chords so formed coincide?

19. A hollow cylinder of length 7 cm. and internal diameter 4 cm. is placed on end on a horizontal plane. On the top of the cylinder is placed a solid sphere of diameter 6 cm., which rests part within the cylinder and part above it. By drawing an accurate diagram, find the height of the centre of the sphere above the horizontal plane.

20. A sphere 1 foot in diameter represents the earth, and is marked with lines of latitude and longitude. A fly starts from the equator and crawls north through 45° of latitude; it then crawls along a line of latitude through 45° of longitude. What is the length of its path on the meridian? By drawing find the radius of the line of latitude and the distance crawled upon it.

21. Q is the centre of a circle of radius 10 cm. and QO is a radius. The seven points A B C .. lie on the circumference and the angles OQA OQB OQC .. have the values 10° 20° 30° 40° 50° 60° 70°. Find, by drawing, the lengths of the chords OA OB OC .., and tabulate the results.

Draw a graph to give the length of a chord of the circle in terms of the angle which it subtends at the centre (for angles up to 70°). Show the chord's actual size, and represent 4 degrees by 1 cm.

From your graph find the length of the chord which subtends an angle of 48°. Make a triangle having one side of this length, and the other two sides 10 cm. long, and therefore having an angle of 48°. Check the accuracy of your drawing by measuring this angle.

22. Explain what is meant by a symmetrical figure. State (without proof) the axes of symmetry for each of the following figures, illustrating your answers by diagrams (which need not be accurately drawn): (i) isosceles triangle, (ii) rectangle, (iii) rhombus, (iv) semicircle, (v) regular pentagon, (vi) regular hexagon.

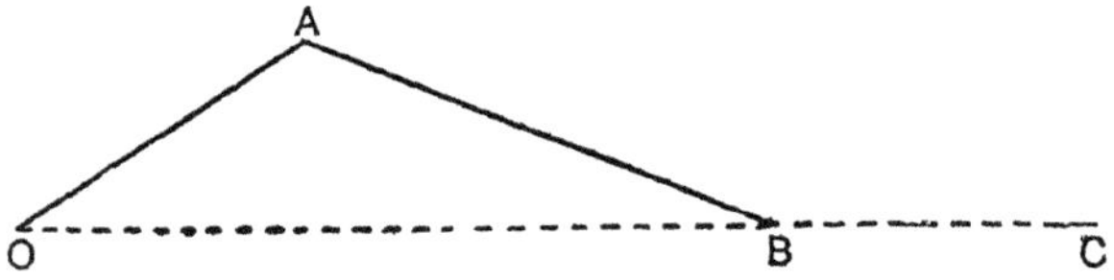

Fig. for Qn. 23.

23. OA and AB are the crank and connecting rod of a steam engine (see the sketch). OA revolves about O, and B can move to and fro along a fixed straight line OC. Taking 2 inches and $3\frac{1}{2}$ inches to represent the lengths of OA and AB, draw the system with OA in different positions, and in each position mark off on OA (produced if necessary) a length OP equal to OB. Draw the complete path of P, and thus represent graphically the variations in the distance OB as OA revolves.

What kind of curve is the graph so constructed?

24. Two rods AB and BC are hinged together at B, as in the figure. The rod AB is fastened to a fixed hinge at A, about which it is free to revolve. AB = 4 cm., BC = 3 cm., and XY is 5 cm.

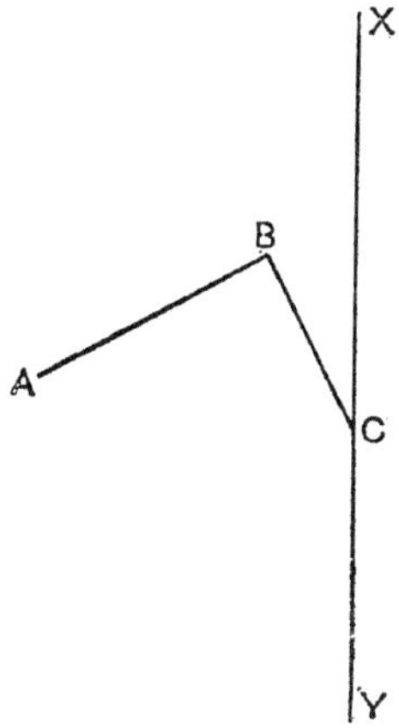

Fig. for Qn. 24.

from A. Draw in your book the path which B is free to take if C is moved along the unlimited line XY, and denote clearly along what portion of XY the point C is free to move.

Section 36. Tangent Properties.

(i.) TANGENTS TO A SINGLE CIRCLE.

1. Construct a regular hexagon about a circle of diameter 6 cm.

Draw also a rectangle equal in area to the hexagon, explaining carefully your method.

2. Draw a circle of radius 2 inches; take a point $4\frac{1}{4}$ inches from the centre of the circle, and from it draw a tangent to the circle.

Prove the correctness of your construction.

Measure the length of the tangent.

Find by calculation what its length should be.

3. A ring is slung on a string, and the ends of the string are then joined together. The string is then hung on a vertical wheel 94 centimetres in diameter so that the ring is vertically below the centre of the wheel and is found to be 47 centimetres from the rim of the wheel. Draw a plan of the apparatus on a scale of 10 centimetres to a metre, and find the length of the string in any way you please.

4. Describe a circle of radius 2·5 centimetres. Take O, the centre, and draw any radius OA. Find a point T in OA produced such that the two tangents drawn from T to the circle are inclined to one another at an angle of 70°.

Explain your construction, and measure the tangents.

5. Prove that all chords of a given circle which are of a constant length touch a fixed circle.

6. Two circles have radii 1·2 in., 2·3 in., and their centres are 0·75 in. apart. Calculate to the nearest tenth of an inch the lengths of the greatest and least chords of the outer circle which can be drawn to touch the inner.

7. AB and BC are two straight portions of a thin bent rod. The rod rotates round a fixed point of AB. Show that BC or BC produced always touches a fixed circle.

8. If a straight line APQB cuts a circle whose centre is O at P and Q, show that the angle OPA is equal to the angle OQB; and if a straight line *touches* the circle at P, show that it is perpendicular to OP.

9. Draw two circles of radii 1·3 and 2 in., which have a common chord of length 1 in. Measure the distance between their centres, and the angle at which the tangents at one of the common points intersect.

10. Show, with proof, how to construct a circle of given radius to touch two intersecting straight lines. Show how many solutions there may be to this problem.

11. Draw a circle of one inch radius. About it describe a square and then an equilateral triangle having one of its sides coincident with a side (sufficiently produced) of the square.

(i) Measure the height of the triangle.

(ii) Measure the length of one of its sides; also deduce this measurement from the properties of a right-angled triangle, to two decimal places.

(iii) Compare the areas of the square and the triangle, giving the lesser as a decimal of the greater, to two places.

12. Circumscribe a regular polygon of eight sides to a circle of radius 5 cm. Hence, by measurement or otherwise, show that the circumference of the circle is less than 3·4 times the diameter.

13. What are the area and the diameter of the largest circle that can be cut from a rectangular sheet of metal, 9 inches by 7 inches? If the circle touches three of the sides of the rectangle, what is the area of the largest square that can be cut from the remaining piece?

Draw the rectangle, scale $\frac{1}{2}$ inch to 1 inch, and show by dotted lines exactly where the cuts are to be made.

How many square inches of the metal are wasted?

14. Draw an isosceles triangle with base BC equal to 3·5 cm. and the sides AB AC each equal to 5 cm.

Describe a circle to touch AB at A and to pass through the point C.

Measure the radius of this circle.

Write out your construction and justify it.

15. If the external angles of a triangle ABC be bisected, prove that the bisecting lines form a triangle such that A B C are the feet of the perpendiculars from its vertices on the opposite sides.

16. By drawing a number of tangents of fixed length to a fixed circle from points on its circumference, find out the locus of their extremities.

Draw this locus accurately for the case in which the radius of the fixed circle is 4 cm. and the fixed length is 4·2 cm.

If from any point in your locus a tangent is drawn to the fixed circle, prove that it is of the required length.

17. A point P is required to be equidistant from two given points A and B, and the length of the tangent drawn from it to a given circle is also given; show how to find P as the point of intersection of two loci. How many positions for P can be found?

18. Prove that the two tangents drawn from an external point to a circle are equal.

AB and AC sides of a triangle are equal, and D and E are points on AB and AC respectively, such that DE is equal to BD + EC. Prove that DE touches the circle to which AB and AC are tangents at B and C.

19. Draw two lines cutting at an angle of 57°; and describe a circle of radius 1·3 inches which touches both lines. Draw also two parallel tangents to the circle each forming together with the two lines a triangle of angles 57°, 68°, 55°. Measure their lengths.

20. Draw a triangle with sides 9 cm., 11 cm., and 14 cm. Inscribe a circle in the triangle, and measure the radius of the circle.

21. Describe a triangle whose sides are 2·6, 2·8, and 3 inches respectively. Measure the greatest angle.

Inscribe a circle in the triangle, and measure its radius.

22. Show how to find a point inside a triangle equally distant from the three sides: and show why your construction is correct.

Also show how to draw three semicircles within the triangle, each touching two sides of the triangle and having its diameter lying along the third side.

23. Show how to inscribe a circle in a given triangle.

A circle is inscribed in a right-angled triangle. Show that the diameter is equal to the difference between the sum of the two sides and the base.

24. Describe a circle touching one side of a triangle and the other two sides produced.

If this circle touches the side BC of a triangle, find the distance of its point of contact with AB from A, given that BC = 12 feet, CA = 14 feet, AB = 16 feet.

25. The sides of a triangle are 5, 7, and 9 centimetres in length. Construct the escribed circle touching the shortest side, and justify your construction by general reasoning. Measure the radius of the circle.

26. Describe a circle with centre A and with radius AB of 1 inch; produce AB to C so that BC = 1/2 inch; draw CD perpendicular to

AC. Find a point P in CD such that the tangents drawn from P to the circle may contain an angle of 60°. Measure the length of PC.

27. Draw a triangle with its side 3″, $3\frac{1}{8}$″, and 5″ long respectively. Inscribe a circle in the triangle, and circumscribe a circle about the triangle. Also draw a circle to touch the 3″ side and the other sides produced. Find the area of the triangle. Measure the radii of the three circles.

28. ABCD is a quadrilateral whose angles at A and B are right angles. The lengths of the sides AB BC AD are 4·5, 8, and 2·1 centimetres respectively. Draw the quadrilateral, and describe two circles to touch AB at its middle point and CD (produced if necessary). Justify your construction of the circles, and measure their radii.

29. Draw a triangle ABC whose sides AB BC CA are respectively 7, 4·4, 4·2 centimetres in length. Produce AB and AC to D and E. Describe a circle whose centre shall be in CE to touch BD and BC; measure its radius.

Draw another tangent to this circle from A. Produce BC to meet it in F. Prove, by measuring and then by general reasoning, that the sum of the sides of the triangle ABF is equal to the sum of the tangents from A to the circle.

30. The four sides of a quadrilateral touch the same internal circle. Prove that the sum of one pair of opposite sides is equal to the sum of the other pair.

31. A piece of wire 16·78 centimetres long is bent into a quadrilateral which circumscribes a circle whose radius is 1·1 centimetres. Calculate the area of the quadrilateral.

32. Draw any triangle ABC having $\angle B = 45°$, and draw AD perpendicular to BC, and prove that $AC^2 = BD^2 + DC^2$.

By this or other geometrical construction divide a line 4 inches long into two parts such that the sum of the squares on them shall be 12 sq. inches.

BC being 4 inches long, and E a movable point in it, state, with reasons, what is the least possible value of $BE^2 + EC^2$, and give the corresponding values of BE and EC.

33. Show, without proof, how to draw regular hexagons inscribed and circumscribed to a given circle, and prove that the area of one is three-quarters of the area of the other.

34. Construct a regular octagon, four of whose sides lie along the four sides of a square of side 10 cm., proving that your method is correct; and measure the side of the octagon.

35. Draw carefully an equilateral octagon, with each of its eight sides distant 1·5 inch from a given point. State the steps of your construction, but do not give a proof of their correctness. Measure the length of a side of the octagon.

(ii.) COMMON TANGENT TO TWO CIRCLES.

36. Two circles of radii 3 and 5 cm. have their centres 9 cm. apart. Draw and measure a common tangent.

37. Draw two circles with radii equal in length to 3 and 1 centimetres respectively, and having their centres 45 millimetres apart.

Construct two unequal common tangents to these circles. Measure the lengths of these tangents and calculate the rectangle contained by them.

38. Construct and measure the common tangents to two circles, radii 31 mm. and 22 mm., whose centres are 45 mm. apart. Also measure the angle between the two common tangents.

Justify your construction.

39. Two circles touch externally at a point A. A common tangent to the two circles touches them at B and C. Prove that BAC is a right angle.

40. Two circles, whose radii are 2 and 3 centimetres respectively, touch externally. Construct the common tangents, and calculate their lengths to two places of decimals.

41. Show, without proof, how to construct the internal common tangents of two non-intersecting circles.

Also calculate the lengths of these tangents, having given the radii a b of the two circles and the distance d between their centres.

42. PQ and RS are the direct common tangents to two circles, P and R being the points of contact with one circle, Q and S the points of contact with the other. Prove that PR and QS are parallel.

43. Two wheels, whose radii are 3·4 centimetres and 2·3 centimetres, have their centres 9 centimetres apart and a strap is drawn tightly round the two wheels.

Draw a plan of the apparatus and measure the length of the strap.

44. Show how to draw an external common tangent touching two given non-intersecting circles, and calculate the length of the tangent, when c is the distance between the centres and a and b the lengths of the radii.

Hence show that if the line joining the centres of the circles cuts the circumferences in the points A B C D in this order, the square on the tangent is equal to the rectangle AC.BD.

(iii.) MUTUAL CONTACT OF CIRCLES.

45. Show how to construct a circle passing through three given points, and prove that your construction is correct.

The triangle ABC has AB = AC = 2·3 in., BC = 1·4 in., and a circle is drawn with centre A and radius 1·2 in. Construct the two circles, each to pass through B and C and touch this circle, and measure their radii. Each step of the construction must be explained, but no proof is required.

46. Draw a circle of radius 1·5 inches, touching each of two equal circles, of radius 0·8 inch, whose centres are 2 inches apart.

Show that six circles can be drawn satisfying the given conditions.

47. Draw two circles of radii 2 and 2·5 inches, and with centres 2 in. apart. Draw one circle of radius 1 in. to touch them both. How many such circles may be drawn?

48. Given a circle and two tangents to it, show how to construct a circle which touches both the given circle and the given tangents, and prove that your method is correct.

49. Describe a circle of $1\frac{1}{2}''$ radius touching a straight line. Describe another of $\frac{3}{4}''$ radius touching the circle and the line. Describe a third circle of $1''$ radius touching both circles.

50. Construct an octagon of $1\frac{1}{2}''$ side, and inscribe four equal circles each to touch one side and two other circles. Protractor not to be used.

51. Draw a circle of 3·5 centimetres radius, touching internally a circle of 2 centimetres radius and externally one whose radius is 1·5 centimetres, the distance between the centres of the latter two circles being 4 centimetres.

Prove that your construction is correct.

52. The triangle ABC has sides BC = 2·4 in., CA = 2·9 in., AB = 1·8 in.; construct three circles having their centres at A, B, C respectively, each of which touches the other two, and calculate or measure their radii. The method of construction must be explained.

53. Two fixed intersecting circles being given, find how many positions there are for a circle of given radius in order that it may touch them; consider different sizes for the touching circle from very small to very large.

54. The figure represents a doorway over which are a circular and two semicircular windows, the curve ACB being also a semicircle. If the radius of each of the two smaller semicircles is R inches and the radius of the circular window r inches, find, by calculation, the ratio of r to R.

Taking AB as 50 inches, make a drawing to scale of the part above AB, representing 10 inches by 1 cm. Give the radius of the circular window, and state the construction and the geometrical principles assumed.

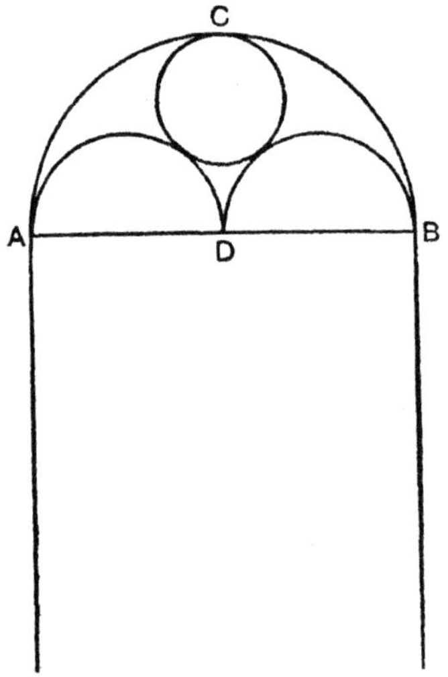

Fig. for Qn. 54.

Section 37. Literal Equations.

Preceding sections on linear and quadratic equations will be found to give most of the educational value which is to be had from the study of equations. The present section, on equations with literal coefficients, contains little that is educative, and is included here only in deference to custom and the requirements of examinations. The section should be omitted by those who are free to do so. (See page V.)

(i.) LINEAR EQUATIONS WITH LITERAL COEFFICIENTS.

1. Solve the following equations, verifying all your results:

(i) $a(x-a) = b(x-b)$;

(ii) $\frac{1}{24}(x-\frac{1}{2}) - \frac{1}{12}(x-\frac{3}{2}) = \frac{1}{8}(x-\frac{5}{2}) - \frac{1}{3}(x-\frac{7}{2})$;

(iii) $(x+1)(x-2)-(x-3)(x+4) = (x-2)(x-3)$.

2. Solve the equations:

(i) $0{\cdot}605x - 0{\cdot}8 = 0{\cdot}75(0{\cdot}5x - 0{\cdot}3)$;

(ii) $\dfrac{x-a}{b} = \dfrac{x+b}{a} - 2$;

(iii) $x^2 + 6x = 20$, to 3 decimal places.

3. Solve: (i) $\dfrac{10-x}{2} - \dfrac{x-3}{4} = \dfrac{x-1}{4} - \dfrac{13-x}{6}$;

(ii) $(a+b)(x-a-b) = (a-b)(x-a+b)$.

4. Solve: (i) $\dfrac{x}{8} - \dfrac{x-8}{16} = \dfrac{x-1}{5} - \dfrac{x-7}{7}$;

(ii) $\dfrac{a(x-a)}{b} + \dfrac{b(x-b)}{a} = x$;

(iii) $\dfrac{x}{6} = \dfrac{y}{2} + 4$,

$5(x-7) + 11(y-14) = 9$.

5. Solve the equations:

(i) $\dfrac{2}{x-5} = \dfrac{1}{3} - \dfrac{x}{3x+4}$;

(ii) $x - c - \dfrac{a-c}{x-b} = x - a$.

6. Solve the equations:

(i) $\frac{x-1}{4}-\frac{15-x}{2}-5 = \frac{x-3}{5}-\frac{5(x-1)}{12}$

(ii) $3{\cdot}7x-1{\cdot}2y = 4{\cdot}6 \qquad 2{\cdot}3x+7{\cdot}2y = 1{\cdot}4$

(iii) $\frac{b-c}{x+c}+\frac{a-b}{x+a} = \frac{a-c}{x+a+b}$

7. What is meant by an equation, and what is meant by solving an equation?

Solve the equations:

(i) $\frac{x-a+b}{x+a-b} = \frac{1-a+b}{1+a-b}$;

(ii) $\frac{x^2-6x+2}{x^2+6x-2} = \frac{1-6+2}{1+6-2}$;

(iii) $x-ay+b = 1, \qquad ax-y+b = 1.$

8. Solve the equations:

(i) $\frac{x}{5}+\frac{2x-7}{3} = \frac{x-5}{8}+2$;

(ii) $(x^2-1)(x-2) = x^3+2$;

(iii) $x+y = 3(a+b), \qquad bx+2(a+b)y = 2(a+b)x+ay.$

Prove the correctness of your answer to (ii).

9. Solve the equations:

(i) $\frac{6}{x+1}-\frac{5}{y-1} = 9, \qquad \frac{7}{x+1}-\frac{2}{y-1} = 5$

(ii) $0{\cdot}5x+y = 2{\cdot}25, \qquad 3{\cdot}4x+0{\cdot}02y = 5{\cdot}13$

(iii) $(a+b)x+(a+c)y = a+b, \qquad (a+c)x+(a+b)y = a+c$

(iv) $\frac{1}{x+1}-\frac{2}{2-x} = \frac{6}{5}$

10. Solve the equations:

(i) $x-1+\frac{5(x+1)}{2x+3} = \frac{2x^2-3}{2x+3}+\frac{17}{(x-1)(2x+3)}$;

(ii) $\frac{5x-3y}{2} = 4x+y = \frac{8}{25}$;

(iii) $\frac{x}{a}+\frac{y}{b} = \frac{a+b}{2}, \qquad \frac{x}{b}-\frac{y}{a} = \frac{a-b}{2}.$

11. (i) Solve the equation $\frac{3}{x}-\frac{2}{x-5}=4$, working out the values to 2 significant figures.

(ii) Solve also the equations

$$px-(p-q)y = q^2-pq, \qquad (p+q)x-py+q^2+pq = 0.$$

12. Solve the equations:

(i) $$\frac{2x^2-7x-4}{x-3} = \frac{2x^2+5x-16}{x+3}$$

(ii) $$\frac{ax+by}{2(a+b)} = c = \frac{ab(x-y)}{b^2-a^2}$$

13. Solve: $$\frac{a}{x}+\frac{b}{y}=2, \qquad \frac{b}{x}+\frac{a}{y}=\frac{a^2+b^2}{ab}$$

14. Solve the equations:

(i) $$\frac{60-x}{14}-\frac{3x-5}{7} = 3\tfrac{1}{2}-\frac{3x-24}{4};$$

(ii) $$\frac{x}{a}+\frac{y}{b} = \frac{2}{a-b} \qquad \frac{b}{x}+\frac{a}{y} = \frac{a^2+b^2}{(a-b)xy};$$

(iii) $$\frac{20}{8-x}-\frac{21}{x-4} = 13.$$

Prove the correctness of your answer to (iii).

15. Solve the equations:

(i) $$\frac{x+4}{3}+\frac{x+7}{5} = \frac{x+6}{2};$$

(ii) $$\frac{a}{a+x}+\frac{a-2x}{a-3x} = \frac{2}{3};$$

(iii) $$x-3y=2, \qquad x^2-xy=12y^2.$$

(ii.) QUADRATIC EQUATIONS WITH LITERAL COEFFICIENTS.

16. Solve the following equations, giving fractions and surds to two significant figures:

(i) $$\frac{x}{5}-\frac{y}{4} = \frac{1}{14}, \qquad \frac{x}{3}-\frac{3y}{14} = \frac{1}{5}$$

(ii) $$\frac{1}{x-2a}-\frac{4}{x-a} = \frac{1}{x+2a}-\frac{4}{x+a}$$

(iii) $$\frac{1}{x-2}+\frac{1}{x-3} = \frac{7}{12}$$

17. Solve the equations:

(i) $x+2y = -2, \qquad x^2+2xy+4y^2 = 19;$

(ii) $\dfrac{x+a}{x}+\dfrac{x}{x+a} = 2\tfrac{1}{2}.$

18. Solve the equations:

(i) $(b-c)x^2+(c-a)x+(a-b) = 0;$

(ii) $x+y+z = 12, \quad x^2-y^2 = 5, \quad x^2+2xy+y^2+z^2 = 74.$

19. Solve the equations:

(i) $3x^2-4y^2 = 2(x+2y), \qquad 2x-y = 3;$

(ii) $\sqrt{x-a}+\sqrt{x-b} = \sqrt{a-b}.$

20. Find the value of a which will make the statement $2x=3-a/x$ true when $x=1{\cdot}7$.

If a has this value, what other value of x will also make the statement true?

21. Given that $x=2$ is a solution of the equation

$$a^2x^2+3ax-4=0,$$

find the value of a.

22. Replace a and b by numerical values so that the equations

$$(x-2y)(2x-y)=a \qquad 2x+y=b$$

may be satisfied by $x=4 \quad y=3$, and find the other common solution of these equations.

23. Solve: $\quad x^2+x(p^2/q-q^2/p)-pq = 0.$

24. Solve the quadratic equation:

$$(x-5)(x-1)-a(x-3)(x-2) = 0.$$

Find values of a for which the two roots of the equation are equal, and state the value of the roots; giving the roots and the values of a to two decimal places. Draw the graph of $(x-5)(x-1)$ from $x=-5$ to $x=5$, and with the same axes draw the graphs of $a(x-3)(x-2)$ for the two special values of a.

Section 38. Angle Properties of the Circle.

1. On a piece of tracing paper make an angle of 37°, vertex at C. In your book take two points A and B 6·3 cm. apart. Move the tracing paper about so that the legs of the angle always pass through A and B; and prick through the path of C. Consider only points on one side of AB.

Show, with proof, how the path of C could be drawn with protractor, ruler, and compasses.

2. A and B are fixed points, 2 inches apart; P is a variable point which moves so that the angle APB is 30°. Construct the complete locus of P, and prove that the method of your construction is correct.

3. Draw a circle 3 inches in diameter and cut off a segment containing an angle of 50 degrees. Measure the length of the chord on which the segment stands.

4. Show how to construct on a given line a segment of a circle containing a given angle, and indicate clearly in a diagram whereabouts a point must be if the angle subtended at it by the line is greater that this given angle.

The base AB of a triangle is 2·5 in., the area is 4 sq. in., and the vertical angle is greater than 30°. Whereabouts must the vertex lie?

5. If ABCDEFG is a regular heptagon, calculate to the nearest minute the angles of the figure, the angles of the triangle ACF, and the angles of the triangle formed by AC and DF produced and the perpendicular at G to FG.

6. Prove that the angle in a semicircle is a right angle.

If a parallelogram be inscribed in a circle, the diagonals of that parallelogram must meet in the centre of the circle.

7. Prove that the angle at the centre of a circle is double of an angle at the circumference standing on the same arc.

If two unequal circles cut one another, the common area consists of a segment of each. Show that of these segments that belonging to the smaller circle has the smaller angle in it.

8. The triangle ABC has the sides AB and AC equal, and D and E are the middle points of these sides. If the circle through D E A touches BC, prove that the triangle is right angled.

9. (i) Given the base of a triangle and its area, what is the locus of its vertex?

(ii) Given the hypotenuse of a right-angled triangle, what is the locus of the opposite angular point?

Make use of these two loci in solving the following problem: Describe a right-angled triangle whose hypotenuse is 3·4 inches and area 2 square inches.

Measure the lengths of the sides containing the right angle.

Would it be possible to describe on the same hypotenuse a triangle of area 3 square inches? Give reasons for your answer.

10. On a base 2·5 inches long describe a triangle with vertical angle 65° and area equal to 2 square inches.

Describe and justify your construction.

11. ABCDE is a regular pentagon; AE and CD meet in F. Prove that the straight lines drawn from D E F parallel respectively to AE AD AB meet in a point.

12. Suppose that an arc of a circle crosses the paper, the centre not being on the paper; show how to draw (i) the tangent at any point on this arc, (ii) the tangent at either point where it meets the edge of the paper. The drawing may not be continued on another sheet of paper.

13. Two circles touch externally at A, and PAS and QAR are straight lines cutting one circle in P Q and the other in R S. Prove that PQ is parallel to RS.

14. In a circle of 5 cm. radius inscribe a quadrilateral ABCD, having the angle ABC equal to 115°, the angle BCA to 30°, and the angle ACD to 60°. Measure the lengths of the sides and prove that the method of your construction is correct.

15. Prove that two chords of a circle cannot bisect each other unless they both pass through the centre of the circle.

Show how to draw a chord through a point A on a circle which shall be bisected by a chord BC. Is this always possible?

16. AB is a chord dividing a circle into two unequal parts. P is any point on the major arc. The straight lines bisecting the angles PAB and PBA meet at a point Q. Find the locus of Q as P moves along the arc.

17. Prove that the angles in the same segment of a circle are equal.

Two circles intersect in A and B. APQ is any straight line drawn through A cutting the circles in P and Q. Show that the angle PBQ is constant.

18. ABC is a triangle inscribed in a circle and the angle A is bisected by the straight line AX. Show that AX makes equal angles with BC and the tangent at A to the circle.

19. If two angles at the circumference of a circle be equal, prove that the arcs on which they stand are equal.

AB AC AD are three thin rods rigidly joined at A. P and Q are two thin pins inserted in a flat table. If the rods be laid on the table, so that AB touches P, and AC touches Q, show that A must lie on a certain circle, and that AD passes through a fixed point in the circumference of that circle.

20. A triangle ABC is inscribed in a circle, the angle A is 65°, the side AB subtends an angle of 115° at the centre, and is 1·5 inches long. Construct the triangle, and measure its remaining angles and sides.

21. A and B are two fixed points on a circle centre O. OC is perpendicular to AB. R is the middle point of any chord of the circle through A. Show that ARC is a constant angle.

22. Define a tangent to a circle. From your definition prove that the angle between the radius drawn to the point of contact and the tangent is a right angle.

ABC is a triangle, right angled at A; AL is the perpendicular from A to BC, and LM LN are the perpendiculars from L to AB AC; prove that MN is a tangent to the circles described on BL LC as diameters.

23. Show, with proof, how to describe a circle about a given triangle.

In a triangle ABC the angle at A is 60°; the bisector of this angle meets the circumscribed circle at D. With D as centre, and DA as radius, a circle is described cutting AB again in E. Prove that the circle circumscribing the triangle DAE is touched by the line CA, and has a radius equal to DA.

24. A straight line OA is drawn at right angles to the diameter MON of a circle centre O, and cutting the circumference in the point A. With centre A a circle is drawn through M and N and cutting OA produced towards A in B. With centre B a circle is drawn through M and N. A straight line MPQR is drawn cutting the circles in P Q R. Show that the triangles QRN PQN are isosceles, and calculate the magnitude of the angle R.

25. AB is a chord of a circle, C a point on the smaller arc, O the centre. Prove that the angle OBC equals the angle between AC and diameter bisecting AB.

26. A B C are three points on the circumference of a circle; BD bisects the angle ABC and meets the circumference at D. Through D the chord DE is drawn parallel to AB. Prove that CE is parallel to BD.

27. Draw a circle of radius 1·2 inches; construct a triangle of which each side touches the circle, and of which two angles are 65° and 41°. State carefully each step of the construction.

28. Construct a right-angled triangle, being given the length of the hypotenuse (7 cm.), and of the perpendicular from the right angle upon the hypotenuse (2·5 cm.).

29. Take a chord AB 2 inches long in a circle of radius $1\frac{1}{2}$ inches. Take a third point C on the greater of the two arcs into which A and B divide the circumference. Join AC and CB, and produce AC to P so that CP = CB. Do this for several positions of the point C, and draw the locus of P as C moves round the circumference.

Measure the angle APB for several positions of P, and discuss the nature of the locus by general reasoning.

30. AB and AC are two chords of a circle; BD is drawn perpendicular to AC, meeting it at D and the circle at L. CE is drawn perpendicular to AB, meeting it at E and the circle at M. Prove that DE and LM are parallel.

31. In the acute-angled triangle ABC the bisector of the angle A and the perpendicular from A on BC meet the circumcircle at L and M respectively. Prove that the arc LM is equal to half the difference of the arcs AB and AC. Each arc is supposed to be less than half the circumference.

32. A rectangular piece of paper ABCD (see the figure, which is not drawn to scale), 20 centimetres by 16 centimetres, is folded in such a way that A falls on BC and the crease passes through D. Draw a figure half size, and on it a line to represent the crease thus formed.

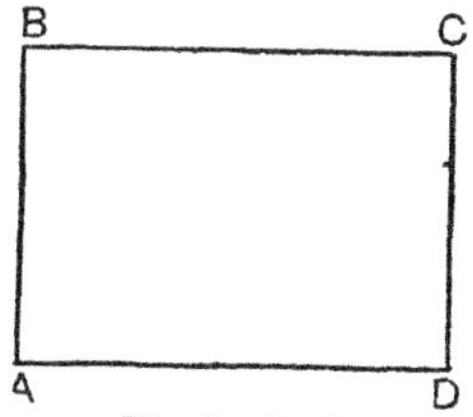

Fig. for Qn. 32.

Draw also the creases through C B A, formed when the corners B C D are similarly folded over so as to come on AD AD BC respectively.

Prove that the quadrilateral formed by the creases is a parallelogram with equal sides.

33. The triangle ABC has the angles at B and C equal to 70° and 75° respectively; the triangle PQR is formed by the tangents at A B C to the circle round ABC, and the triangle XYZ is formed by the tangents at P Q R to the circle round PQR. Calculate the angles of the triangles PQR and XYZ, and verify by measurement from a carefully drawn figure.

34. Explain how to construct a triangle having given the base, area, and vertical angle.

If the base is 15 yards long and the vertical angle is 37°, find, by drawing to scale, the greatest possible area.

35. If a triangle is to have its base 3 inches long, and its vertical angle 110°, find the largest area it can possibly have, using drawing, measurement and calculation in any proportions you please.

Draw the triangle when its area is half this maximum, and measure its sides.

36. Prove that the line joining the centre of a circle to the middle point of any chord is perpendicular to that chord.

Draw a circle of radius 2·7 in. and take a point P which is 2·2 in. from the centre O. Construct a chord of the circle passing through P, and having its middle point 1·4 in. from P, and measure its length.

37. Prove that if a line AB subtends equal angles at the points C and D upon the same side of it, then a circle drawn through any three of the points A B C D will also pass through the fourth point.

38. A B C D are four points *in order* along the circumference of a circle, B and C being opposite ends of a diameter. All possible chords joining any two of these points are drawn, and BA and CD are produced to meet at O. Give as many instances of equal angles as you can find in the figure.

39. Prove that the opposite angles of a quadrilateral inscribed in a circle are supplementary.

In a quadrilateral ABCD of this type the angles CAD, BDA, BDC are 15°, 65°, 35° respectively. Find the values of the angles of the quadrilateral. Construct it when AC = 1·7 in., and measure the length of BD.

40. Show that if a quadrilateral inscribed in a circle has one of its sides produced, the exterior angle so formed is equal to the interior and opposite angle of the quadrilateral.

K is the middle point of the arc BC of the circle circumscribed to a given triangle ABC. KL and KM are drawn perpendicular to AB and AC. Show that LB = MC.

41. If A B P Q are any four points on a circle, prove that the angles APB AQB are either equal or supplementary.

If one pair of opposite sides of a cyclic quadrilateral are equal, prove that the other pair are parallel.

42. ABCD is a quadrilateral inscribed in a circle of which AC is a diameter; AB = 6·3 cm., BC = 5·85 cm., CD = 7·9 cm. Calculate the length of AD.

43. Show that a parallelogram inscribed in a circle is necessarily a rectangle.

44. A and B are the common points of two circles; C D are points in the same arc AB of one of these circles. AC AD meet the other circle again in G H; BC BD meet it again in E F. Prove that the angles EAG FBH are equal or supplementary.

45. Two circles have a common chord AB, and PAQ is a double chord drawn through A. If the tangents at P and Q meet in T, prove that a circle can be described about the quadrilateral TPBQ.

46. Show, with proof, how to describe a circle about a given square.

A circle circumscribes a rectangle ABCD. Another circle is described with centre D and radius DA cutting the circle ABCD in E. Prove that CE is parallel to BD and that BE is equal to CD.

47. Prove that the opposite angles of a quadrilateral inscribed in a circle are together equal to two right angles.

Enunciate and prove the converse of this proposition.

48. Describe a circle of radius $1\frac{1}{2}$ inches and to it circumscribe a quadrilateral ABCD with the angles A, B, C, D equal to 100°, 120°, 70°, 70° respectively.

49. A B C D E F are six points on a circle such that AB is parallel to DE and BC is parallel to EF. Prove that CD is parallel to FA.

50. The angles of a quadrilateral are bisected, each bisector meeting the two adjacent bisectors and so forming another quadrilateral. Prove that a circle can be described about this latter quadrilateral.

51. Two circles intersect in A and B. APQ is any straight line drawn through A cutting the circles in P and Q. Show that the angle PBQ is constant.

52. Prove that if A B P are three points such that APB is a right angle, then P lies on the circle described on AB as diameter.

Given the middle points of two adjacent sides of a rectangle, prove that the loci of its vertices and of the intersection of its diagonals are all circles.

53. A B C D are four points on a circle such that the chords AC BD intersect at right angles inside the circle. Prove that the tangents at A B C D form a quadrilateral which can be inscribed in a circle.

54. A quadrilateral is inscribed in a circle; and has also a circle inscribed in it, which touches its sides taken in order at the points P, Q, R, S. Prove that PR is at right angles to QS.

55. If A and B are fixed points, and the point P moves so that APB is a right angle, prove that its locus is the circle described on AB as diameter.

ABC is an isosceles triangle having the vertical angle A equal to 48°, and P is a point on the minor arc BC of its circumcircle such that the arc BP is one-quarter of the arc BC. If PL PM PN are drawn perpendicular to BC CA AB (produced if necessary), show, by calculating the angles MLC BLN, or otherwise, that the points L M N are collinear. The figure need not be drawn to scale.

56. P is any point on the circumcircle of the triangle ABC; X Y Z are the feet of the perpendiculars drawn from P to BC CA AB. Prove that X Y Z are collinear.

57. Prove that three straight lines drawn from the vertices of a triangle respectively perpendicular to the opposite sides meet in a point.

58. If AD BE CF be the perpendiculars drawn from the angular points of a triangle to the opposite sides, prove that the perpendiculars bisect the angles of the triangle DEF.

59. Inscribe in a circle of 5 cm. radius a triangle equiangular to a triangle whose sides are 5, 6, 8 cm. respectively, and find the area of the inscribed triangle.

60. P Q R are any points in the sides BC CA AB of a triangle. Prove that the circles circumscribing the triangles AQR BRP CPQ meet in a point.

61. Show, with proof, how to draw a triangle of given angles circumscribed to a given circle.

Also show how to draw a triangle of given angles so that the given circle touches the shortest side and the other two sides produced.

62. The base of a triangle is given and the centre of the circumscribed circle. Show that the line which joins the vertex of the triangle to the centre of the inscribed circle passes through a fixed point.

63. A and B are two fixed points whose distance apart is 5 inches. Draw (i) the curve which is the locus of a point whose distances from A and B differ by 2 inches; (ii) the curve which is the locus of a point at which AB subtends an angle of 40°.

Hence obtain a triangle PAB, whose vertical angle is 40° and the difference of whose sides is 2 inches. Measure the lengths of the sides PA and PB.

State how such a triangle could be constructed without drawing any other loci than straight lines and circles.

64. A semicircle is described on AB as diameter and a point C taken on the circumference such that BAC is one quarter of a right angle. If BD be drawn meeting AC at D and making the angle DBA equal to the angle DAB, prove that $AD^2 = 2DC^2$.

65. If a pair of opposite angles of a quadrilateral are together equal to two right angles, its vertices are concyclic.

ABCD is a square. P is a point in AD. Within the square an equilateral triangle AOB is described on AB. PO produced meets BC in Q. OR bisecting PQ at right angles meets AB in R. Prove that PRQ is an equilateral triangle.

66. Two circles intersect in A and B. A straight line CPQD cuts the circles in C P and Q D respectively. If BP cuts AD in F and BQ cuts AC in E, prove that a circle can be drawn through A E B F.

67. The diagonals of a quadrilateral inscribed in a circle intersect at right angles, and from the point of intersection a perpendicular is drawn on one of the sides. Prove that this perpendicular, if produced backwards, will bisect the opposite side.

68. In equal circles prove that angles at the circumference which stand on equal arcs are equal.

Having given the base of a triangle, the size of the vertical angle, and a point in the bisector of the vertical angle, construct the triangle, stating your construction without proof.

Draw a figure for the case in which the base BC is 4″, the vertical angle BAC is 75°, and the point P on the bisector is such that, P and A being on the same side of BC and PN the perpendicular to BC, PN = 0·6″, BN = 3″.

Measure the angle B, and deduce the value of C.

69. Two circles of radii 8 and 3 cm. have their centres 13 cm. apart. P and Q are the points of contact of a direct common tangent, P being on the greater circle. The line of centres cuts the greater circle in the points A B and the smaller in C D. AP and BP are drawn and produced to cut DQ and CQ (both produced in the necessary directions) in R and S respectively. Prove that PRQS is a rectangle, with RS perpendicular to the line of centres.

Measure its sides and find its area.

70. Two circles intersect at A and B, and from any point P on the one straight lines PA PB are drawn and produced to cut the other again in Q and R. Show that the length of QR is constant for all positions of P.

71. A straight line PQ of constant length slides with its extremities P and Q upon two lines AB AC containing an acute angle. At P and Q lines at right angles to AB AC are drawn, intersecting at R. Find the locus of R.

72. Any point P is taken on the base BC of the isosceles triangle ABC, and circles are described round the triangles APB, APC. Prove that their radii are equal, and also that if D and E are their respective centres, then BD and CE meet on the circle round the triangle ABC.

73. Prove that in equal circles chords of the same lengths subtend equal angles both at the centre and at the circumference.

ACB is a straight line in which $AC = 2{\cdot}4''$, $CB = 1{\cdot}6''$; P is any point such that the circles circumscribing PAC and PBC are equal. If Q be the centre of the circle circumscribing PAC, R that of the circle circumscribing PBC, find and construct (i) the locus of Q, (ii) the locus of R, (iii) the locus of P; stating your construction clearly, with short reasons for it.

74. A and B are fixed points on a fixed circle and O is a point within it. AO and BO meet the circle again at P and Q. If the point O be chosen so that the length of PQ is always the same, prove that for all the positions of O the circle circumscribing the triangle OPQ has the same size.

75. ABCD is a quadrilateral inscribed in a circle, AB being parallel to CD. AD and BC meet at X, and the tangents at A and C intersect at Y. Show that XY is parallel to AB and CD.

Section 39. Algebraic Manipulation.

This section has little educative value, and should be omitted when circumstances allow that course. See page v. Sufficient manipulative skill may be acquired from the study of portions of greater intrinsic value.

(i.) SIMPLE MANIPULATION.

1. From $3x^3-7x^2+9x-16$ subtract x^3-4x^2+4x+9, and divide your result by $2x-5$.

2. Find the remainder when $2x^4+3x^3-4x^2-5x-6$ is divided by $x+2$.

3. Divide $2x^4+27y^4-5x^3y-24x^2y^2$ by x^2-9y^2-3xy.

4. Divide x^2-x+1 into $6x^7+3x^6+2x^4+4x^3+4x^2+2$.

Multiply x^3+x-1 by $x^5-x^3+x^2+x-2$.

5. Multiply $2x^3-6x-5$ by x^2-3x+3.

Divide $x^4-24x+55$ by x^2-4x+5.

6. Find the quotient when x^2-4x+3 is divided into the product of x^4+6x^2-3x-4 and x^4-6x^2-6x-9.

7. Multiply $x^2+11x+28$ by $x^2-11x+30$, and divide the product by x^2-x-20.

Write down the equation, connecting your answer with the three expressions given in the question, and verify this equation when $x=2$.

8. Find the value which c must have, in order that the expression x^3-3x+c may be divisible by x^2-2x+1 without remainder.

9. Find the remainder when $6x^4+5x^3-9x^2-ax-b$ is divided by $2x^2-x-6$. Find the values of a and b for which the divisor is a factor of the dividend.

10. Find the remainder when x^4+4x^2+ax+b is divided by x^2+2x+3, and hence find the values of a and b which make x^4+4x^2+ax+b a multiple of x^2+2x+3.

11. Write the expression

$$(3+7x-x^2)^2-2(4x-1-2x^2)(2+x-x^2)$$

in ascending powers of x, and verify your work by calculating the values of the expression and your result when $x=-3$.

12. Show that the statement $(1+x)^5=1+5x+10x^2$ is correct to 5 places of decimals, if x is less than 0·005.

Hence find the value of $(0{\cdot}998)^5$ correct to 5 places of decimals.

(ii.) FACTORIZATION.

13. (i) Find the values of a and b for which x^2+ax+b will be divisible by x^2-3x+2 for all values of x.

(ii) Factorize the expressions:

$$x^3-9x^2-9x+81;\quad x^4+x^2+25;\quad x^3-3x^2+4.$$

14. Find the factors of the expressions:

(i) $36+7x^2-4x^4$; (ii) $x(3x-2)-p(3p+2)$.

15. Show that any common factor of A and B is also a factor of $mA+nB$.

Resolve into factors:

(i) x^2+x-2;

(ii) $2x^2-x-1$;

(iii) $(2a+b)x^2-(a-b)x-(a+2b)$;

(iv) $201x^2-99x-102$.

16. Resolve into factors:

(i) $2x^2-11x-21$; (ii) x^2-x-a^2+a.

Use your results to solve the equations:

(iii) $2x^2-11x=21$; (iv) $x^2-x=a^2-a$.

17. Factorize: (i) $14x^2+193xy-42y^2$;

(ii) $(px+qy)^2+(qx-py)^2$.

18. Find the factors of the expression $x^2y-p^2q+x^2q-p^2y$, and prove that $(x^2+1)^4=(x^4-6x^2+1)^2+16(x^3-x)^2$.

19. (i) Simplify the following expression, and express the result as a product of factors:

$$(a+b)^2-\{9ab-5(a+b)(a-b)\}-[b(b-3a)-\{3ab-5(a-b)^2\}].$$

(ii) Find, correct to three decimal places, the value of

$$\frac{1}{1-3x}-(1+3x+9x^2) \quad \text{when} \quad x=0{\cdot}09.$$

20. Factorize: (i) $24x^2+119x-48$;

(ii) x^4+x^2+1;

(iii) $b^2x+a^2x+ab+bc+acx+abx^2$.

(iv) Show that $x-2$ is a factor of $x^3-10x^2+23x-14$, and find the other two factors.

21. Express $4x^8+17x^4y^4+4y^8$ as a product of 4 factors, and employ the result to find the factors of 400170004.

22. Simplify and resolve into factors: $(3x^2+1)(3x^2+3x+1) \times (3x^2-3x+1) - (x^2+3)(x^2+3x+3)(x^2-3x+3)$.

23. If x^3+ax^2+bx+1 is exactly divisible by x^2+cx+1, what must the quotient be?

A certain algebraical expression is exactly divisible by $x-2$, the quotient being x^2-x-6. Show that the expression is divisible by $x+2$, and find the quotient.

24. Factorize: (i) $a^2-b^2-c^2-2bc$;

(ii) $4x^2-8x-5$.

25. Factorize: (i) $x^2-13x-30$;

(ii) $x^2-y^2+2yz-z^2$;

(iii) $3x^2+10xy-8y^2$.

(iv) Use the identity $a^2-b^2=(a+b)(a-b)$ to find the difference of the squares of 62357 and 47643.

26. Find the factors of

(i) $xy-ab-b(x-y+b)+ay$;

(ii) $pq(x-pq)(x+pq)-x(p^2-q^2)(p^2+q^2)$.

27. Find the factors of

$$6x^2+5xy-4y^2-x+17y-15.$$

28. Assuming that $b-c$ is a factor of

$$bc(b-c)+ca(c-a)+ab(a-b)$$

give a reason for $c-a$ $a-b$ being also factors. Do not multiply the expression out, if you can answer the question without doing so.

29. Find the factors of

(i) $9b^2 - 6bc + c^2 - 16$;

(ii) $a^3 - b^3 - (x^2 - ab)(a - b)$;

(iii) $24(x^2y^2 - 1) + 14xy$;

(iv) $a^2(b - c) + b^2(c - a) + c^2(a - b)$.

30. Factorize the expression $ab(1 + c^2) - c(a^2 + b^2)$, and find what the value of k must be if the expression $9x^2 - 4x - k$ contains $x + 1$ as a factor. What is the other factor?

31. Find the factors of

(i) $ac^2 + bd^2 - ad^2 - bc^2$;

(ii) $x^2 - 2(p + q)x - (p^2 - 1)(q^2 - 1)$.

(iii) What number must be added to $x(x + 1)(x + 2)(x + 3)$ in order to make it a perfect square?

32. (i) If $x - 2$ is a factor of $120x^3 - 167x^2 - ax + 56$, find what the numerical value of a must be, and then find the two remaining factors of the expression.

(ii) Find the square root of

$$3(x^2 - x + 1)^2 + (x^2 + x + 1)^2 - \tfrac{9}{2}(x^2 - x + 1) - \tfrac{5}{2}(x^2 + x + 1) + 4.$$

33. For what value of m is $4x^4 + 24x^3 + mx^2 + 60x + 25$ a perfect square?

Find the square root of $4x^6 + 28x^4 + \frac{4}{3}x^3 + 49x^2 + \frac{14}{3}x + \frac{1}{9}$.

34. (i) Simplify:

$$\frac{1}{256} \times \left\{\left(x + \frac{1}{x}\right)^2 - \left(x - \frac{1}{x}\right)^2\right\}^4 \times \left\{\left(x + \frac{1}{x}\right)^2 + \left(x - \frac{1}{x}\right)^2\right\}.$$

(ii) Apply the formulae $\left(x \pm \dfrac{1}{x}\right)^2 = x^2 + \dfrac{1}{x^2} \pm 2$ to find the square roots of $731\frac{1}{729}$ and $254\frac{1}{256}$.

35. (i) Find the value of x which will make

$$x^4 + 4x^3 - 2x^2 - 11x + 7$$

a perfect square.

Solve the equations:

(ii) $2\sqrt{x + 1} - \sqrt{x - 6} = 5$,

(iii) $x + y = 18 \qquad x^{1/2} + y^{1/2} = 6$.

(iii.) COMMON FACTORS AND COMMON MULTIPLES.

36. Prove that x^2-3x+2 is a common factor of x^3-7x+6 and $3x^3-7x^2+4$ and x^4-3x^3+6x-4. For what values of x do these three expressions simultaneously vanish?

37. Find the L.C.M. of

$$x^3-6x^2+11x-12 \quad \text{and} \quad x^4+x^3-x^2+5x+6.$$

38. Resolve into factors and find the L.C.M. of x^2-5x+6, $3x^2-4x-4$, and $3x^2-7x-6$.

39. Factorize:

$$56x^2+x-56, \qquad 56x^2-79x+18, \qquad 63x^2+58x-16,$$

and write down the L.C.M. of these expressions.

40. Find the H.C.F. and L.C.M. of x^2+2x-8, $x^2+8x+16$, and x^3+64.

41. (i) Find the H.C.F. of

$$x^4+6x^3+9x^2+3x-1 \quad \text{and} \quad x^4-3x^3-9x^2-6x-1.$$

(ii) Find the L.C.M. of x^2-y^2, x^3-y^3 and $(x^2+y^2)^2-x^2y^2$.

42. Write down the factors of

(i) $1-x^3-y^3-3xy$ and (ii) $ab+4bc-2ac-2b^2$.

(iii) Find the H.C.F. of

$$6x^3-19x^2-x-21 \quad \text{and} \quad 10x^3-21x^2-55x+21.$$

43. Write down in factorial form all the expressions that have $(x-1)^3(2x+3)(2x-5)$ as their common multiple.

44. If $A=B+C$ where A B C are integral expressions in powers of x, prove that the G.C.M. of A and B is also the G.C.M. of B and C.

Find the G.C.M. of $x^5-3x^4+x^3-5x-12$ and $x^3-5x-12$.

Section 40. Extension of Pythagoras' Theorem.

1. ABC is an acute-angled triangle and AD is drawn perpendicular to BC. Prove in any way you please that

$$AB^2 = BC^2 + AC^2 - 2BC.CD.$$

The sides of a triangle are 6·5 cm., 7·2 cm., and 3·6 cm. Find, by calculation, whether the triangle is right angled, acute angled, or obtuse angled.

2. The sides of a triangular plot of ground are 22, 26, 37 yards in length. Ascertain without drawing it, whether it is obtuse angled.

Draw a plan of it, taking 5 mm. to represent one yard, and find its area by any method you please.

3. Given four straight lines of lengths 3, 4, 5, 6 inches, find which of the triangles formed with three of them are obtuse angled.

4. In the triangle ABC the angle A is equal to that of an equilateral triangle. Prove that the sum of the squares on AB and AC is equal to the square on BC together with the rectangle contained by AB and AC.

5. If APQ is a triangle and R the middle point of PQ, prove that the squares on AP and AQ are together equal to twice the sum of the squares on AR and RP.

6. Show that if the square on one side of a triangle is greater than the sum of the squares on the other two, the angle opposite that side is obtuse.

ABC is any triangle, D a point in the side BC. Show that if AC^2 is greater than $AD^2 + DC^2$, then AB^2 is less than $AD^2 + DB^2$.

Show also why if D is the middle point of BC, the excess in the one case is equal to the deficiency in the other.

7. Prove that the rectangle contained by the sum and the difference of two straight lines is equal to the difference of their squares.

ABC is an isosceles triangle and D is a point in the base BC. Prove that the difference of the squares on AB and AD is equal to the rectangle BD.DC.

8. The side AB of a triangle ABC is bisected at F. Prove that the sum of the squares on CA and CB is equal to twice the sum of the squares on FA and FC.

Two points K L are taken in the side AB of a triangle ABC so that AK is equal to BL. Prove that the sum of the squares on CA and CB is greater than the sum of the squares on CK and CL by twice the rectangle AK.KB.

9. Two rods, AB of length 8 cm. and AC of length 6 cm., are hinged together at A, and the ends B and C are joined by an elastic string. Find by drawing, and show in a table, the lengths of the string BC when the two rods make angles of 0°, 20°, 40°, 60°, 80°, 100°, 120°, 140°, 160°, 180°.

Also mark off along a straight line distances 2, 4, 6, . . cm. to represent 20°, 40°, . . ; representing 10° by 1 cm. At the distance 2 cm. that represents 20°, mark off a distance perpendicular to the first straight line and equal to the length that BC has when the angle BAC is 20°. Do the same for each angle, and through the ends of the perpendiculars draw freehand a curve or graph to show how the length of the string BC varies as the angle between the rods increases. Show the two lines from which you measure, and graduate them.

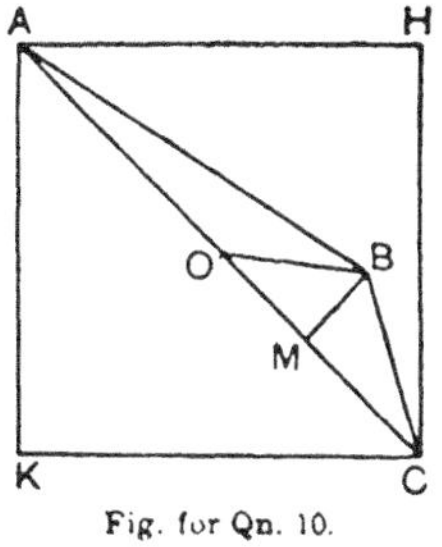

Fig. for Qn. 10.

10. ABC being any triangle, O the middle point of AC, and BM a perpendicular from B on AC (see the figure), write down an expression for BC^2 in terms of AB, AC, and AM. Also express AB^2, BC^2, and AB^2+BC^2 in terms of OA, OB, OC, OM.

AHCK being a square 5 cm. in the side, and B a point such that $BA^2+BH^2+BC^2+BK^2=58$ sq. cm., calculate the distance of B from O (to two decimal places), and sketch a figure showing the locus of positions that are possible for B.

11. D is the middle point of the base BC of a triangle ABC. Show that the sum of the squares on AB and AC is double the sum of the squares on AD and half of BC.

A and B are two fixed points. Find the locus of a variable point P such that $PA^2 + PB^2$ is of constant magnitude.

12. In the figure the rectangular gate ABCD is formed of four bars, AB and CD being each 9 ft. long, BC and AD each measuring 4 ft. It is kept in shape by a fifth bar forming the diagonal AC. Calculate the length of this bar to the nearest tenth of an inch.

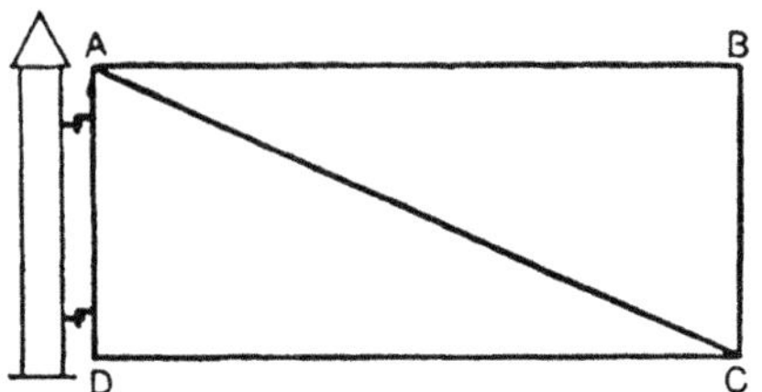

Fig. for Qn. 11.

If the length of AC is increased by 1 inch, so that the gate is strained out of its proper shape, the remaining bars being of the same length as before, through what vertical distance (to the nearest tenth of an inch) will the corner C fall ?

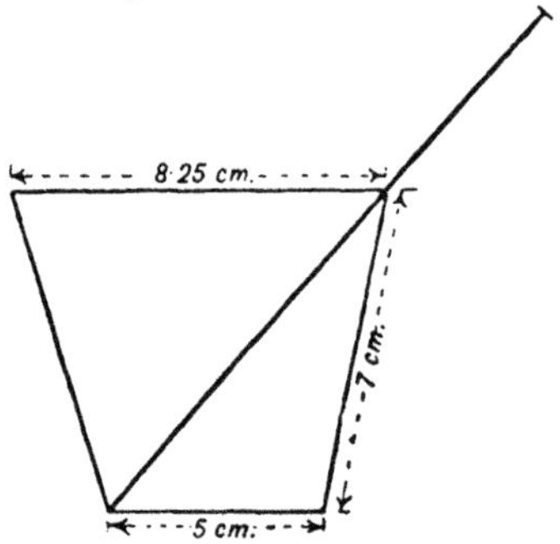

Fig. for Qn. 12.

13. A cylindrical coffee cup measures 6 cm. in diameter and 8·5 cm. in height. A spoon is placed in the cup, touching the top and bottom at opposite points. Calculate, to the nearest millimetre, the length of the spoon which is inside the cup, treating it as a straight line.

Work out the corresponding result in the case of a teacup with slanting sides measuring 7 cm. from rim to base, the diameters of the base and rim being 5 cm. and 8·25 cm. respectively (see the sketch-figure).

14. If ABC is a triangle and AM is a perpendicular on the base BC, prove that $AB^2 - AC^2 = BM^2 - CM^2$.

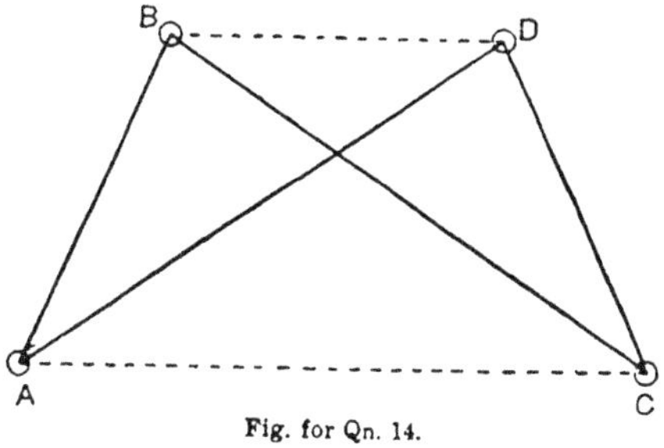

Fig. for Qn. 14.

The figure represents a framework ABCD formed of four rods jointed at A B C D, such that

$$AB = CD = a \quad \text{and} \quad AD = BC = b.$$

The lengths of AC and BD being denoted by x and y, prove that for all positions xy is constant and equal to $b^2 - a^2$.

15. ABCD is a quadrilateral whose diagonals meet at O. Prove that the angle AOB is obtuse if the sum of the squares on AB and CD is greater than the sum of the squares on AD and BC.

16. The sides of a triangle have lengths n, $n+1$, $n-1$, where n is a whole number: show that the triangle is acute angled save in the cases $n = 1, 2, 3, 4$. Discuss these cases.

17. P being any point on the circumference of a circle circumscribing an equilateral triangle ABC, show that

$$PA^2 + PB^2 + PC^2 = 6R^2,$$

where R is the radius of the circle.

18. A point O is taken inside a polygon. Prove that the sum of the squares on the straight lines joining O to the vertices of the polygon exceeds the sum of the squares on the straight lines joining O to the mid-points of the sides by one-quarter the sum of the squares on the sides.

Section 41. Further Manipulation.

This section has little educative value, and should be omitted when circumstances allow. *See page v.*

(i.) TRANSFORMATIONS.

1. If $a=1,\quad b=2,\quad x=-2,\quad y=3,$ find the value of

$$\left(\frac{a^2}{a+x}+\frac{b^2}{b+y}\right)\left(\frac{a}{a-x}+\frac{b}{b-y}\right)\{2a+b-(x-y)\}.$$

2. Prove the identities

(i) $$\frac{1}{x+1}-\frac{2}{x+2}+\frac{1}{x+3}=\frac{2}{(x+1)(x+2)(x+3)};$$

(ii) $$\{a^2+(a+b)^2+b^2\}^2=2\{a^4+(a+b)^4+b^4\}.$$

3. Simplify:

$$\frac{x^2-4xy-21y^2}{x^2-9xy+20y^2}\div\frac{x^2-13xy+42y^2}{x^2+3xy-40y^2}.$$

4. Simplify:

$$\frac{1-x}{(2-x)(3-x)}-\frac{28x}{3(3-x)(1-x)}+\frac{22x+4}{3(1-x)(2-x)},$$

and verify your result by substituting for x the value -2 in the separate fractions and in the final form.

5. Simplify:

$$\left(x+\frac{1}{x}\right)^2+\left(y+\frac{1}{y}\right)^2+\left(xy+\frac{1}{xy}\right)^2-\left(x+\frac{1}{x}\right)\left(y+\frac{1}{y}\right)\left(xy+\frac{1}{xy}\right).$$

6. Simplify: (i) $$\left(\frac{x}{2}-\frac{2}{x}\right)\left\{\frac{1}{x+2}-\frac{1}{2-x}\right\};$$

(ii) $$\frac{x}{x-3}-\frac{x}{x+3}-1.$$

Find, to two decimal places, the values of x for which the expression (ii) is equal to $\frac{2}{3}$.

7. (i) Resolve into factors:

$$ab(x^2+y^2)+(a^2+b^2)xy.$$

(ii) Simplify:

$$\left[\left(x^2+\frac{a^2}{4}\right)^2\left\{x^2+\frac{a^2}{4}-3ax\right\}-a^2x^2\left\{ax-3\left(x^2+\frac{a^2}{4}\right)\right\}\right]$$
$$\div\left\{(x^2-ax)^2+\frac{a^2}{2}(x^2-ax)+\frac{a^4}{16}\right\}.$$

(iii) Solve the following equation, giving the roots to three significant figures:

$$\frac{1}{3x-8}+\frac{1}{3(x-2)}+\frac{1}{3(x+2)}+\frac{1}{3x+8}=0.$$

8. Simplify: $\dfrac{1}{x-4}-\dfrac{x-4}{x^2+4x+16}-\dfrac{7x+20}{x^3-64}.$

9. If $z=\dfrac{y-a}{y+1}$ and $y=\dfrac{x+a}{x-1}$ express z in terms of x and a. What is the value of a if z and x can have the value 2 at the same time?

10. If $x=\dfrac{a+2}{a-2}$ and $a=\dfrac{2b+1}{2b-1}$ find the value of $x(a+b)+(a-b)$ in terms of a, expressing the result as a single fraction in its lowest terms.

11. In each of the following, y is given in terms of x. In each case find x in terms of y.

(i) $y=mx+c$;

(ii) $y=\dfrac{2-3x}{x+3}$;

(iii) $y=\dfrac{-1\pm\sqrt{3x-2}}{3}$.

12. Given that $y=\dfrac{3-2z}{4z+1}$ and $z=\dfrac{5+4x}{3x-2}$ find the values of y in terms of x, and of x in terms of y.

13. Find the value of $\dfrac{ab^2}{b+x}-\dfrac{a^2b}{a+x}$ when $x=\dfrac{ab}{a+b}$ expressing the result as a single fraction in its simplest form; and verify your work by taking $a=-1$, $b=4$.

(ii.) IDENTITIES AND INEQUALITIES.

14. (i) Factorize $25x^2+55x-42$, and show that if x is a multiple of 7, the expression is a multiple of 14.

(ii) Solve the equations:

$$\frac{a}{x+a-1}+\frac{1}{x-a+1}=\frac{a}{x-1}; \qquad 6x^2+23x+20=0.$$

(iii) Show that $\dfrac{a}{1+a}+\dfrac{b}{1+b}+\dfrac{c}{1+c}=1$ if

$$\frac{x}{y+z}=a, \quad \frac{y}{z+x}=b, \quad \frac{z}{x+y}=c.$$

15. Show that $x^2+y^2+z^2-xy-yz-zx$ is not altered in value by increasing x y z each by the same quantity a.

16. Prove

$$(ax^2+2bx+c)(ay^2+2by+c)-[x(ay+b)+by+c]^2 = (ac-b^2)(x-y)^2.$$

17. Prove that

$$(a^2+b^2+c^2)(x^2+y^2+z^2)-(ax+by+cz)^2 = (bz-cy)^2+(cx-az)^2+(ay-bx)^2$$

18. Prove that the sum of the three fractions

$$\frac{b-c}{1+bc} \qquad \frac{c-a}{1+ca} \qquad \frac{a-b}{1+ab}$$

is equal to their product.

19. Prove that $\dfrac{a(b^3-c^3)+b(c^3-a^3)+c(a^3-b^3)}{(a-b)(b-c)(c-a)}=a+b+c.$

20. If $a=x^2-yz$, $b=y^2-zx$, $c=z^2-xy$, prove that $a^3+b^3+c^3-3abc$ is the square of $x^3+y^3+z^3-3xyz$.

21. Having given $a^2=b^2+c^2-2bcQ$, calculate $1+Q$ and $1-Q$ in terms of a b c, and show that

$$1-Q^2=\frac{1}{4b^2c^2}(a+b+c)(-a+b+c)(a-b+c)(a+b-c).$$

22. Find the necessary restrictions on the value of x if

$$\frac{2x+1}{x-3}+\frac{4x-5}{x-2}$$

is greater than 6.

23. The expression $7x^2-5x-1$ is to be written in the form:

$$A+B(x-1)+C(x-1)(x-2).$$

Find the values of A B C which will make the latter identically equal to the former expression.

24. Express $x^3-7x^2+18x-14$ in the form

$$A(x-3)^3+B(x-3)^2+C(x-3)+D.$$

25. Show that $x=y=z$ if x y z are real quantities and $(x-y)^2+(y-z)^2+(z-x)^2=0$.

26. Prove that $\frac{pa+qc}{pb+qd}$ lies between $\frac{a}{b}$ and $\frac{c}{d}$, all the symbols representing positive quantities; and find the ratio of p to q in order that the first fraction may be midway between the others.

27. The percentages of passes in an examination among the boys and girls were m and n respectively, and x per cent. of the total number of candidates passed. Find the total number of candidates if y boys entered for the examination.

Prove that m x n are in order of magnitude, and the number of girls will be equal to the number of boys if the difference between m and x is equal to the difference between x and n.

(iii.) SIMPLIFICATION OF FRACTIONS.

28. (i) Find the L.C.M. of a^3-a^2b, $a^2b^2+ab^3$, $a^6b-a^3b^4$.

(ii) Simplify: $$\frac{b^2(a-b)}{a(a^2-b^2)}-\frac{a(a-b)}{b(a+b)}+\frac{a^2-b^2}{ab}.$$

29. Simplify: (i) $$\frac{a^2b^2}{a^2-b^2}\left(1+\frac{a}{b}\right)\left(\frac{a}{b}-1\right);$$

(ii) $$\frac{a^2-a-6}{a^3+4a^2+5a+2}.$$

30. Simplify:

(i) $\left(\frac{a-b}{a+b}-\frac{a^3-b^3}{a^3+b^3}\right)\times\left(\frac{a+b}{a-b}+\frac{a^3+b^3}{a^3-b^3}\right)$;

(ii) $\frac{x}{(x-y)(x-z)}+\frac{y}{(y-z)(y-x)}+\frac{z}{(z-x)(z-y)}$.

31. Express $\frac{x-6a+4b}{(x+2a)(x+b)}+\frac{x+3a-2b}{(x+a)(x+b)}$

as a single fraction in its lowest terms, and calculate separately the values of the expression and your result when $x=1$, $a=1$, $b=-2$.

32. Simplify:

(i) $\frac{x(x+y)(2x-3y)}{(x-y)(x^3-3y^3)}-\frac{x^2-3y^2}{x^3-3y^3}$;

(ii) $(a+b+c)\{a(b+c-a)+b(c+a-b)+c(a+b-c)\}$
$-(b+c-a)(c+a-b)(a+b-c)$.

33. Simplify:

(i) $\frac{(a+b)^3-(a-b)^3}{(a+b)^3+(a-b)^3}\times\frac{\frac{1}{a^2+b^2}+\frac{1}{2b^2}}{\frac{1}{a^2+b^2}+\frac{1}{2a^2}}$;

(ii) $\frac{x^3-a^3}{x^2-ax+bx-ab}\times\frac{x^2-b^2}{x^2+ax-bx-a}\div\frac{x^2+ax+a^2}{x^2+ax+bx+ab}$.

34. Simplify the following expressions:

(i) $\frac{1}{x+1}+\frac{2}{x^2+3}+\frac{8}{(x^2-1)(x^2+3)}$;

(ii) $x^3+y^3+z^3+3(x+y+z)(yz+zx+xy)-(x+y+z)^3$.

35. Simplify:

$$\frac{a}{(a-b)a-c)}+\frac{b}{(b-c)(b-a)}+\frac{c}{(c-a)(c-b)}.$$

36. Simplify:

(i) $\left\{\frac{1}{(x-1)(x-2)}+\frac{1}{(x-2)(x-3)}\right\}\times\left\{x+\frac{3}{x-4}\right\}$;

(ii) $\frac{x(x+y)+y^2}{x^2-y(x-y)}\times\frac{x^2(x-y)+y(x^2+y^2)}{x(x^2+y^2)-y^2(x+y)}$.

37. Simplify the expression:

$$\frac{1}{(a-b)(a-c)(x-a)}+\frac{1}{(b-c)(b-a)(x-b)}+\frac{1}{(c-a)(c-b)(x-c)}.$$

Section 42. The Rectangle Property of the Circle.

(i.) CONSTRUCTION OF A SQUARE EQUAL TO A GIVEN FIGURE.

1. Show how to construct a square equal to a given rectangle whose sides are 3 and 5 cm. Measure the side of the square found by your construction, and verify your result arithmetically.

2. Show how to describe a square equal to a rectangle whose sides are 2·87 and 3·19 inches in length respectively, and describe the square.

Measure the side of the square, and verify your result arithmetically.

3. Draw an isosceles triangle having a base 10 cm. long, and each angle at the base 65 degrees.

Construct a square of equal area, stating how you do so, but giving no proof.

From measurements of the side of the square and the height of the triangle, find and write down their areas.

4. Describe a square which shall be equal in area to an equilateral triangle whose side is 10 cm.; and calculate the side of the square to three decimal places.

5. Construct a square equal in area to a triangle whose sides are 2, 3, and 4 inches. Measure the side of the square.

6. Construct a triangle ABC, having given BC = 4 centimetres, CA = 5·8 centimetres, and AB = 7·4 centimetres.

Construct a square equal in area to this triangle, and measure its side.

7. Draw a triangle having an area of 3 sq. in. Construct a square equal to it in area, and measure the side of the square.

8. It is required to ascertain which of the two triangles ABC and DEF in the figure has the greater area. Test the point by transferring the triangles to your book, and constructing geometrically *either* two rectangles having a common base *or* the sides of two squares which are equal in area to the respective triangles.

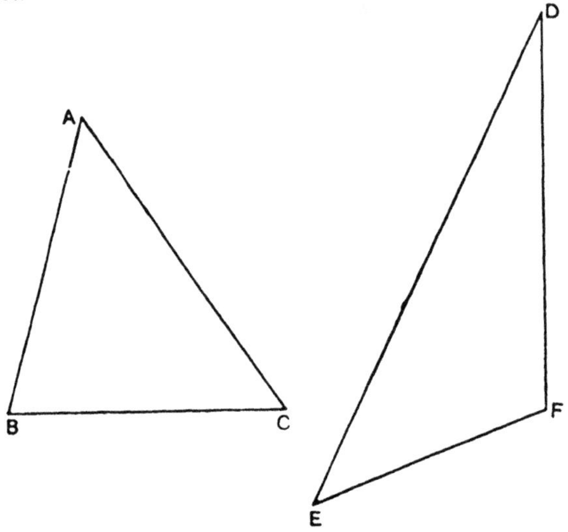

Fig. for Qn. 8.

Check your conclusion by measuring the bases and altitudes of the triangles, in millimetres, and so calculating their areas.

9. Construct an equilateral triangle with each side 4 inches long. Construct (without proof) a square equal in area to the triangle.

Show that the area of the triangle is equal to $4\sqrt{3}$ square inches and hence find the square root of $4\sqrt{3}$ correct to one decimal place by measurement from your figure.

10. ABCD is a convex quadrilateral; AB is 3 in., BC and CD are each 2 in.; ABC and BCD are right angles. Construct a square equal in area to ABCD, and measure its side.

11. Draw a quadrilateral ABCD having AB = 5 cm., BC = 6 cm., CD = 7 cm., DA = 8 cm., AC = 7 cm.; and then construct a square of equal area, and measure its side. Every step of the method you employ must be explained.

12. Show, with proof, how to construct a square whose area is the same as that of a given triangle.

Also show how to construct a square whose area is equal to the sum of the areas of two given triangles.

(ii.) OTHER CONSEQUENCES OF THE RECTANGLE PROPERTY.

13. If AB CD be two chords of a circle which intersect in an internal point E, prove that the rectangles AE.EB and CE.ED are equal in area.

PQ is a straight line 6 inches in length, and O is a point in PQ such that PO = 2·56 inches. If from O a straight line OR be drawn in any direction 3·01 inches in length, calculate, to two decimal places, how many inches RO must be produced to S, in order that the four points P Q R S may lie on a circle.

14. Draw a straight line AB one and a half inches long. Produce AB to C, making BC half an inch in length. Through the points A and B describe a circle of any convenient radius, and draw CD to touch the circle. Measure the length of CD. Quote some geometrical proposition to prove that the measured length is correct. If a number of circles are drawn through A and B, what will be the locus of D?

15. Explain clearly what is meant by the converse of a proposition, and state a converse of each of the following:

(i) If AOB and COD are two intersecting chords of a circle, then AO.OB = CO.OD.

(ii) If a point P lie inside a triangle ABC, then the angle BPC is greater than the angle BAC.

Also discuss the truth of the converses which you state.

16. Show, with proof, how to describe a square equal to a given rectangle.

Show how to describe a rectangle equal to a given square, such that the sum of its sides shall be equal to a given straight line. When is this impossible?

17. AB is a straight line of length five inches. Divide it by a geometrical construction into two parts at the point C so that the rectangle AC.CB may be equal in area to four square inches.

18. Show how to construct a rectangle equal to the square on a line of length a, such that its perimeter may be equal to four times a line of length b.

What limit is there to the size of b in order that this may be possible?

Show that if x be the length of a side of the rectangle, this problem is equivalent to the solution of the equation

$$x^2 - 2bx + a^2 = 0.$$

19. Prove that there cannot be two similar segments of a circle on the same base and on the same side of it.

ABC is a triangle and AD BE are two straight lines meeting BC and AC produced if necessary in D and E so that the angles AEB ADB are each 45°. Show that the rectangle AO.OD is equal to the rectangle BO.OE, where O is the intersection of AD and BE.

20. AOB is a straight line of unlimited length and OC is perpendicular to AOB, and of length a. A circle of radius r has its centre in AB, passes through C, and cuts AB in P and Q. Show that if $OP = x$ then $x + a^2/x = 2r$, and find a geometrical interpretation for the other root of the equation.

21. AB is a chord of a circle which is produced to P, and PT is the tangent from P. Prove that $PT^2 = AP.BP$.

Find a point G in the base EF of a triangle DEF such that $DE^2 = EF.EG$.

22. ABCD is a square whose side is 2 inches in length. A circle with centre A is described to pass through B and D, and CEF is a straight line cutting the circle in E and F.

Show that $x = 4/y - y$ where $x = EF$ and $y = CE$.

Find CE when EF is 3 inches.

23. Prove that if two chords of a circle intersect internally the rectangles contained by the segments are equal.

Thence, if the chords are OR and PQ intersecting at S, and if O is the end of a diameter bisecting PQ, prove that the square on OP is equal to the rectangle contained by OR and OS.

24. The tangent at A to the circle ABC meets the diameter BC at D, and AL is drawn perpendicular to BC. Prove that

$$DB.DC - LB.LC = DL^2.$$

25. The figure represents a circular arch spanning a canal. From measurement give the total width AB of the canal inclusive of the towing path, and the height CD of the arch, the figure being drawn on a scale of 1 inch to 10 feet.

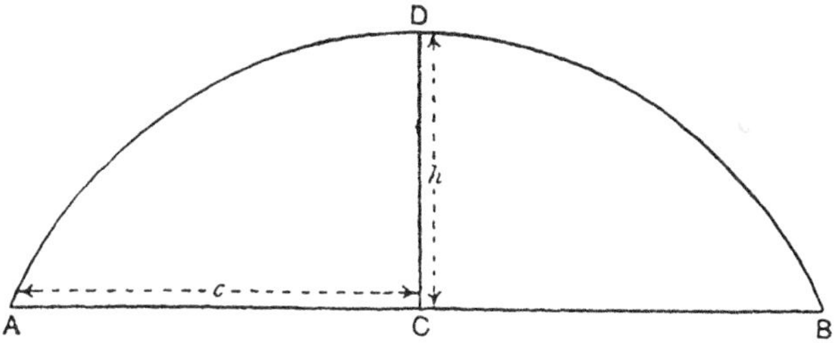

Fig. for Qn. 25.

Denoting AC by c, CD by h, and the radius of the arch by r, show that the radius can be found from the formula

$$r^2 = (r - h)^2 + c^2.$$

Substitute the values of h and c, and find, to the nearest foot, the radius of the arch.

Check, by drawing a figure in which each line in the given figure is represented by a line double its length.

26. Draw two straight lines OX OY inclined at an angle 45° to each other; along OX measure OA = 4 cm., along OY measure OB = 7·5 cm.; then construct a circle touching OY and passing through A and B. State the steps of your construction.

27. Show, with proof, how to describe a circle passing through two given points and touching a given straight line.

28. A B C D are four points in a straight line such that AB is equal to BC. Two semicircles are drawn on AC and BD meeting in E. Prove that a circle with centre D and radius DE will cut at right angles every circle which passes through A and C.

29. Two circles touch externally at O, and POQ is a chord of both drawn through O; a tangent is drawn from each of the points P Q to the circle on which the other lies. Prove that the sum of the squares on these tangents is equal to PQ^2, and hence find for what position of PQ this sum is greatest.

30. In the figure A and B are two points on the same side of a fixed straight line XY. Give, with proof, a geometrical construction for drawing two circles to pass through A and B and to touch XY. Transfer the line and points to your book and carry out your construction. Measure the radii of your circles, and prove that the points of contact are the points of XY at which AB subtends the maximum angles.

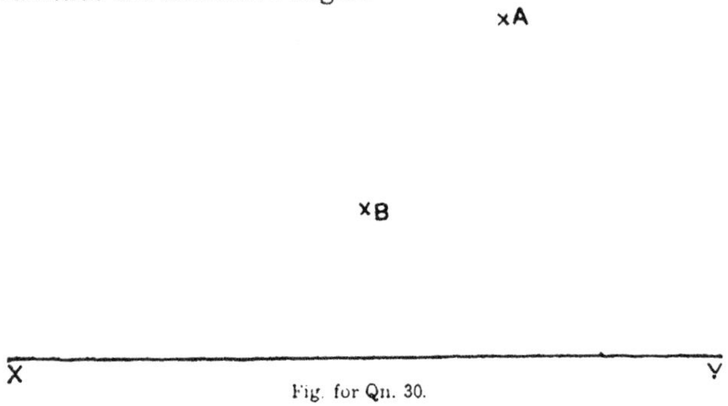

Fig. for Qn. 30.

31. A steeple 30 feet high stands on a tower 29 feet high; at what distance from the foot of the tower should an observer, whose eye is supposed to be 5 feet from the ground, stand in order that the steeple may subtend the greatest angle at his eye?

32. Describe a triangle ABC so that AB BC CA are 3·5 7·1 and 9·6 centimetres in length respectively. Describe two circles to pass through the middle point of BC and touch AB and AC, produced if necessary.

Justify your construction by general reasoning.

33. Find the locus of a point from which the tangents drawn to two intersecting circles are equal.

34. Describe, with proof, how to draw a tangent to a circle from a point outside it.

If the tangent from a point T touches at P a circle whose centre is C, and PN is the perpendicular from P on CT, prove that

$$CP^2 = CN . CT.$$

35. If a tangent OT and a chord OPQ are drawn to a circle from an external point O, prove that $OT^2 = OP.OQ$.

The triangle ABC has the sides AB AC equal, and a circle is drawn to touch AC at C and BA produced at Q. If this circle cuts BC produced at P, prove that $BC.BP = 4AB^2$; and deduce that $BC.CP = 4AD^2$, where AD is drawn perpendicular to BC.

36. (i) An oak beam rests in a horizontal position upon two supporting knife-edges at its ends 2 metres apart. When a weight is attached to its middle point the beam is bent in such a way that the depression y measured vertically downwards from a point situated x centimetres along the horizontal from the nearer supporting edge is given in centimetres by the equation $y = 0{\cdot}3x - x^3/10^5$. On squared paper draw a diagram representing the shape of one half of the loaded beam.

(ii) A diameter of a circle is at right angles to a chord which is not a diameter. Without assuming the rectangle property of the circle, show that the square of the length of the semi-chord is equal to the product of the lengths of the two segments of the diameter.

When a loaded beam rests upon two supports, in the manner described in (i), the depression of its middle point for varying distances between the supports is proportional to the cube of the distance between the supports. If the depression is so small that its square may be neglected, the shape of the beam is approximately an arc of a circle. Show that the radius of this circle is inversely proportional to the distance between the supports.

Section 43. More about Equations.

The non-specialist may with advantage omit this Section. The mathematical pupil should (if possible) omit it for the present; at a later stage, when he must deal with such questions, they will be easier to him and take less time. See page v.

(i.) EQUATIONS OF HIGHER DEGREE THAN THE SECOND.

1. Make a graph of x^3 from $x = 0$ to $x = 3$.

On the same figure and to the same scale draw the graph of $15 - 5x$.

From your graph find to one decimal place a value of x which makes $x^3 = 15 - 5x$.

Calculate the value of $x^3 + 5x - 15$ for this value of x, and also for $x = 1$ and $x = 2$.

2. Draw the graph of $10y = 3(x-3)(2x+3)(x-1) - 4$ from $x = -2$ to $x = 3\frac{1}{2}$, and hence show that the roots of the equation $3(x-3)(2x+3)(x-1) = 4$ are nearly equal to 1, 3, and $-1\frac{1}{2}$. Find the values of the roots, correct to 2 per cent., if necessary by drawing on a larger scale the small portions of the graph in the neighbourhood of the roots.

3. Three numbers are such that the second is the double of the first, and the third the double of the second. Find the numbers if their product is fifty-six times their sum.

4. Find all the real roots of the equations:

(i) $12\left(x + \frac{1}{x}\right)^2 - 56\left(x + \frac{1}{x}\right) + 65 = 0$;

(ii) $\begin{cases} x^3 - y^3 = 91, \\ x^2(x - y) = 36. \end{cases}$

5. Solve the equations:

(i) $7x^2 - 15x - 52 = 0$;

(ii) $\begin{cases} x - 2y = 6, \\ x^3 - 8y^3 = 936. \end{cases}$

6. Solve the equations:

(i) $3x^2 + 8x = 4$ (to two decimal places);

(ii) $x^2 - 2xy - y^2 = 7, \quad xy + 2y^2 = 12$.

7. With a unit of 10 cm. trace the curves $2x^2 + 5y^2 = 1{\cdot}3$ and $x^2 + y^2 - 1{\cdot}3y + 0{\cdot}11 = 0$; and determine by calculation or by measurement the coordinates of their points of intersection.

8. Trace the curves $x^2 - y^2 + 0{\cdot}5 = 0$, $x^2 + 4xy + y^2 = 0$, and give the coordinates of their points of intersection to 3 significant figures.

9. Solve the equations:

(i) $(a+b)(x+a+b) + (a-b)(x+a-b) + 2b(x+2a) = 0$;

(ii) $\sqrt{x-2} + 3\sqrt{x-3} = \sqrt{4x-5}$;

(iii) $\frac{x}{y} = 4\frac{1}{2} - x^2; \quad \frac{y}{x} = 8 - y^2$.

10. Solve the equations:

(i) $$\frac{\sqrt{x}+\sqrt{x-1}}{\sqrt{x}-\sqrt{x-1}} = \frac{\sqrt{x}-\sqrt{x-1}}{\sqrt{x}+\sqrt{x-1}};$$

(ii) $$\frac{5}{x+2a}+\frac{5}{x-2a} = \frac{8}{x+a}+\frac{8}{x-a};$$

(iii) $$x+y+1 = x^3+y^3-12 = 3.$$

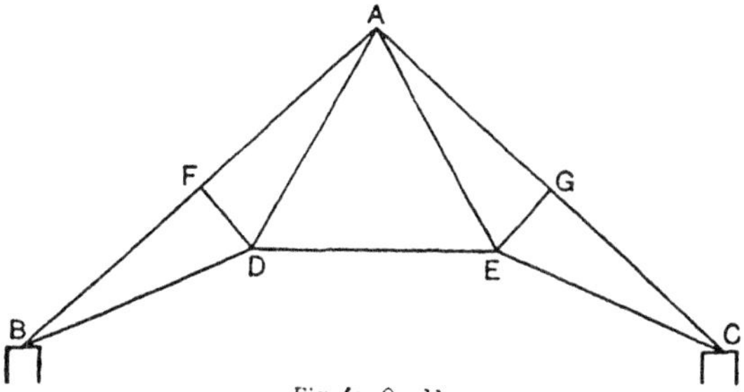

Fig. for Qn. 11.

11. The diagram shows a roof truss. The span BC is 32 feet, the height of A above BC is 8 feet, and the height of DE above BC is 2 feet. DF and EG bisect the rafters AB and AC at right angles. Prove that if the length of BD is x feet, then

$$\sqrt{x^2-36}+\sqrt{x^2-4}=16.$$

Find from this equation the value of x, and find also the lengths of the other bars of the truss.

12. Two poles whose lengths are 10 feet and 17 feet are upon opposite sides of a wall; each has one end resting on the ground and one end just reaching to the top of the wall. The distance between the bottom ends of the poles is 21 feet. Find the height of the wall by calculation, and verify by means of an accurately drawn figure.

13. A triangle is to be drawn in accordance with the directions: side AB 20 cm., side AC 24 cm., side BC 30 cm. Ascertain by calculation whether the triangle so constructed is right angled, acute angled, or obtuse angled. Verify, by drawing a figure half-size and drawing a line through C at right angles to AB. Calculate the distance of the foot of this perpendicular from A.

14. $x76$ is a number of three digits such that its square ends in the digits x 7 6; find x.

15. A bill of £1. 10s. was paid in shillings, florins, and half-crowns. The number of coins used was 17, and the number of shillings was one more than the number of half-crowns. Find how many there were of each.

16. Find, by trial or otherwise, positive integral values of x and y which satisfy the equation $4x + 5y = 61$.

(ii.) CHIEFLY THEORY OF QUADRATIC EQUATIONS.

17. A person living 4 miles from the bank of a straight reach of river, and wishing to catch a boat travelling at 6 miles an hour along the river, sets out three-quarters of an hour before it passes the point nearest to his house. If he walks at 4 miles an hour, find by calculation what point of the river he should make for in order to catch the boat in the shortest possible time.

Is there any other point of the river which he could reach simultaneously with the boat by travelling in a straight line at 4 miles an hour?

18. Reduce the equation

$$\frac{x+y}{a+b} = \frac{x}{a} + \frac{y}{b}$$

to its simplest form as a relation between x and y, and deduce that, if all the quantities are real and not zero, x and y must have different signs.

19. (i) Simplify: $a^2(a+7)^2 + ab(a^2+b^2-10) + b^2(b+7)^2$, where a b are the roots of the equation $x^2 + 7x - 9 = 0$.

(ii) The two equations

$$x^2 + 81x - 1666 = 0, \qquad 2x^2 - 49x + 255 = 0,$$

have a common root. Find it.

20. (i) Resolve into factors:

$$x^2 - 5x + (3{\cdot}7 - a)(1{\cdot}3 + a).$$

(ii) Simplify:

$$\frac{2x^2 - 9xy + 12y^2}{x^2 - 5xy + 6y^2} - \frac{4x^2 - 14xy + 12y^2}{x^2 - 4xy + 3y^2} + \frac{2x^2 - 5xy + 4y^2}{x^2 - 3xy + 2y^2}.$$

(iii) Find the value of $x + y$ and $x - y$ when

$$50{\cdot}5x + 49{\cdot}5y = 169, \qquad 49{\cdot}5x + 50{\cdot}5y = 143.$$

21. A body is originally rising in the air with a velocity v feet per second, and it is known that if, t seconds later, h is the height of the body in feet above its original position, the quantities h and t are connected by the formula $h = vt - 16t^2$.

Taking $v = 100$, give a graph of the height of the body in terms of the time elapsed, and find when (if ever) the body is at the heights 100 feet and 200 feet above its original position, and when it is 100 feet below its original position.

Treating $h = vt - 16t^2$ as a quadratic equation to find t when h is given, write down the conditions that the roots of this quadratic may be (i) imaginary, (ii) real and equal, (iii) real and unequal, both being positive, (iv) real and unequal and of opposite signs; and interpret the results.

22. Solve the equations :

(i) $(x-2a)(3x+a)-(x-4a)(2x+a) = (x+2a)^2-2a^2$;

(ii) $\dfrac{3x-4y}{6}-\dfrac{2x-3y}{2} = 5, \qquad \dfrac{x}{5}+\dfrac{y}{3} = 6$;

(iii) $\dfrac{x}{x+2}+\dfrac{x+2}{x} = \dfrac{25}{12}$.

(iv) Find a value of k which will make the roots of the following equation equal: $3x^2+2(3k-4)x+3k^2+4=0$.

23. Two quantities x and y are unequal and are such that:

$$x^2-117x+47=0, \qquad y^2-117y+47=0.$$

Find the values of $x+y$ and xy.

24. If m and n are different numbers, and

$$am^2+bm+c=0, \qquad an^2+bn+c=0,$$

prove that $$a = \frac{-b}{m+n} = \frac{c}{mn};$$

and find m^3+n^3 in terms of a b c.

25. If $x+y=a$ and $xy=b$, find the values of x^3+y^3 and x^4+y^4 in terms of a b.

26. Find the factors of $x^2-px+q-1$, having given that:

$$100-10p+q=(1{\cdot}5)^2-(1{\cdot}5)p+q=1.$$

27. You are informed that the roots of the equation

$$2x^2 - 11{\cdot}7x + 15{\cdot}91 = 0$$

are $3{\cdot}7$ and $2{\cdot}15$. Construct the equation which results from writing $y-1$ in place of x in the above, and find its roots.

28. Show that x^2+px+q must always be positive if $p^2<4q$. Form the equation whose roots are the squares of the reciprocals of the roots of the equation $x^2+px+q=0$, and verify your result by actual solution of both equations for the case $p=-5$, $q=6$.

29. Construct a quadratic equation whose roots are $a-2b/a$ and 2.

One root of the equation $2x^2-10{\cdot}283x+12{\cdot}566=0$ is 2. Find the other root.

30. Show that if the roots of the equation $x^2-bx+32=0$ are p and q, then $p+q=b$ and $pq=32$. Draw a graph of b as a function of p, taking 5 mm. as unit and taking values of p from 0 to 32. Show how to use your graph for finding the roots of the equation $x^2-21x+32=0$ and for determining the value of b that would give equal roots. Check your results by calculation.

31. (i) Show that x^2+px+q can always be expressed as the sum or difference of two squares.

(ii) Solve for x: $\dfrac{2x+3}{4x+3}+\dfrac{3x+2}{4x+2}=\dfrac{5(x+2)}{4x+5}$.

32. Prove the formula which gives the roots of the quadratic equation $ax^2+bx+c=0$, and apply it to the solution of the equation $abx^2-(a-b)^2x=(a-b)^2$.

(iii.) THE REMAINDER THEOREM.

33. (i) Show, preferably without actual division, that $x+3$ is a factor of $15x^3+16x^2-101x-42$, and find the remaining factors.

(ii) Find the square root of $a^4+b^2+16+8b-8a^2-2a^2b$.

34. Find without actual division the remainder which is left when $3x^4-4x^2+x+2$ is divided by $x-3$. Explain why the method adopted gives the correct result.

35. If $f(x)=\dfrac{x^2+4x+5}{(x-2)(2x-1)}$, find $f(0)$ and $f(-2)$.

36. Find the value of a in order that $x=-3$ may satisfy the equation $2x^3-x^2+ax+15=0$. Substitute this value of a, and then by inspection (or otherwise) find the other values of x which satisfy the resulting equation.

37. Find a value of a such that $x^3+4x^2+ax+4a$ may have $x-2$ for a factor.

What are the other factors?

38. If any integral function of x is divided by $x-a$, prove that the remainder is the expression formed by substituting a for x in the function.

If 3 is one root of the equation $x^3-49x+a=0$, find the other roots.

39. An algebraic expression contains terms in x^2 and x, but none in x^3 or higher powers of x. Three values for x can be found for which the expression equals zero. What can you infer from this?

Show that the expression

$$(b-c)(x-b)(x-c)+(c-a)(x-c)(x-a)$$
$$+(a-b)(x-a)(x-b)+(a-b)(b-c)(c-a)$$

equals zero if a be written for x, and make what inferences you can.

40. When is an algebraic expression involving the letters x and y said to be symmetrical in x and y?

An algebraic expression contains no letters other than a and b, is homogeneous, of the third degree, and symmetrical in a and b, and $a-2b$ is a factor. What can you infer as to the nature of the expression?

Section 44. Coordinates.

1. Draw the triangle whose vertices lie at the points (1, 2) (8, 3) (6, 7). Take 1 cm. as unit.

Find the length of each side by measurement, and also by calculation.

Circumscribe to the triangle a rectangle whose sides are parallel to the axes of coordinates, and so find its area. Also find the area from the "base-height" formula with the aid of measurement.

2. Two opposite corners of a square have coordinates

$$(-3, -2) \quad \text{and} \quad (2, -1).$$

Find the coordinates of the other two corners by plotting the figure on squared paper, unit 1 inch.

3. Plot the line $y=x$ and the points $(3, 2)$ $(3, -2)$. Draw a circle to pass through the points and touch the line.

Justify your construction.

4. Solve graphically or otherwise the equations:

$$2(x-0{\cdot}8)^2 - 5y(x-0{\cdot}8) + 2y^2 = 0, \qquad 5x + 9y = 22{\cdot}45.$$

5. A circle passes through three points whose coordinates with reference to two rectangular axes X′OX Y′OY are $(12, 17)$ $(-12, 17)$ $(0, 25)$. By drawing find the points at which the circle cuts the axis X′OX.

6. Plot the points A B C D E F whose coordinates in millimetres are $(-44, 27)$ $(0, 46)$ $(42, 19)$ $(30, -8)$ $(6, 0)$ $(-33, -26)$. These points are to be connected in the order named by a curve made up of circular arcs, so arranged that consecutive arcs do not meet at an angle but have a common tangent at the point of junction. Draw the curve, indicating clearly the method of determining the centres.

State how far the form of the curve is fixed, and how far it is at your disposal.

State two properties of the circle that you use in drawing the curve.

7. A square lies in the positive quadrant and has two vertices at $(5, 0)$ and $(0, 2)$. Give the coordinates of the other vertices, and draw the square, unit 1 cm.

Find the length of a side of the square by measurement and by calculation, and deduce the area.

Also circumscribe a rectangle with sides parallel to the axes, and so find the area.

8. The bisectors of the angles B and C of a triangle ABC intersect at O. Through O a straight line is drawn parallel to BC, meeting AB and AC respectively in D and E. Show that it is possible to draw two circles touching DE at O, of which one touches AB at B and the other touches AC at C.

Plot the points P Q R S whose coordinates are, in centimetres, $(-2{\cdot}4, 0)$ $(0, 7{\cdot}3)$ $(9{\cdot}8, 0)$ $(6{\cdot}0, 6{\cdot}2)$. Join PQ, and join RS. Then draw from Q to S a line consisting of two circular arcs which touch one another, while one of them touches PQ at Q, and the other touches RS at S.

State and justify your construction.

9. A set square ABC, with the right angle at B and BC equal to 2, slides along with BC upon the axis of x and B between O and C; the vertex C is joined to the fixed point D (1, 1), cutting AB in P. Draw the path of P, unit 1 inch.

10. Plot the straight lines given by the equation

$$(x+1)+k(y-2)=0$$

for the values $k=3$, 0·3, $-0·6$. What is the common feature of all these lines? Will it be present in the lines found by giving other values to k?

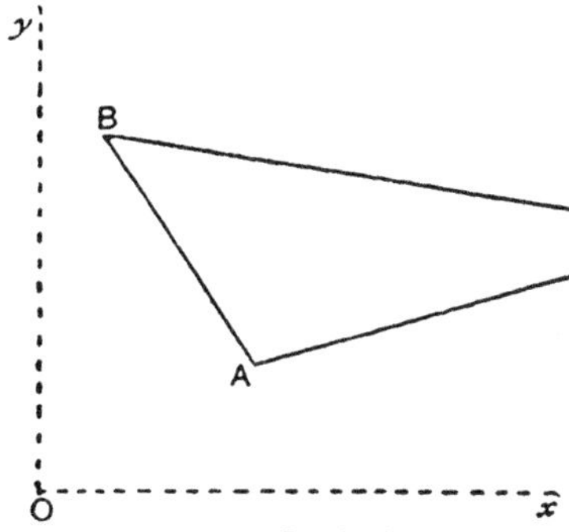

Fig. for Qn. 11.

11. Transfer the given figure to your book, and taking 1 centimetre as unit, find from measurement and drawing, (i) the coordinates of the vertices of the triangle ABC; and (ii) the equations of the sides of the triangle, each in the form $\frac{x}{a}+\frac{y}{b}=1$ with a and b given to two significant figures.

Section 45. Problems depending on the Perimeter and Area of the Circle.

1. A racing track with semicircular ends is represented in the figure, the dotted line showing the middle of the track. If the straight length AB is a yd., BC b yd., and the width MN of the track is w yd., find expressions for the length of one lap (that is, the distance once round) measured (i) along the dotted line, (ii) along the inner boundary, (iii) along the outer boundary.

Give the length of a lap in each case if $a=220$, $b=100$, $w=12$.

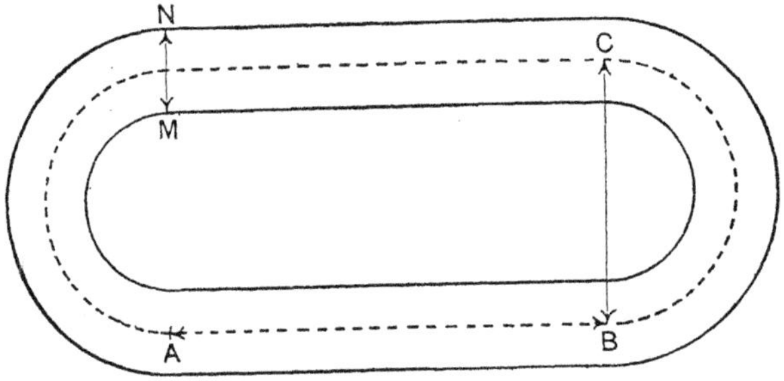

Fig. for Qn. 1.

Two boys, one of whom goes 11 yards while the other goes 10 yards, run a lap, the faster one close to the outer boundary, the slower close to the inner boundary of the track; which wins? Show all your calculations.

2. A triangular piece of cardboard, whose sides are 3 2·6 and 1·7 inches in length, is laid flat on a table; a halfpenny is also placed flat on the table and rolled round the outside of the triangle. Draw accurately the path traced out by the centre of the halfpenny and calculate its length.

A halfpenny is one inch in diameter.

3. A piece of corrugated metal measures 10 ft. by 10 ft. A section taken across the corrugations is a wavy curve of the shape indicated in the figure, the portions between a and b, b and c, c and d, etc., being each one quarter of a circle of radius 2 in. If the piece is made of sheet-metal which, without corrugations,

weighs 2·73 lb. to the square foot, find the weight of the corrugated piece.

Fig. for Qn. 3.

4. Semicircles are described on AB AC CB (see the figure). Given that the radius of AKB is r cm. and that of ALC x cm., show that the area of the unshaded portion of the figure is $\pi x(r-x)$ sq. cm.

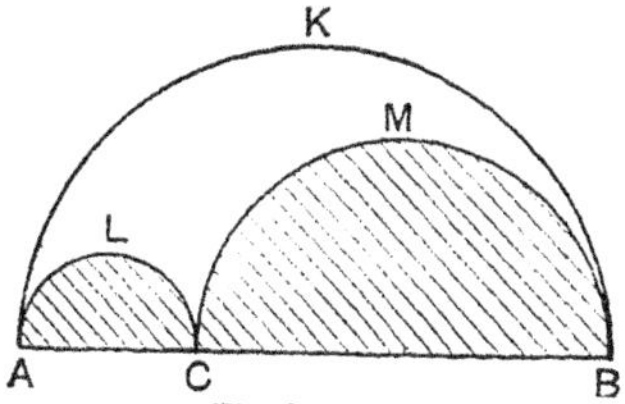

Fig. for Qn. 4.

With 9 as the value of r, calculate the area of the unshaded portion for the following values of x: 0, 1, 2, . . . Tabulate your results, and draw a graph to show the variation in the area, showing x full size, and representing area on the scale of 5 mm. for π sq. cm. What is the position of C when the area of the part between the circles is a minimum? a maximum? Compare the maximum area with that of the semicircle AKB.

5. Some of the following statements are obviously wrong; indicate any errors that you can find, explaining your reasons. You need not correct the mistakes. The last figure in each answer is approximate only.

(i) $(4{\cdot}816)^3 = 111{\cdot}7$.

(ii) $\sqrt{0{\cdot}4816} = 0{\cdot}2194$.

(iii) The area of a circle whose radius is 81·90 feet is 5268 square feet.

(iv) A regular seven-sided figure being inscribed in a circle of radius 81·9 feet, the length of each side is 37·8 feet.

6. The rough sketch represents a set of friction wheels in a piece of mechanism. The wheels marked B and b are keyed to the same shaft and so turn with it, also the wheels C and c similarly turn on one shaft. The wheel A revolves, as indicated by the arrow; wheels A and b are held against one another so that they cannot slip over one another but turn together; wheels B and c similarly turn together. The figure gives the length of each circumference. Draw the system to scale, representing the

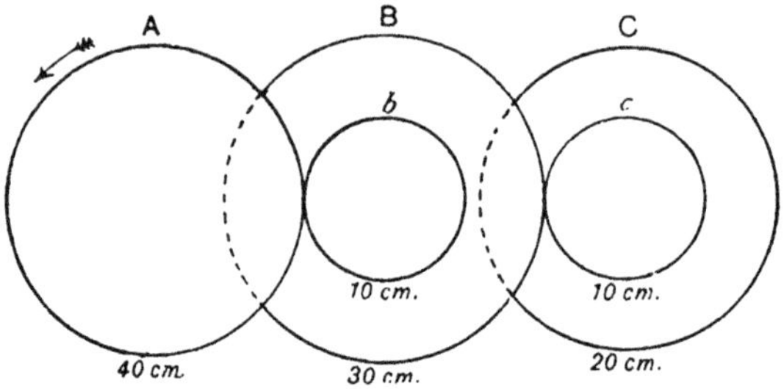

Fig. for Qn. 6.

diameter of the smallest wheels by 2 cm. Indicate by arrows how the various wheels turn, and compare the rates of revolution of A, B, and C. If A makes one revolution in 3 seconds, at what rate does C revolve? Give all the steps of your reasoning.

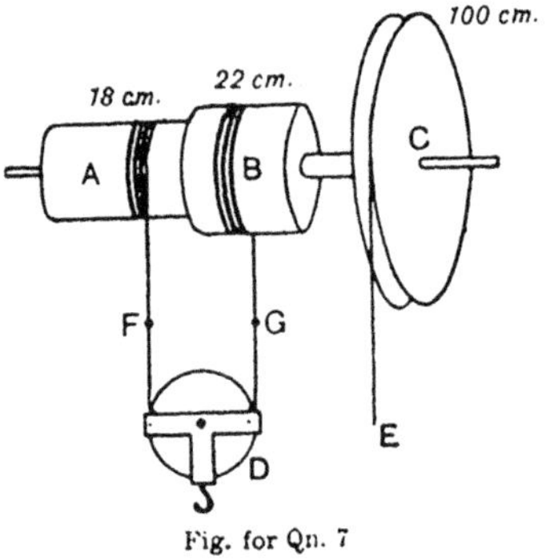

Fig. for Qn. 7

7. The figure shows two drums A and B, and a pulley C, turning together on the same shaft. The drums have diameters of 18 and 22 cm., and the pulley a diameter of 100 cm. Another pulley D is supported by a rope so wrapped on the two drums that rotation of the shaft either way wraps up one end and unwraps the other. Another rope E is wrapped on the pulley C. When the end of the rope E is pulled down one centimetre, through what angle does the shaft turn, how far are the points F and G of the other rope raised or lowered, and how far is the pulley D raised or lowered?

When the diameter of drum A is p cm., of drum B q cm., and of pulley C r cm., give an expression for the distance the pulley D will move when the end of rope E is lowered h cm.

8. The figure represents a triangular piece of cardboard ABC, right angled at C. D is the middle point of BC, and E the middle point of AC. G is the intersection of AD and BE, and GH is the perpendicular let fall from G on AC. Prove that $GH = \frac{2}{3}DC$.

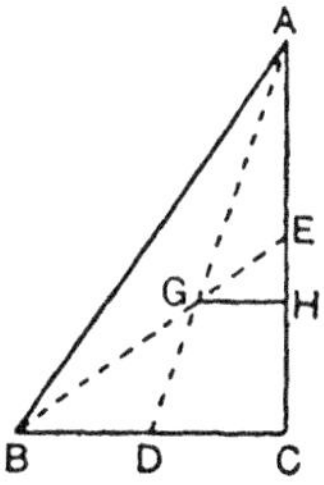

Fig. for Qn. 8.

Assuming that the volume of the cone generated by the rotation of the triangle ABC about AC is equal to the product of the distance travelled by G and the area of the triangle, calculate this volume when $BC = CA = 24$ cm.

Also find an expression for the volume when $BC = a$ and $AC = b$.

9. AOB is the diameter of a semicircular piece of paper, O the middle point of AB. OA is brought into coincidence with OB so that the paper forms a right cone. Given that the length of OA is 7 inches, find the volume of the cone. Assume that the volume of a cone is the area of the base multiplied by a third of the height.

10. The figure shows a fixed circle A of radius 15 cm., and a disc B of radius 5 cm. that touches A and rotates about its own centre C. There is a groove across the disc, through C, in which a compressible rod PQ lies, the ends P and Q lying always on the circle A. Draw a graph showing the length of the rod PQ as a function of the angle turned through by the disc, for one revolution from the position in which PQ passes through the centre of A. State the greatest and least lengths of the rod.

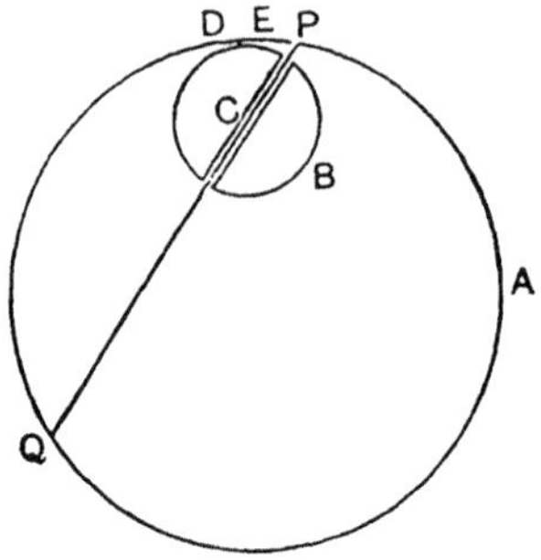

Fig. for Qn. 10.

Consider now a hollow cylinder A of length 40 cm., and radius 15 cm. Inside it there is a solid cylinder B of the same length and of radius 5 cm. that touches A and rotates about its own axis C. The cylinder B is made in two halves, and between these halves there slides a diaphragm that separates the cavity of A into two parts. The diaphragm consists of two overlapping rectangular plates, which are held by springs against the cylinder A at P and Q. This arrangement is used as a fan, the air being drawn in at E and driven out at D. What volume of air is drawn in at E and what volume driven out at D in half a revolution?

Section 46. The Golden Section.

This section may be omitted when circumstances allow. See page v.

The title "Golden Section" indicates the high estimation in which this construction has been held. The construction and its consequences are full of beauty, and access to it should always be made possible for the student who cares to spend time upon it. For practical purposes, however, it is superseded by the protractor, decimal calculation, and mathematical tables; and acquaintance with it should not be made compulsory.

1. Draw a straight line AB 6 centimetres in length, and divide it in H so that the rectangle $AB.HB = AH^2$. Give construction and proof.

Measure AH and HB as accurately as you can, and verify arithmetically whether your construction has given the required result.

2. A line AB is divided in H so that the rectangle AB.BH is equal to the square on AH. Given the points A and H, show how the point B may be found.

3. Show how to divide a line AB at C so that the rectangle AB.BC is equal to the square on AC.

A straight line AB is bisected at D and divided at C as in the first part of the question. Show that the square on the sum of AD and AC is equal to the sum of the squares on AD and AB.

4. Show how to find by geometrical construction a point P in a straight line AB produced, so that the rectangle AB.AP is equal to the square on BP. Give a proof.

Being given that a point Q within AB makes $AB.AQ = BQ^2$, and that a point P in AB produced makes $AB.AP = BP^2$, prove that $AB = BP - BQ$.

5. The angle B of a triangle ABC is a right angle, and AB is divided at D so that the square on AD is equal to the rectangle AB.BD. Prove that AB is equal to twice BC if AC is equal to the sum of AD and BC.

6. Show how to construct an isosceles triangle, each of whose base angles is double the vertical angle.

7. Show how to construct an angle of 72° by means of ruler and compasses only.

8. Explain how, without the use of a protractor, you could construct angles of 18°, 15°, 3°. The constructions need not be accurately performed, but proofs must be given.

9. Construct accurately a regular pentagon whose perimeter is 4 inches. State the steps of your construction, but do not give a proof of its correctness.

When your pentagon is constructed, show how from it you would proceed to make a regular ten-sided figure by cutting off an isosceles triangle at each corner. (This need not be done accurately.)

PART VI.

SECTIONS 47–54. CHIEFLY SIMILARITY AND TRIGONOMETRY.

Section 47. Scale Drawing.

1. To a man standing on the edge of a river, the direction of the top of a tower on the opposite bank makes an angle of 25° with the horizontal. He walks backwards 40 feet, and the top of the tower now makes an angle of 20°.

Determine, by drawing a figure, the width of the river.

2. To find the distance of an inaccessible point P from his position at A, a man measures off AB at right angles to AP and 50 metres long. He then goes back from A to a point C in line with A and P and 100 metres from A. From C he walks at right angles to CP until he reaches a point D in line with B and P. The distance CD being 82·6 metres, find the distance AP by drawing to scale.

3. Two observations taken in one vertical plane at a point A give the elevation of a mountain summit as $7\frac{1}{2}$°, and that of a point B between A and the mountain as 5°. A straight road 3 miles long runs from A to B, and at B the elevation of the summit of the mountain is $16\frac{1}{2}$°. Find its height above the level of A.

4. A man sets off from O in a direction between north and east and making 68° with the east, and walks 1450 yards to A. Then he turns 14° to his right and walks 1600 yards further to B. Draw his path on a scale of 1 inch to 400 yards, and measure the distance and direction of his final position B from his starting-point O. Give also the angle between AB and the east.

From your drawing measure how far A is east of O and how far north, and how far east and north B is of A; measure also how far east and north B is of O.

5. A map is constructed on the scale of twenty-two miles to the inch. Find the area of land represented on the map by an oblong measuring 3·5 in. by 7·4 in.

6. The figure shows a hill path between two places, A and B, 15 miles apart. One scale shows the distance from A of each point on the path. The other scale shows the height in feet of each point of the path. From the figure find the heights above

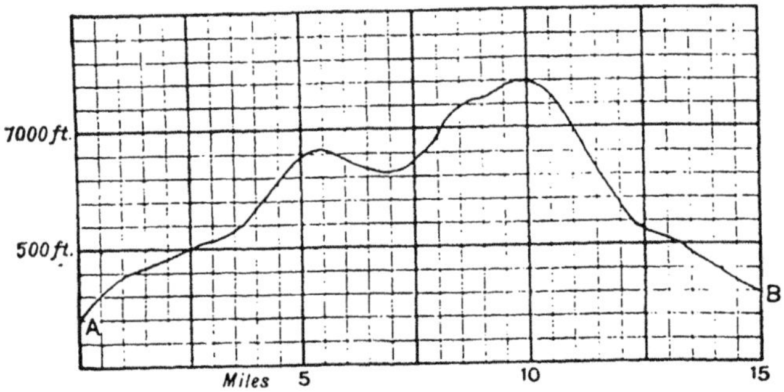

Fig. for Qn. 6.

sea-level of A and of points which are 1, 2, 3, ... 15 miles from A. When the height is not an exact number of hundreds of feet, either put down the nearest hundred or estimate the fraction of a hundred feet. Show your results in a table.

7. The figure is a map of a piece of country, in which an inch represents a mile. Each irregular line (called a contour line) passes through all the points at a certain height above sea-level, and this height is shown against the end of the line. The two numbers against crosses show the heights of those two points.

Draw a line equal in length to AB. Measure the distance from A at which AB cuts each of the irregular lines, mark corresponding points on your line and write against them their heights above sea-level. Also erect perpendiculars at these points to

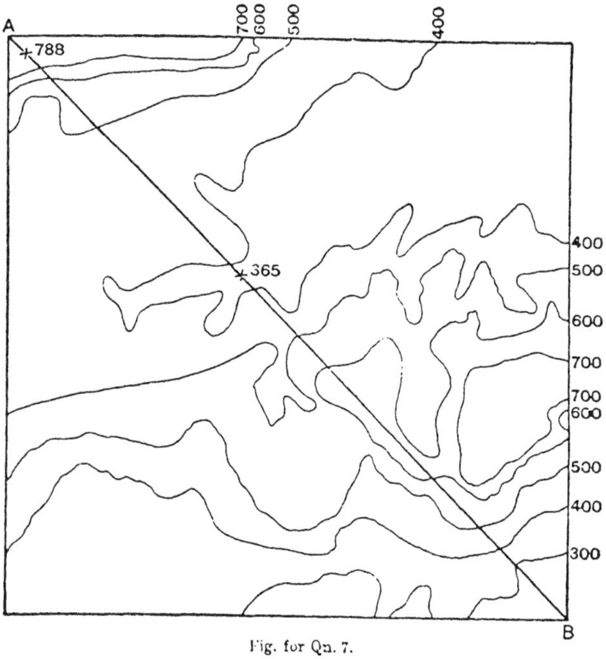

Fig. for Qn. 7.

represent the heights, representing 100 feet by 1 centimetre, and draw a smooth curve through the ends of the perpendiculars.

8. A man walking along a straight road which runs due east observes that at a certain point the bearing of a lighthouse is 50° east of north. After he has walked 2 miles from that point the bearing is 20° east of north.

Draw a figure to scale, taking two inches to represent a mile, and find, by measurement, the distance of the observer from the lighthouse in each position. Also find how much further he must walk along the road in order to come to the point of the road which is nearest to the lighthouse. State your construction, showing how you make use of the data.

9. It is known that two villages A and B are distant 9 and 12 miles respectively from a town C, and that A lies between north and north-west of C, while B lies between east and north-east of C. Find the positions of A and B for which the distance AB is as large and as small as possible. Find the length of AB in these two positions in any way you please.

10. Helvellyn is 9 miles distant from Scafell, bearing N. 54° E. Skiddaw bears E. 78° N. from Scafell and W. 58° N. from Helvellyn. Grasmoor bears W. 71° N. from Scafell and N. 77° W. from Helvellyn.

Find the distance and bearing of Grasmoor from Skiddaw.

11. An aviator thinks that he is flying due east at the rate of 40 miles an hour, whereas, owing to a wind blowing, he is really flying in a north-easterly direction at the rate of 35 miles an hour. By drawing a figure and measuring the distance, find how far he is at the end of 3 hours from the place where he expected to be.

12. It is required to draw the plan of a building 150′ long and 100′ wide upon a sheet of paper measuring 21″ × 15″. The margin should be about 2″ wide. What scale could be employed? Construct it.

13. To find the height of a mountain C above a plain two points A and B are taken in the plain 1000 metres apart (see figure). The angles CAB and CBA are measured and found to be 62° and 84°. Also, CD being vertical and BD horizontal, the angle CBD is measured and found to be 12°. By drawing the triangle ABC to a suitable scale, find the length of BC, and then by drawing the triangle BCD find the height CD of the mountain.

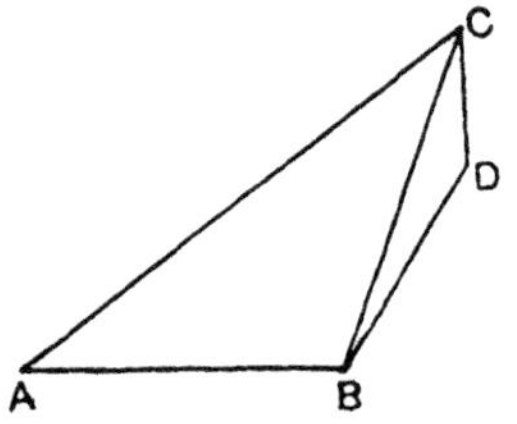

Fig. for Qn. 13.

Section 48. Indices and Logarithms.

A table of logarithms will be found inside the back cover.

1. Find the value correct to five places of decimals of

$$\frac{3}{7}+\frac{3}{7^2}+\frac{3}{7^3}+\frac{3}{7^4}.$$

2. It is known that a triangle, whose sides are a b c units long, has an area (in the corresponding square units) of

$$\sqrt{\left(\frac{a+b+c}{2}\times\frac{b+c-a}{2}\times\frac{c+a-b}{2}\times\frac{a+b-c}{2}\right)}.$$

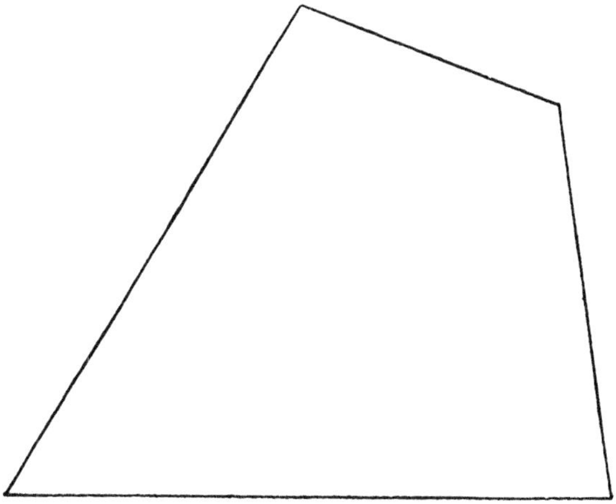

Fig. for Qn. 2.

The figure is a plan of a field on a scale of 1/10000. Divide it into two triangles, and use this formula to find its area. Give the area in hectares and in acres.

Check one of your results by finding the area in some other way.

3. The distance D of the horizon from the top of a mountain is given by the formula $D=\sqrt{h^2+2hr}$, where h = the height in miles of the mountain, and $r=4000$ = the length in miles of the radius of the earth. Find, to five significant figures, the distance of the horizon from the top of a mountain 3 miles high.

4. The area of a sphere is proportional to the square, its volume to the cube, of its radius. When the area is 25 sq. in. the volume is 11·754 cu. in. Find the volume when the area is $30\frac{1}{4}$ sq. in.

5. Find the fifth root of 587 as accurately as is possible by the use of your logarithm tables.

6. The velocity of water in a rectangular mill stream, whose breadth is A feet and depth D feet, is $\dfrac{520 \times R}{0{\cdot}516 + \sqrt{R}}$ feet per minute, where $R = \dfrac{AD}{2D + A}$. Calculate the amount of water which would pass along such a stream when $A = 4$ and $D = 3$, and compare it with the amount when $A = 4$ and $D = 2$.

7. Write down the logarithms of 10·39 and 0·0912. Find the number whose logarithm is $\bar{1}{\cdot}3$.

Evaluate:
$$\frac{\sqrt[3]{7} \times (1{\cdot}83)}{2{\cdot}857} + \sqrt[5]{190}.$$

8. Explain the rules for writing down the characteristic of the logarithm of any number or decimal.

Find in inches, using the tables, the diameter of a hemispherical bowl which will hold 3 gallons of water.

Volume of sphere $= \frac{4}{3} \times 3{\cdot}1416 \times (\text{radius})^3$. 1 cubic foot $= 6\frac{1}{4}$ gallons.

9. What is meant by saying that 10 is the base of the ordinary system of logarithms? What can you infer (i) if the logarithms of two numbers have the same characteristic? (ii) if they have the same mantissa?

Find, by logarithms, the approximate value of
$$\frac{3{\cdot}456 \times 4{\cdot}7816}{8{\cdot}9635}.$$

10. Prove that
$$\log a^m = m \log a,$$
and that
$$\log \sqrt[n]{a} = \frac{1}{n} \log a.$$

Use the table of logarithms to find the approximate values of

(i) $143{\cdot}5 \times 6{\cdot}714$, (ii) $\sqrt[3]{41{\cdot}38 \times (0{\cdot}01674)^4}$.

11. Prove that $\log_x y \log_y x = 1$.

Given that $\log_{10} 5 = x$, express $\log_5 1250$ in terms of x.

12. A man of 34 insures his life as follows: He makes to a society 25 annual payments of £10, the first immediately. In return the society pays £353 at his death. Assuming that m such annual payments of £P will amount n years after the last payment to £$PR^n(R^m - 1) \div (R - 1)$, the rate of interest being $100(R - 1)$ per cent. per annum; calculate what the man's payments amount to at $2\frac{1}{2}$ per cent. per annum 4 years after the last one is made; and give the result to two significant figures, which is all that the tables allow.

13. If a corporation borrows p pounds at the rate of r per cent., the loan and interest can be paid off by n equal annual instalments of a pounds, where $a = p \times R^n \times (R - 1) \div (R^n - 1)$ and $R = 1 + r/100$. What must be the annual payment if a loan of £5000 at 3 per cent. is repaid by 10 equal payments? Give the result as accurately as the tables allow.

14. A sum of £100 is borrowed, to be repaid in 30 equal yearly instalments, the payments covering principal and compound interest at 4 per cent. per annum. However, just after making the nth payment, the borrower wishes to pay off the remainder of the loan. Assuming that the sum he must pay is

$$\frac{(1{\cdot}04)^{30} - (1{\cdot}04)^n}{(1{\cdot}04)^{30} - 1} \times £100;$$

calculate (to the nearest pound) the value of this sum when $n = 15$; and determine the least value of n that gives a less sum to repay than £50.

15. Prove, either from arithmetical or from more general considerations, that a pounds invested at c per cent. compound interest will amount in b years to $a\left(1 + \frac{c}{100}\right)^b$ pounds, the interest being added once a year to the principal.

A man sets aside £800 to be invested for his two sons so that each of them on attaining the age of 21 shall be entitled to the same amount. If one is 13 and the other 17 when the investment is made, and compound interest at the rate of 3 per cent. is reckoned, how must the £800 be apportioned? Give each result to the nearest pound.

16. The present value £P of an annuity of £a per annum to last for n years is given by

$$P = \frac{100a}{r}\left\{1 - \left(1 + \frac{r}{100}\right)^{-n}\right\}$$

when money is worth r per cent. per annum. Find, as accurately as the tables allow, the value of an annuity of £425 to last for 48 years when money is worth 5 per cent. per annum.

17. Explain how a meaning is found for the symbols $x^{2/3}$ and $x^{-1/2}$.

Find, using the tables, the value of $\dfrac{3^{-1/3}+1}{3^{1/3}+1}$.

18. Find the square root of $93 + 28\sqrt{11}$ to three significant figures.

Simplify $\sqrt[8]{a^4} \times \sqrt[5]{a^3} \times a^{-1/10} \div a^{2/7}$, expressing your result with a fractional index.

19. The difference in feet between the heights of two places is given approximately by the formula $x = 60360\,(\log p_1 - \log p_2)$, where p_1 is the atmospheric pressure, observed with a barometer, at the lower station, and p_2 that at the upper one; the base of the logarithms being 10.

The pressure at the top of a mountain is found to be $\frac{2}{3}$ that at the bottom. What is the difference in height?

What is the pressure at the top of Mt. Blanc (15,775 ft. above sea-level) when that at the sea-level is 15 lb. per sq. in.?

20. In steam engines the mean pressure is $(1 + \log_e r)/r$ times the initial pressure, where r is a constant depending on the form of the engine. Calculate to two decimal places (*i.e.* to the nearest hundredth), the ratio of the mean pressure to the initial pressure when $r = 2$. Take 2·718 as the value of e.

21. Show that in addition to $x = 0$, another value of x which satisfies the equation $10 \log_{10}(1 + x) = 5x^2 + 2x$ lies between $x = 0$ and $x = 1$. Find the value of this root correct to two significant figures, preferably by plotting on a large scale the graphs of $y = 10 \log_{10}(1 + x)$ and $y = 5x^2 + 2x$, between $x = 0$ and $x = 1$, and finding their point of intersection. Check your result by putting

$$\log_{10}(1 + x) = \left(x - \frac{x^2}{2} + \frac{x^3}{3}\right)\log_{10} e$$

and solving the resulting approximate equation in x. Take $e = 2{\cdot}718$.

22. Make a list or table of the integral powers of 2 from index -2 to index $+4$. Use your values to draw a curve the ordinates of which show the logarithms of numbers between 0·3 and 16·0 to base 2; use as unit for the numbers 1 cm., and for the logarithms 5 cm. From your curve determine as accurately as you can $\log_2 1{\cdot}7$ and $\log_2 5{\cdot}4$.

Use your curve to graduate two strips of paper for use as a slide rule and fix the two together so as to show the product of 1·7 and 5·4

23. If a heavy chain of length $2k$ feet is suspended from two points in the same horizontal line, $2x$ feet apart, and the sag at the centre is z feet, it is known that

$$2k = c(e^{x/c} - e^{-x/c}) \qquad \text{and} \qquad 2(c+z) = c(e^{x/c} + e^{-x/c}),$$

in which c is a constant dependent on the circumstances and e has the value 2·718.

Prove that $k + z + c = ce^{x/c}$ and $k^2 = z^2 + 2cz$.

Also given the length of the chain to be 100 feet and the sag at the centre 20 feet, find the value of c and the span, each correct to 3 significant figures.

Section 49. Similarity of Triangles.

1. Prove that a parallel to the base of a triangle cuts the two sides in the same ratio.

On a base BC 10 cm. long draw a triangle ABC having AB = 16·84 cm. and AC = 11·02 cm. Parallel to BC and at distances of 4 cm., 6 cm., and 9 cm. draw three straight lines cutting the sides of the triangle in D and E, F and G, H and K respectively. From measurements of the figure calculate to three significant figures each of the ratios AD/AE, AF/AG, AH/AK.

From the data, and without any measuring, calculate the true value of each ratio to four significant figures.

2. Draw any triangle abc and outside it take a point O. Join aO, bO, cO, and produce the lines through O to A, B, C respectively, so that OA = 2 Oa, OB = 2 Ob, OC = 2 Oc. Compare the sides, angles, and area of $\triangle$ ABC with the sides, angles, and area of $\triangle\, abc$, and justify your conclusions.

3. Construct a triangle whose angles are in the ratio 7 : 5 : 3, and whose greatest side is of length 6 centimetres. Bisect the greatest angle and measure the length of the bisector taken to its intersection with the opposite side.

4. A man stands at some distance from a straight road, and notices a row of telegraph posts along the far side of it. The first post is known to be 242 yd. from the last one. He watches a motor-car, and finds that it appears in line with the last post exactly 15 seconds after it was in line with the first. Walking in a straight line to the first post, he crosses the track of the car after 60 yd., and reaches the post after walking 14 yd. more. Assuming that the car runs parallel to the far side of the road, draw a figure to scale, and from it find the speed of the car.

5. To divide a set of numbers by 5·44 I use a straight line graph (OP in the rough sketch). Draw the graph on the scale

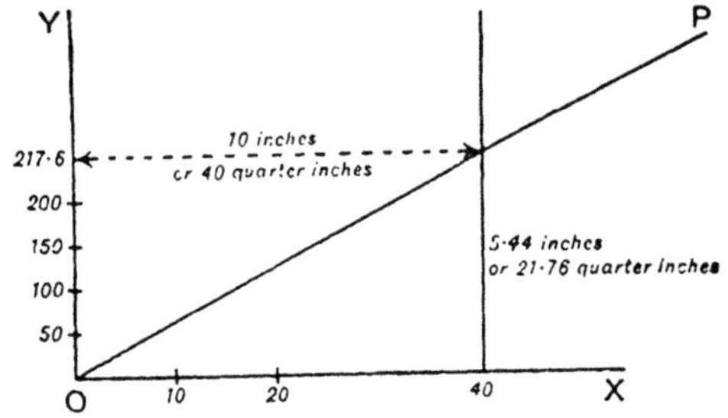

Fig. for Qu. 5.

indicated or any other fairly large scale. Obtain from your graph the quotients of 60 and 218 by 5·44, and verify by calculation with the tables.

Explain why the graphical method gives the result.

6. The frame of a travelling crane is represented in plan in the figure. AB and CD represent two fixed parallel girders forming opposite sides of a rectangle, EF a movable girder which travels upon them so that the straight line EF remains always parallel to AC, and P is the crane which can be moved from one end to the other of the girder EF.

Initially EF coincides with AC, and P coincides with F. If the parts are set in motion in such a way that the distance P has separated from F is always proportional to the distance by which EF has separated from AC, what sort of path does P describe? Give a proof of your answer.

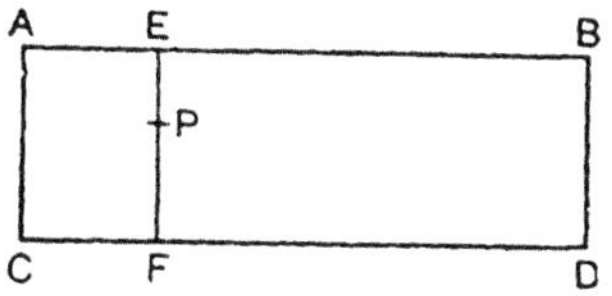

Fig. for Qn. 6.

If AB is a feet and AC b feet long, and P moving in the way described arrives in due course at B, find the distance BP at the moment when the distance CF is a/n feet.

7. To find without instruments the height H of an inaccessible object AB on level ground (see figure) the following method has been employed: Two staffs CD and EF of equal height h are set up in line with the foot of the object at a distance d apart. Let

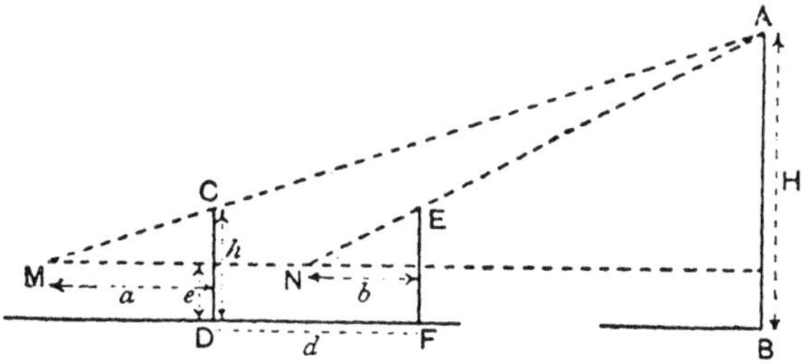

Fig. for Qn. 7.

M be the point where the observer's eye must be placed so as to see C and A in the same straight line, let N be the point from which E and A appear in the same straight line, and let a and b be the distances from M and N to the staffs CD and EF respectively. Let e be the height of the observer's eye from the ground.

Then $\quad H = \dfrac{d(h-e)}{a-b} + h.\quad$ Prove this relation, and apply it to find H when $a = 15$ ft. 5 in., $b = 13$ ft. 4 in., $d = 20$ ft., $e = 5$ ft. $4\frac{1}{2}$ in., $h = 10$ ft.

8. Draw a fairly large triangle whose sides are in the ratio

3 to 4 to 5; and another whose angles are in the ratio 3 to 4 to 5. Determine, by measurement, whether the two triangles are similar.

9. The sides of one triangle measure 6, 7, and 9 inches, and those of another triangle 18, 14, and 21 inches. Show that, corresponding to each pair of sides of one triangle, there is a pair of sides of the other which have the same ratio. Are the triangles similar? Give reasons, and prove the theorem upon which your test is based.

10. ABCD is a quadrilateral. Its diagonals AC BD intersect at a point O, such that O is a point of the trisection of each. Prove that two of the triangles into which the quadrilateral is divided by its diagonals are equal in area. Find also the ratio of the areas of the two remaining triangles.

11. DE is a line 2 inches long, DX and EY two parallel lines perpendicular to DE. A triangle ABC is laid across the parallel lines so that its sides cut them as in the sketch. The following measurements are given (in inches):

DP = 3	DR = 6	DT = $6\frac{1}{2}$
EQ = $1\frac{1}{2}$	ES = 3	EU = $8\frac{1}{2}$

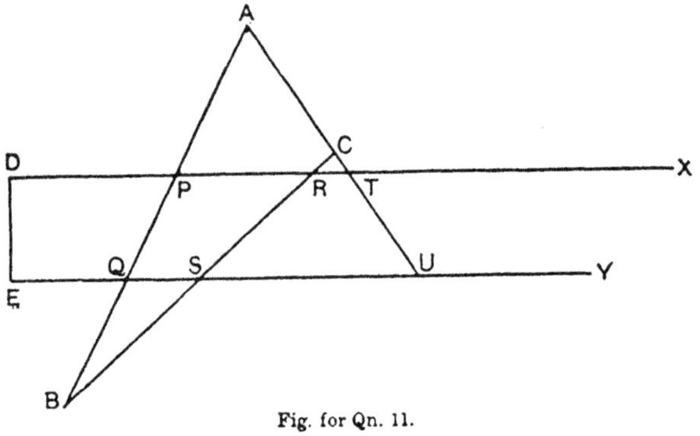

Fig. for Qn. 11.

(i) Draw the figure to scale and measure AB.

(ii) Without making use of your drawing, calculate the length of PQ, and also the ratios of AP and BQ to PQ, and thus find AB.

12. Show that one diagonal of a quadrilateral divides the other diagonal in the same proportion as it divides the quadrilateral.

13. XYZ is a straight line, and XYAB XZCD are equal rectangles on the same side of the line XYZ. Prove that AC and DY are parallel.

If XY = 4 inches, XZ = 6 inches, and the area of each rectangle is 18 square inches, find, by means of a geometrical proof and without actual measurement, the perpendicular distance between AC and DY.

14. Through a given point P between two intersecting straight lines OX OY draw a straight line APB terminated by the given lines, so that AP : PB = 1 : 2. Prove your construction.

15. From F, a point in the side PQ of the triangle PQR, a straight line is drawn parallel to QR, meeting PR in G. Prove that

PF : FQ :: PG : GR.

A given point D lies between two given straight lines AB and AC. Find a construction for a line through D terminated by AB and AC, such that D is one of its points of trisection. Prove also that there are two such lines.

16. Assuming that the image of an object appears as far behind a plane mirror as the object itself is in front, discuss the movement of the image in each of the following cases. In the sketch O is a fixed object 10 inches from the mirror and P its image; MM is the mirror.

(i) Suppose the mirror moved parallel to its first position and away from O at the rate of 10 inches per second; discuss the rate at which the image P moves, and give its distance from O in *m* seconds, starting from the position represented.

(ii) Suppose the mirror rotated about X through a right angle in the direction shown by the arrow. Sketch a few positions to get a rough notion of the movement of the image, then make an exact statement about its path, and prove your statement.

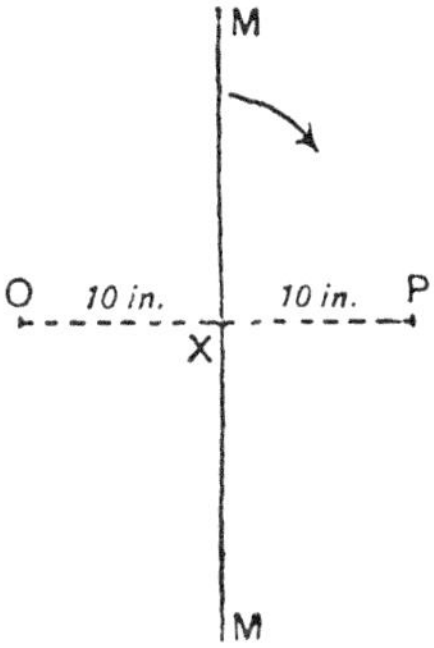

Fig. for Qn. 16.

17. OCA and OBD are two intersecting straight lines. P is

a point on AB, and BC and AD are both bisected by OP or OP produced. Prove that CPD is a straight line divided at P in the same ratio as AB.

18. If two triangles ABC PQR have the angles A and P equal, and CA : AB = RP : PQ, prove that the angles B and Q are equal.

If two similar triangles have their corresponding sides parallel, prove that the lines joining corresponding vertices meet in a point.

19. In the figure OC is a square, each side of length c cm. Imagine a rod of indefinite length to turn about C in the plane of the paper. Denoting its intercept OA on OX always by a and

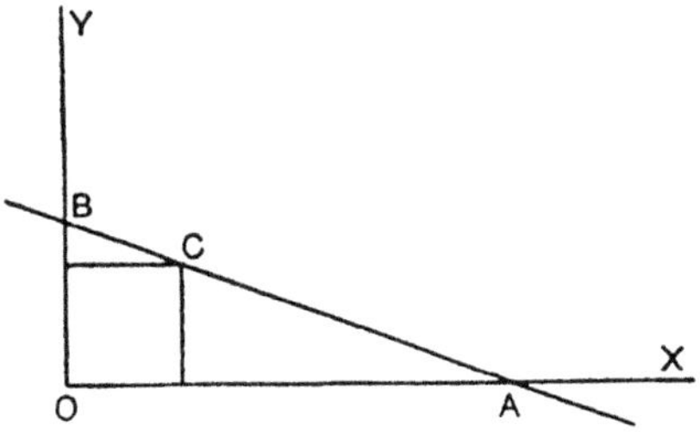

Fig. for Qn. 19.

its intercept OB on OY by b, verify by drawing and by calculation that the relation $1/a + 1/b = 1/c$ is true when $c = 1{\cdot}4$ in. and $a = 2{\cdot}35$ in. Then prove that it is true generally.

20. ABCD is a square. Through C any straight line QCP is drawn to meet AD produced and AB produced in Q and P. Prove that $1/AP + 1/AQ = 1/AB$.

In the particular case in which the triangle APQ is isosceles, find the ratio of the area of the triangle APQ to that of the square ABCD.

21. A triangle ABC being given, inscribe a square in it, so that one side of the square may lie on BC, and a vertex on each of the sides AB and CA.

If x be the length of the side of this square, a the length of BC, and p the perpendicular of BC from A, prove that $x = \dfrac{ap}{a+p}$.

22. A triangular field ABC, right angled at B, with AB equal to 42 chains, and BC equal to 35 chains, has in it a circular pond of radius 5 chains whose centre is 12 chains from AB and 15 chains from BC.

It is desired to run a straight fence across the field from AB to BC so as to divide it into two equal parts, each part containing half the pond. If the distances of the ends of the fence from B are x and y chains respectively, find two equations connecting x and y, and calculate their numerical values.

Draw the field and pond on as great a scale as your paper allows, and show one possible position of the fence.

23. The interior of a glass vessel has the form of the frustum of a cone, slant height 10 cm., top radius 6 cm., bottom radius 4 cm. Assuming that the volume of the frustum of a cone is given by the formula $\frac{1}{3}\pi h(a^2+ab+b^2)$, where a and b are the radii of the ends and h the true height (the perpendicular distance between the ends), find the capacity of the vessel and show that when the vessel is filled to a slant height of x cm., the ratio of the volume of liquid to the capacity of the vessel is given by the expression $\frac{1}{180}(12x+0{\cdot}6x^2+0{\cdot}01x^3)$.

Draw a graph showing the value of this fraction for values of x between 0 and 10.

Supposing a scale is to be placed on the vessel which will graduate it in tenths of its capacity, find from the graph the distance of each graduation mark from the bottom of the interior of the vessel, measured on the slant side.

24. From an external point O a perpendicular ON is drawn to a straight line XY. Any other straight line OA is drawn from O to XY, and on it a point B is taken such that $OA . OB = ON^2$.

Draw a figure with ON equal to 5 cm., and find the locus of B. Give a geometrical proof that your result is correct, and extend that proof to determine the locus of B (a) when $OA . OB = 2ON^2$; (b) when $2OA . OB = ON^2$.

25. AOB is the diameter of a circle of radius x inches. OD is a radius drawn at right angles to AB. In OD a point P is taken such that $OP = y$ inches. Join AP and produce it to meet the circumference at C.

Express the lengths of the lines AC and CB in terms of x and y, and hence find an expression for the area of the triangle ACB.

Section 50. Trigonometrical Ratios.

1. Draw an angle the cosine of which is 2/3. Measure its tangent.

2. Draw a right-angled triangle with an hypotenuse of length 4 in. and one angle 60°. Deduce by measurement the sine and cosine of 60°.

3. Draw an angle whose sine is 0·3, and find the value of the angle, of its cosine, and of its tangent.

4. An angle has a cosine of 0·6. By drawing find the angle, its sine, and its tangent.

5. The tangent of an angle is 3. By drawing find the angle, its sine, and its cosine.

6. With your protractor make an angle of 42°. By drawing find the values of the sine and tangent of the angle.

Tabulate beside these the values given in trigonometrical tables, and account for any discrepancies.

7. Draw four right-angled triangles in which one side is 0·2, 0·4, 0·6, 0·8 times the hypotenuse. From these find the angles whose sines are 0·2, 0·4, 0·6, 0·8, and verify your results by reference to a book of tables.

8. Draw the angle x and measure its sine

(i) when $\tan x = 2{\cdot}3$; (ii) when $\sec x = 1{\cdot}4$.

Verify your results by using the tables.

9. Can the cosine of an acute angle be equal to 2? What values can the cosine have? and the sine? and the tangent?

10. From the tables take, to two decimal places, the values of the sine, cosine, and tangent of angles of 20, 40, 60, 80 degrees.

For each of these angles verify that the tangent is the quotient of the sine by the cosine.

11. A certain tower is known to be 150 feet high. Its angle of elevation as seen from a point at the same level as the base of the tower is 35·4°; calculate the distance from this point to the base of the tower, and verify your answer by a carefully drawn diagram.

12. A and B are directly opposite to each other on the banks of a stream; C is on the same bank as A and is 30 yards distant; A and B subtend an angle 47° 30′ at C. How broad is the stream?

13. Calculate from the tables the length of that chord in a circle of radius 5 inches which subtends at the centre an angle of 42°.

14. Construct an angle whose sine is double its cosine. Determine the angle and its sine and cosine by measurements of your figure, and compare the results with those in your tables.

15. A ladder 20 ft. long leans against a wall, making with it an angle of 20°. Calculate how far the top will descend if the foot of the ladder is pulled out 4 ft. farther from the wall?

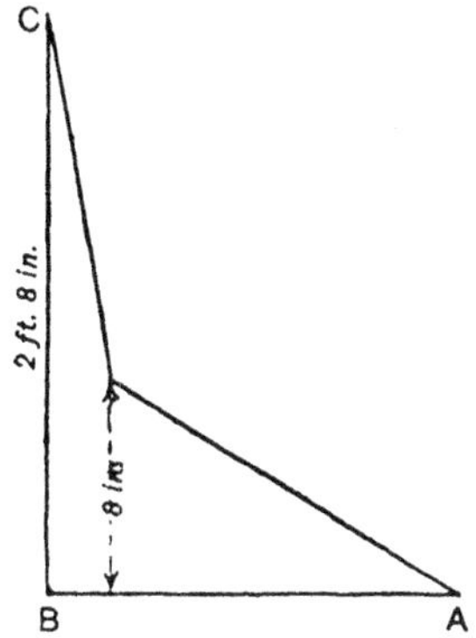

Fig. for Qn. 16.

16. In the figure AB is horizontal and 2 ft. long, BC is vertical and 2 ft. 8 in. long. A rod 2 ft. long has one end fixed at A, and the other end is supported from C by a string 2 ft. 8 in. long. The string (still fixed at C) is now shortened until the free end of the rod is raised 8 inches. Find the new length of the string, either by drawing or calculation. Find also the angle which the rod now makes with the horizontal.

What is the shortest length the string could have, and what is the angle which the rod would then make with the horizontal?

17. A straight rod AB 3 ft. 9 in. long is held under water, A being 2 ft. 6 in. and B 9 in. below the surface. Calculate (i) the distance below the surface of a point C on the stick which is 12 inches from A, (ii) the angle which the stick makes with the surface of the water.

If a parallelogram is held under water, show that in every position the sum of the depths of the four corners is 4 times the depth of the point of intersection of the diagonals.

18. The ancient Greeks measured the latitude of a place by setting up a vertical rod and comparing its length with the length of its shadow. Supposing observations taken at midday at the equinox (when the sun is vertical at the equator) to give 5/3 as the ratio of rod to shadow at Alexandria, and 4/3 as the ratio at Carthage, find the latitude of each place.

19. In the figure BCA and BDC are right angles. How many right-angled triangles are there in the figure? In terms of the lengths $a \quad b \quad p \quad m \quad n$ shown on the figure, state the relation between the three sides of each triangle.

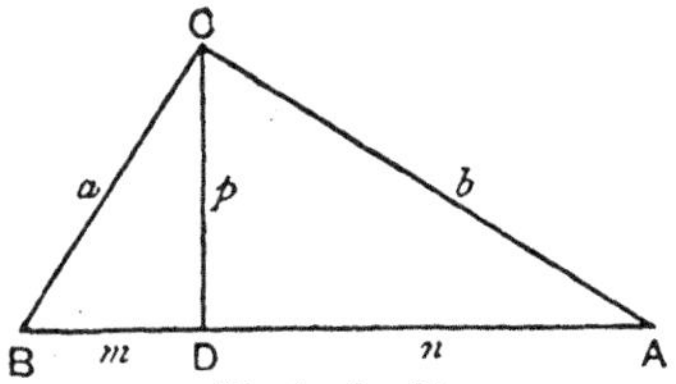

Fig. for Qn. 19.

Prove the rectangle property of a circle, namely, that O being a fixed point, and OPQ a line through O meeting a fixed circle in P and Q, then OP.OQ is the same for all positions of OPQ; take O inside or outside the circle as you please.

In how many ways can you choose a set of three points from the four A B C D of the figure? Make such a figure in your book, and, as far as possible, draw a circle through each set of three points. How does the construction for a circle fail in one case? Show that the rectangle property of a circle gives for each circle a relation among the lengths $a \quad b \quad p \quad m \quad n$, and write down these relations.

You have now six relations among $a \quad b \quad p \quad m \quad n$. Select three independent relations, and deduce the remaining 3 relations from them. How many of the lengths $a \quad b \quad p \quad m \quad n$ must be known in order that the rest may be calculated?

20. A point moves so that at any time t seconds its coordinates x and y are given by $x = \sin t$ and $y = \sin 2t$, the angles being measured in degrees. Find the values of x and y for a number of values of t and draw the curve, using 5 cm. as unit.

How long does the point take to describe the curve once?

21. Take a rectangular sheet of paper 20 cm. by 12 cm., and fold it along a diagonal AC, as in the figure, so that B comes to b. Letter your sheet of paper as in the given figure, and then answer the following questions:

(i) Describe as accurately as you can the triangle marked OAC in your figure, adding short reasons for your statements.

(ii) Measure AC and the height-line OM on your sheet, and thence obtain the area of the triangle OAC. Then obtain the true area *entirely by calculation* from the given length and breadth of the sheet.

(iii) Find, to two significant figures, what the ratio of the sides of the original rectangular sheet of paper must be to make it possible to fold down ODC and ObA so as just to cover up OAC. Justify your statement.

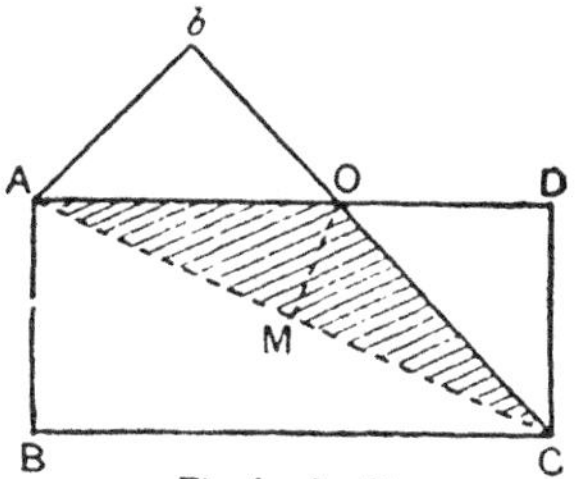

Fig. for Qn. 21.

22. The side of a regular 12-sided polygon inscribed in a circle is one inch in length; employ any method you please to find the radius of the circle to two decimal places.

23. Upon a tower 55 feet high stands a vertical pole. If the tower and the pole subtend angles of 45° and 12° respectively at a point level with the base of the tower, find the length of the pole.

24. There are two knots A B in a rope which hangs from the top of a flagstaff. When A is on the ground and that part of the rope above it taut the inclination is 55°, and when B is on the ground and all the rope above it taut the inclination is 40°. Given that $AB = 15$ feet, find the height of the flagstaff.

25. A flagstaff 10 ft. high is placed on the top of a tower. At a certain spot on the ground the elevation of the top of the tower is 45°, at a spot 90 ft. further from the tower the elevation of the top of the flagstaff is 30°. Find the height of the tower.

26. I take measurements to determine the air space of a rectangular hall: Length 18·4 metres, breadth 11·8 metres, inclination to floor of diagonal of side wall 31·8°, of diagonal of end wall 44°. Calculate the air space.

More measurements were taken than were necessary. Check the measurements by deducing one of them from the other three.

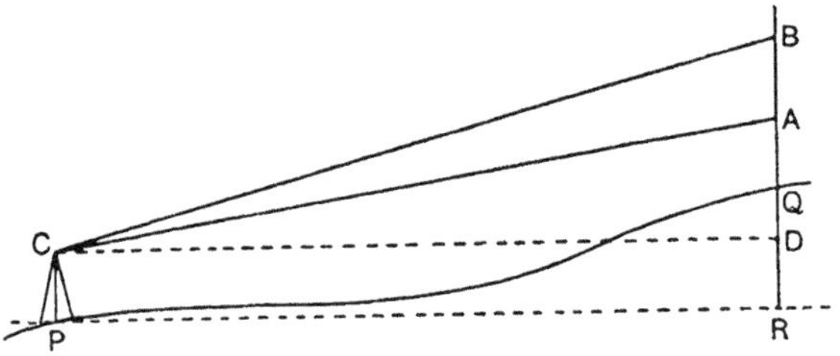

Fig. for Qn. 27.

27. The figure shows a method of determining the horizontal distance PR, and the difference of level QR between two points P and Q. A rod with fixed marks A B upon it is held vertical at Q, and the elevations ACD (equal to α) and BCD (equal to β) of these points are read at C by a telescope and divided circle, the axis of the telescope being at a distance CP (equal to h) above the ground at P. If $QA = b$ and $AB = s$, write down expressions for PR and QR.

Find PR and QR when $\alpha = 6° \ 10'$ and $\beta = 7° \ 36'$, the values of h b s being 5 feet, $2\frac{1}{2}$ feet, and 5 feet respectively.

28. A railway is to be driven through a cutting at a depth of 80 ft., the width at the level of the track being 30 ft. and the greatest allowable width of the top of the cutting 65 ft. The soil is such that the inclination of the sides must not exceed 75°.

Show, by drawing or calculation, that retaining walls must be built at each side of the track; and find their height, supposing them vertical.

29. A girder to carry a bridge is in the form of a circular arc; the length of the span is 120 feet and the rise of the arch (*i.e.* the height of the middle above the ends) is 25 feet. Find the angle subtended by the arc at the centre of the circle, and the radius of the circle.

If the girder expands 0·00001 of its length for a rise of 1° C. of temperature, what rise of temperature will increase the span by 1 inch?

30. AOB is the diameter of a circle whose centre is at O. At C in the circumference a tangent CE is drawn. OE parallel to AC meets this tangent in E. Prove that EB touches the circle, and find an expression for the distance EB, in terms of r the radius of the circle, and θ the angle COA.

31. Determine how many apple trees will be required to plant an orchard in the form of a square containing 10 acres, the trees to be placed 20 feet apart, so that any three neighbouring trees are at the corners of an equilateral triangle, and no tree is to be within 9 feet of a boundary of the orchard.

32. A regular pyramid has a square base. The faces are isosceles triangles having the base angles equal to 70°. Suppose such a pyramid made of cardboard, the base measuring 4 cm. in the side. Suppose it slit along the slant edges and spread out flat, and draw this flat figure.

Find the inclination of the faces of the original pyramid to the base by drawing or by calculation.

33. The Great Pyramid is 481 feet high, and stands on a square base 756 feet in the side. Find the slant height of a face, and the length of an edge. Find also the inclination of a face to the horizontal.

34. A chasm in level ground is bounded by parallel vertical sides. A is a point at the top of the chasm, and the depth of a point B at the bottom, vertically below A, is wanted. It being impossible to take measurements from C the point opposite to A, a point D 50 yards along the side from C is chosen. The angle ADB is 43° and the angle ADC is 52°. Find the depth AB.

Section 51. Three-dimensional Problems.

1. The curved surface of a right circular cone whose semi-vertical angle is 45° is made by cutting out a sector from a circular sheet of copper, the diameter of the sheet being 56 cm. Determine the angle of the required sector to the nearest minute.

If the metal is 1·4 mm. thick, what will be the weight of the cone in kilograms, assuming that its base is made of metal of the same thickness and that 1 cubic centimetre of copper weighs 8·94 grams?

2. A tool clamp for a lathe is shown by plan and elevation in the figure.

Draw the views full size, and add another elevation as seen when looking in the direction indicated by the arrow A.

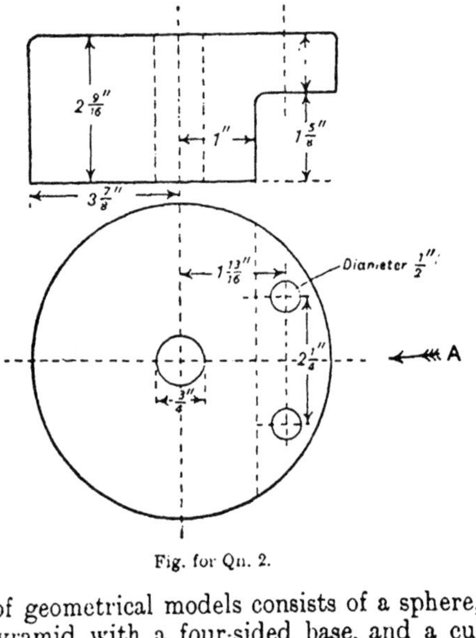

Fig. for Qn. 2.

3. A set of geometrical models consists of a sphere, a cone, a cylinder, a pyramid with a four-sided base, and a cube. How many boundary surfaces, plane angles, and edges, has each of these? Which of the surfaces are plane?

4. A cube with each edge an inch in length is placed on a table. A right circular cone just covers the cube and rests on the table. Find the height of the cone, if the length of the sloping side is equal to the diameter of the base.

5. Draw a development of the piece of tinned plate which would make the corner box shown by the isometric sketch in the figure. The bottom of the box is 3″ × 3″. Lap, for jointing, should be provided where necessary. Scale full size.

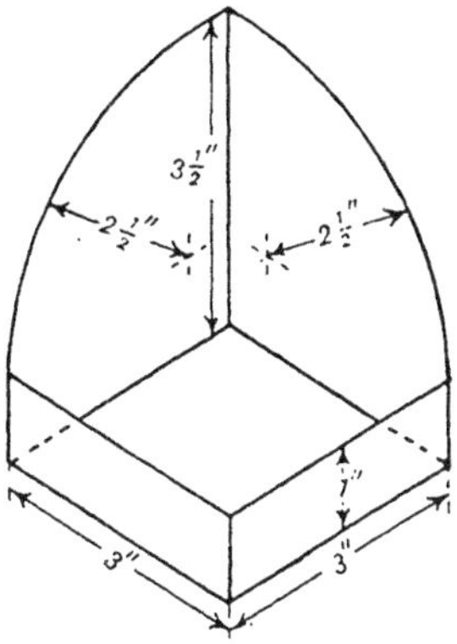

Fig. for Qn. 5.

6. A sphere of radius 2 inches rests on three horizontal wires which form a plane equilateral triangle of side 6 inches. Find the height of the top of the sphere above the plane of the wires.

7. ABCD is a rectangle whose diagonals meet in E, the length of each diagonal being 1·5″. Through E a straight line EF, 1·8″ in length, is drawn perpendicular to the plane of the rectangle. Find the centre of the sphere passing through A B C D F, and calculate the length of its radius.

8. (i) A camera-stand is a tripod with legs a metre long. When the 3 feet of the tripod form an equilateral triangle 70 cm. in the side, find, by drawing or calculation, the distance of the centre of the triangle from a foot of the tripod, and the height of the tripod.

(ii) Assuming the height y cm. of the tripod to be given in terms of the distance x cm. between two feet of the tripod by the equation $y=\sqrt{10000-x^2/3}$, find, by calculation from this equation or by drawing, the values of y corresponding to the values 0 40 80 120 160 of x, and the value of x that makes $y = 0$. Tabulate these values, and draw a graph of y as a function of x on a scale of 1/10, showing clearly your axes and your scales.

(iii) Prove the given relation between x and y.

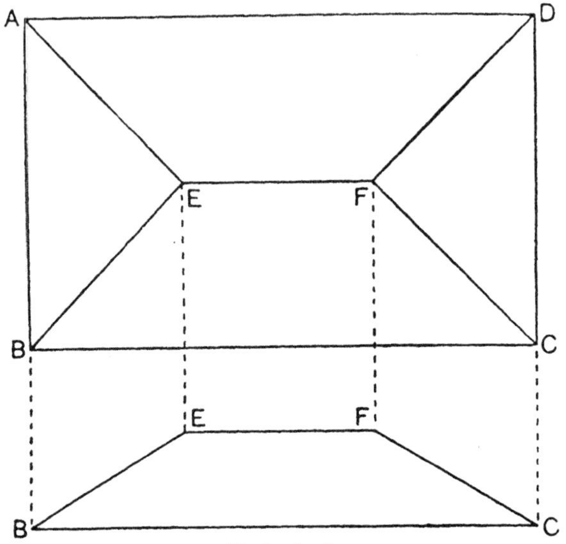

Fig. for Qn. 9.

9. The figure shows the plan and elevation of a slated roof, consisting of four sloping parts, which run up to the horizontal ridge EF. The upper diagram shows the lengths AB BC CD DA EF on a scale of 1 inch to 20 feet, and the lower diagram shows the height of EF above ABCD on the same scale. Find in any way you please the length of the edge AE of the roof and the slated area.

10. A cord is stretched in a room between two points in space, and the three planes of reference are (i) the north wall of the room, (ii) the east wall, and (iii) the floor. If the co-ordinates of the two points, between which the cord is stretched tightly, are $(5, 2, 4\frac{1}{2})$ and $(4, 5\frac{1}{2}, 1\frac{1}{2})$, unit 1 foot, calculate the length in feet of this cord, and the points on the floor and east wall where the cord, if prolonged, would intersect them.

Find all these results also by drawing.

11. The sketch shows an elevation of two roofs, which run into each other at right angles. The slopes of roof A make angles of 60°, and those of roof B angles of 45° with the ground. The heights of A, B, and C above the ground are 18 feet, 16 feet, and 8 feet respectively, and CD is 8 feet long. To a scale of 2 feet to 1 centimetre draw the given elevation, the elevation at right angles to it, and the plan, and determine the angle of slope of the line of intersection of the two roofs. Also calculate this angle from the given data.

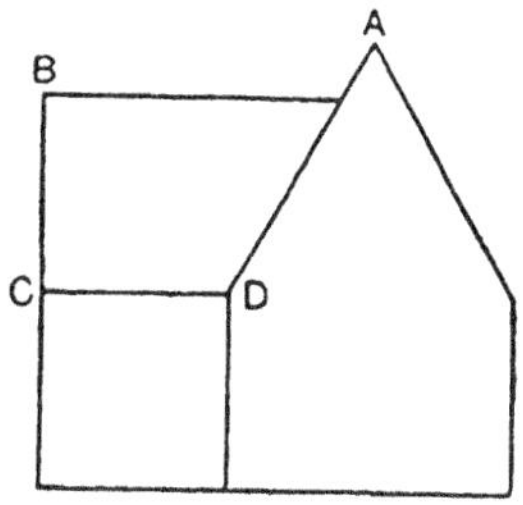

Fig. for Qn. 11.

12. A line is 3″ long, and its plan measures 2·5″; find the inclination of the line to the horizontal plane. How much higher is one end of the line than the other? Draw the two projections of the line when its plan makes an angle of 30° with the ground line.

13. AB and AC are two straight lines which meet at an angle of 45°. AB is 10 cm. in length. At B BD is drawn perpendicular to the plane BAC and 2·5 cm. in length. Calculate the length of the perpendicular from D to AC.

14. Of the three coordinate axes, that of x points out of the paper towards you, that of y to the right, and that of z is vertically upwards; the paper on which you are working being supposed in a vertical plane. The coordinates of the four vertices A B C D of a triangular pyramid are, in millimetres, (70, 52, 80) (45, 7, 0) (102, 60, 0) (20, 100, 0). Draw the pyramid in plan and elevation, that is, draw its projections on the planes $z=0$ and $x=0$.

Draw the three slant faces of the pyramid and find the area of each in sq. cm. Find the volume of the pyramid in c. cm.

Find the inclination of the slant face ABC to the horizontal. The pyramid being turned about the edge BC until the vertex A lies in the horizontal plane, show the pyramid in plan and elevation.

15. A rectangular block of wood measuring 12 by 12 by 2 inches has a bevel taken off the four edges of the upper square face, the bevel making 45° with the face, and being 1 inch broad. Draw a section of the block by a plane containing a diagonal of the upper and of the lower square faces, and by a plane parallel to this plane and 2 inches from it; the scale to be half full size.

16. The plan of a square is a rhombus whose diagonals are 3·5″ and 2″ long. Determine the length of the side of the square, and the inclination of the plane of the figure to the horizontal.

17. The depth of a coal bed at A is 525 feet; at B, which is 300 yards east and 160 yards north of A, it is 472 feet; and at C, which is 400 yards due east of A, it is 416 feet. At what depth may coal be expected at a point D 600 yards east and 200 yards north of A? The points A B C D are all in the same horizontal plane, and the coal bed may be taken to be an inclined plane.

18. Three shafts are sunk at three places at the corners of an equilateral triangle a mile apart on flat country, and the depths of a seam of coal are found to be 400, 500, and 600 feet. If the surface of the seam is a plane, draw a diagram showing the line where it reaches the surface of the ground, and find the perpendicular distances of this line from the places at which the shafts are sunk.

19. Prove that the volume of a tetrahedron is one-third of the volume of a prism on the same base and with the same height.

20. ABCD is a plane quadrilateral, and O is a point outside the plane. OA OB OC OD are divided in P Q R S so that

$$\frac{OP}{OA}=\frac{OQ}{OB}=\frac{OR}{OC}=\frac{OS}{OD}.$$

Prove that PQRS is a plane quadrilateral, and that it is similar to ABCD.

21. A B C D are any four points in space. Prove that the straight line which joins the mid point of AB to the mid point of CD intersects the straight line which joins the mid point of AC to the mid point of BD, and that both lines are bisected at their point of intersection.

Show that the straight line joining the mid points of BC and AD is also bisected at that point.

Section 52. Similarity, and Calculation of Figures.

1. A distance of 52 miles is represented in a map by a length of 12 inches. Construct a length representing 15 miles on such a map, and measure it.

2. Prove that a straight line drawn parallel to the base of a triangle, and cutting the sides, divides them proportionally.

Express the resulting proportion algebraically, assigning values to the various lengths.

Show a simple practical method of ascertaining geometrically the approximate value of the product of 3·9 and 1·7, and obtain the product.

3. Show, with proof, how to find a mean proportional between two given straight lines.

Find approximately by a geometrical construction the square root of the product of the two numbers 3·75 and 1·75.

4. D is any point in the side AB of a triangle ABC; DE, parallel to AC, meets BC in E; EF, parallel to CD, meets AB in F. Prove that $BF.BA = BD^2$.

5. Draw a square ABCD, each side measuring one inch in length. Divide each side into four equal parts, and name the points of section (and vertices) taken round the figure A E F G B H J K C L M N D P Q R. Join AN EM FL GC RH QJ PK.

Prove that this figure enables you to measure distances less than an inch expressed in sixteenths of an inch, and show how to find in your figure lines equal to 5/16 and 11/16 of an inch.

6. Draw a triangle ABC (to scale or actual size), given $AB = 7{\cdot}4$ cm., $BC = 4{\cdot}2$ cm., $CA = 8{\cdot}8$ cm. Bisect the angle ACB by a line CD meeting AB in D.

Measure AD DB and then calculate the values of the ratios AC/CB and AD/DB.

7. The angle A of a triangle ABC is bisected by a straight line AD meeting BC in D. Prove that the four lengths AB AC BD DC are in proportion.

The sides BC CA AB of a triangle are 5 6 4 centimetres in length, and D and E are points in CB and CA such that $BD = \frac{2}{5}\,BC$ and $AE = \frac{4}{9}\,CA$. AD and BE meet in O, and CO produced meets AB in F. Calculate the lengths AF FB.

8. If the vertical angle A of a triangle ABC be bisected internally and externally by straight lines cutting the base in the points D and E respectively, prove that the ratios $BD : DC$ and $BE : CE$ are each equal to the ratio $AB : AC$.

If O be the mid point of BC, prove that $OD.OE = OB^2$.

9. In a triangle ABC, D is the middle point of the side BC. DE and DF are drawn bisecting the angles ADB ADC and meeting AB and AC in E and F respectively. Show that EF is parallel to BC.

10. It is known that if the vertical angle of a triangle be bisected by a straight line which also cuts the base, the segments of the base have the same ratio which the two sides of the triangle have to one another. Make use of this proposition in order to obtain a construction for dividing a line into two parts in the ratio of $4 : 3$.

11. Define a circle, a segment of a circle, a diameter of a circle.

Draw a circle with centre O and a radius of one inch. On any diameter mark off the two points P and Q midway from the centre to the circumference.

Now from P draw any straight line PA to the circumference and produce it to B so that PB is double PA. Draw another such straight line PC and produce it to D so that PD is double PC.

Measure the distances QB and QD.

What do these results suggest?

Briefly prove the truth of your inference.

12. From a fixed point A inside a circle a straight line is drawn to meet the circle in any point P; find the locus of the mid point of AP.

13. AB BC CD DA are four rods forming a parallelogram with joints at the four vertices. The rod BC is continued beyond C to a point E.

Show that the straight line AE will always cut CD in the same point F at whatever angle AB is inclined to AD.

Prove that AE bears a constant ratio to AF.

Show that by means of this property the instrument may be made of practical use in drawing to scale.

14. A sector of a circle of 3″ radius contains an angle of 80°. Draw it and inscribe a square symmetrically in it.

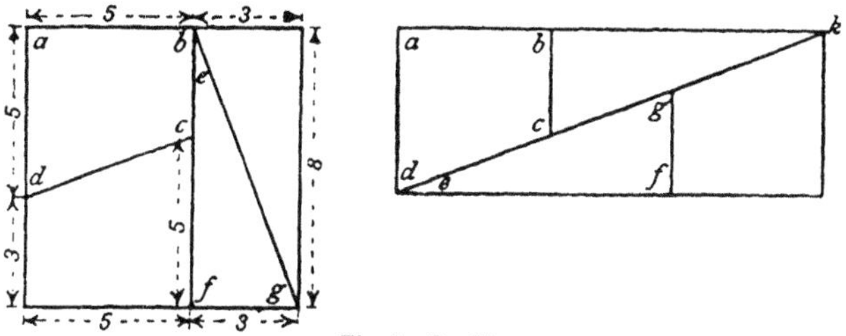

Fig. for Qn. 15.

15. In the figure the first diagram shows a square, 8 inches in the side, divided by straight lines into four parts whose forms are indicated in the figure. Calculate the angles *adc* and *feg*, and the lengths *cd* and *eg*, as accurately as four-figure tables allow.

The square is cut up into its four parts and fitted together again into the form shown in the second diagram. As the square measures 8 inches by 8 inches and the second diagram measures 13 inches by 5 inches, it appears that we have converted an area of 64 sq. inches into an area of 65 sq. inches. Show that in the second diagram what appears to be a straight line *dcgk* is really a parallelogram; find the values of one of its angles and two adjacent sides; hence calculate the area of the parallelogram, and compare with what you already know the area to be.

16. Two sides of a triangle are 13·6 cm. and 15·4 cm., and the included angle is 46°. What would be the increase in area if each of the two sides was lengthened by 0·3 cm.? State, with reasons, whether the new triangle would be similar to the old.

17. Prove by general reasoning, the accuracy of the following graphical method of determining the area of a quadrilateral ABCD: "Join BD; through C draw CE parallel to BD meeting AB, produced if necessary, in E; with centre E and radius equal to twice the unit of length, describe a circle; from A draw a tangent to this circle, to meet DX, which is parallel to AB, in X. Then the number of units of length in AX is the number of units of area in ABCD." Use it to find the area of the quadrilateral when: BD = 2 in., AB = 1·6 in., BC = 1·8 in., CD = 1·6 in., DA = 1·3 in.

18. A triangular piece of ground ABC, the sides of which are AB = 120 feet, AC = 100 feet, BC = 50 feet, has to be divided into 3 plots of equal area by fences parallel to the shortest side BC. Find, by calculation or drawing, the length of the frontage which each plot will have on a road running along the side AC (i.e. the lengths of the segments which the dividing lines intercept on AC). Do the lengths of the frontages on AC depend on the lengths of the other sides of the triangle?

What length of fencing is required to make the divisions? Give the cost of the fencing at 8 shillings a yard.

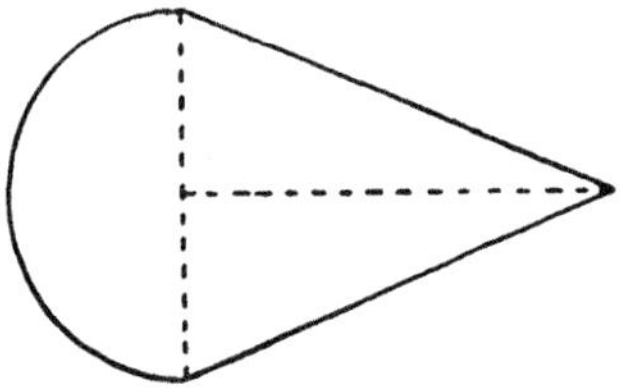

Fig. for Qn. 19.

19. Find, in sq. ft., the area of the kite shown in the figure on a scale of an inch to a foot.

Find a formula for the area of a kite of the same shape whose breadth is a feet; and state what the area would be of a kite of the same shape and breadth $3a$ feet.

20. Show that an equilateral triangle described on one of the sides of an isosceles right-angled triangle is (in area) half of an equilateral triangle described on the hypotenuse.

21. Squares are described on the sides of a triangle, and a second triangle is formed whose sides are equal to the diagonals of the squares. Prove that the two triangles are equiangular, and compare their areas.

22. Draw an isosceles triangle whose altitude is 10·5 centimetres, such that the angles at the base are each 74°. Construct geometrically an equilateral triangle equal to it, and measure its side. Justify your construction.

23. AB and CD are two chords of a circle intersecting in O. Prove that if AC BD be joined and the triangles AOC and BOD are equal, the two chords are equal.

24. If an angle of a triangle be bisected, and the bisector be produced to meet the opposite side, prove that the square on the bisector, together with the rectangle contained by the segments of the side, is equal to the rectangle contained by the sides containing the angle.

Section 53. Series.

SUBSECTION (i). DIRECT QUESTIONS ON SERIES.

1. Find to two significant figures the value of

$$3^9 - \frac{18}{1}3^8 + \frac{18.16}{1.2}3^7 - \frac{18.16.14}{1.2.3}3^6.$$

2. Find the sum of n terms of an arithmetical progression of which the first term is a and the last is z.

Find the total distance run by a competitor in a potato race. There are 12 potatoes placed 3 feet apart in a straight line, and the first is 8 feet from the starting point, to which each is to be brought separately.

3. A man started saving in 1890, and during that year he saved £100. During each succeeding year he saved £20 more than in the preceding year. Find how much he saved during 1908, and how much he saved in all by the end of that year, taking no account of interest.

4. A triangle ABC, right angled at B, has side AB 6 inches long and side BC 4 inches long. AB is divided into a number of equal parts, on which rectangles are constructed as in the figure. Find the sum of the areas of the rectangles (i) when 9 in number, AB being divided into 10 parts, (ii) when 99 in number, AB being divided into 100 parts, (iii) when $n-1$ in number, AB being divided into n parts.

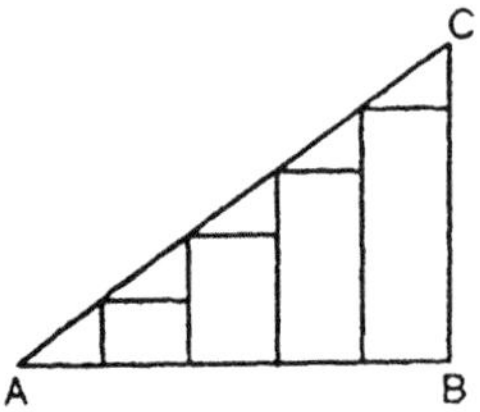

Fig. for Qn. 4.

In each case find the difference between the sum and the area of the triangle.

5. Sum to 50 terms the arithmetical series:

$$1{\cdot}5 + 1{\cdot}51 + 1{\cdot}52 + \dots.$$

6. Prove that the mean of an odd number of terms in arithmetic progression is equal to the middle term.

If the nth term of a series is given by the formula $5n+6$, find the sum of n terms.

7. An open square has like houses built at equal intervals all round it, and the houses are numbered 1, 2, 3, ..., beginning at an end of one side of the square, proceeding along that side, and going all round the square in order. If the house numbered a is exactly opposite to the house numbered b, how many houses are there round the square?

In what possible circumstances would a definite answer to this question be impossible?

8. A town council repays a loan of £70,000 in half-yearly instalments of £1000, together with simple interest at 5 per cent. paid at the time of each repayment. If the first instalment be paid at the end of a year from the date of the loan, how much will the council pay altogether?

9. The growth of a tree in any year is 9/10 of its growth in the preceding year; in January 1901 it was 10 ft. 6 in. high, and a year later it was 12 ft. high. Find the year in which its yearly growth will first be less than 0·1 in., its height at the beginning of that year, and the maximum value of the height; having given that $(9/10)^{50} = 0{\cdot}00515$.

10. Along the face of the wall of a quay there is built a pathway which slopes uniformly down to the shore. On the pathway 100 vertical rods are erected at equal intervals; at high water the rods at the two ends of the row have respectively 35 feet and 13 feet of their length immersed. How deep is the 28th rod, counting from the lower end of the row, immersed? Also what is the cost of sheathing all the rods with copper up to high-water mark, the price of the sheathing being half-a-crown per foot?

11. Twenty boys make a snow man; each boy starts with a ball of snow weighing 4 lb., which on rolling picks up in each yard one-fifth of the weight of snow which it had at the beginning of the yard. The first boy rolls his ball 2 yards, the next 3 yards, the next 4 yards, and so on. What is the whole weight collected?

[$1{\cdot}2^{10} = 6{\cdot}19$ approximately.]

12. If a person 35 years old pays a single premium of £50. 7s. 6d. into the Post Office Savings Bank, he assures that the sum of £100 will be paid on his death.

Assuming that the interest is reckoned at $2\frac{1}{2}$ per cent., and is added annually to the capital, find, to the nearest year, what is the expectation of life at that age, that is, find the number of years in which the sum paid will amount to £100.

13. A man invested £100 on 1 January 1905. On the 1st of January in each succeeding year he invests £100, and also the interest at the rate of 4 per cent. on the money he has previously invested. How much money, to the nearest shilling, will he have in this investment immediately after 1 January 1908?

14. A circle of radius r is divided into 8 equal sectors by four diameters. Through the extremity A of one of the diameters a perpendicular AM_1 is drawn to an adjacent diameter; then M_1M_2 is drawn perpendicular to the next diameter, M_2M_3 to the next, and so on. Find how far M_n is from the centre of the circle, and show that the length of the spiral formed is not greater than $2{\cdot}42r$ however large n may be.

15. Prove that the amounts at the ends of successive years of a sum of money put out at compound interest form a series in geometric progression.

Explain the phrase "sum to infinity" in the case of a decreasing geometric progression.

How many terms must be taken of the series $1+1/2+1/4+\ldots$ in order that the sum may exceed $1{\cdot}998$?

16. Without quoting any formula, find to three decimal places the sum of the series

$$1-\frac{4}{5}+\frac{16}{25}-\frac{64}{125}+\ldots$$

to eight terms.

Also find the limiting value to which the sum approaches as the number of terms increases indefinitely.

17. (i) What is the error made in taking the sum of the infinite series 1, $0{\cdot}2$, $0{\cdot}04$, $0{\cdot}008$, $\ldots$ as being $1{\cdot}248$?

Find to three decimal places, the sum of the square roots of the terms of this series (ii) taken as all positive, (iii) taken as alternately positive and negative.

18. State and prove the rule for expressing a recurring decimal as a vulgar fraction, taking as an example $\dot{}abc\dot{d}$, where a b c d are any of the digits 0 1 2 $\ldots$ 9.

Give reasons why the decimal obtained in extracting the square root of 2 cannot recur.

19. Define a geometric progression, and show how to sum a given geometric progression to a given number of terms.

Show that the quotient of $x^{12}-y^{12}$ by $x-y$ is a geometric progression. Give the number of terms and the common ratio of the progression; and find in its simplest form the sum of the first six terms.

20. Obtain a formula for finding the sum of n terms of an arithmetical progression.

Sum the series

(i) $a+3a+5a\ldots$ to 24 terms.

(ii) $1-\frac{2}{3}+\frac{4}{9}\ldots$ to 8 terms and to infinity.

SUBSECTION (ii). RIDDLE QUESTIONS.

This Subsection should be omitted when possible. *See page* v.

21. A workman who saved every year half as much again as he saved the previous year had in seven years saved £514. 15s. 0d. How much did he save the first year?

22. The output of a machine in its first year of working was valued at £800. Every year this output fell off by the same number of pounds, and the machine was cast away when it was found that the value of the output for the preceding year had fallen to £200. The value of the total output of the machine during its lifetime was £8000; find how many years it was used.

23. (i) Find the nth term and the sum to n terms of the arithmetical progression whose first two terms are a and b. (No formula may be quoted unless it is proved.)

(ii) Deduce that one term in the progression will be zero if $a/(a-b)$ is a positive integer.

(iii) If the sum of n terms of a series be $5n^2+6n$, find the nth term.

(iv) The sum of n terms of a series is $1+3n+n^2$. Write down the series.

24. Find, without quoting any formula, the sum to infinity of the geometric progression whose first two terms are a and b, where a is greater than b and both are positive. Also find the sum to infinity of the series formed by taking alternate terms of this series commencing with the first.

25. The first and third terms of a geometrical progression are 2/3 and 3/8. Find the sum of the series to 12 terms, giving the result to four decimal places.

26. The first and fourth terms of a geometric progression are 27 and -8. How many terms must be taken so that their sum may differ from the sum to infinity by less than 1/100?

27. If 7·5 be the first and 29·4 the seventh term of a geometrical progression, find the thirteenth term to the nearest integer.

28. (i) Prove that the ratio of the sum to the difference of the arithmetic and the geometric means between $(a+b)^2$ and

$(a-b)^2$ is equal to the square of the ratio of the sum to the difference of $a+b$ and $a-b$.

(ii) The first term of an arithmetic progression is 3, the last term is 9, and the sum of all the terms is 36; find the second term.

29. (i) Prove that the arithmetical mean of two positive quantities is greater than their geometric mean.

(ii) The sum of four numbers in geometric progression is 60, and the arithmetic mean of the first and last is 18; find the numbers.

30. The first, fifth, and eleventh terms of an arithmetical progression whose first term is a form the first three terms of a geometrical progression. Find the common difference of the arithmetical progression and the common ratio of the geometrical progression, and show that only one more term of the latter series is also a term of the former.

31. (i) Continue the series $1, \; 1\frac{1}{2}, \; \ldots$ in harmonic progression, and find the nth term.

(ii) Show that if
$$\frac{a}{1+ak} \qquad \frac{b}{1+bk} \qquad \frac{c}{1+ck}$$
are in harmonic progression for any value of k, they are so for all values of k.

32. If y is the arithmetic mean between x and z, and z is the geometric mean between x and y, prove that x is the harmonic mean between y and z.

33. Show how to find (i) the arithmetic mean, (ii) the geometric mean, between two given numbers.

If $1 \; a \; A$ are in arithmetic progression, $1 \; a \; G$ in geometric progression, and $1 \; a \; H$ in harmonic progression, show that $4G=(A+1)^2$, and find H in terms of A.

34. Prove that the geometric mean of a and b is also the geometric mean of the arithmetic and harmonic means of these quantities.

The arithmetic mean of two numbers is $45\frac{1}{2}$, and their harmonic mean is $38\frac{10}{13}$. Find the numbers.

35. Prove that the arithmetic mean of two quantities cannot be less than their geometric mean, and that the harmonic mean cannot be greater than the geometric mean.

The arithmetic mean of two quantities is $1\frac{3}{8}$, and the geometric mean is $1\frac{1}{3}$; find the harmonic mean.

36. If a b c be three given numbers, find an expression for x in terms of a b c such that $a+x$ $b+x$ $c+x$ may be in harmonic progression.

Verify your result when $a=1$ $b=2$ $c=4$.

37. Three numbers are in harmonic progression. If the sum of the numbers be 37, and the sum of their reciprocals be $\frac{1}{4}$, find the numbers.

38. Find the sum of n terms of an arithmetical progression whose first term is a and whose second is b.

If a_1 a_2 ... a_n are in harmonical progression, prove that

$$\frac{a_1a_2 + a_2a_3 + \ldots + a_{n-1}a_n}{a_n - a_1} = \frac{a_1a_2}{a_2 - a_1}.$$

39. (i) In an arithmetical progression the common difference is 2 and the 117th term is 3. Without assuming any formulae, find the sum of the terms from the 101st to the 133rd, both inclusive.

(ii) The middle term of an arithmetical progression of $2n+1$ terms is m. Without assuming any formulae, show that the remaining terms may be arranged in pairs each equal to $2m$.

(iii) If p q r s be in geometric progression, show that

$$(p-q+r-s)^2 = (p-q)^2 + 2(q-r)^2 + (r-s)^2.$$

Section 54. Laboratory Questions.

THESE QUESTIONS REQUIRE FURTHER APPARATUS THAN MATHEMATICAL INSTRUMENTS.

1. Clamp the metre stick along the edge of your table over a sheet of paper. Cut out of cardboard a circular disc of 6 cm. diameter. Make a small semicircular cut P in the edge of the disc just sufficient to take the point of your pencil.

Roll the disc along the metre stick, tracing out with your pencil the path taken by P, and read off two consecutive points along the metre stick where your pencil touches it. Using a piece of string, measure the length of the curve traced out by P between those two positions.

Repeat the experiment using circular discs of 5, 4, 3 cm. diameter, always cutting the smaller disc from the larger.

Draw a graph showing the relation between the length of the path traced out by P and the distance rolled by the corresponding disc.

2. Roll the given cone on a sheet of squared paper, till every point of the curved surface has been in contact with the paper; draw the boundary of the area thus traced out, and estimate the area of the curved surface of the cone.

Trace out on another portion of the squared paper the base of the cone, and by counting the squares estimate also the area of the base.

Check both your results by measuring d the diameter of the base, and h the height of the cone, and using the following formulae:

$$\text{area of circle} = 0{\cdot}7854d^2,$$

$$\text{area of curved surface} = 1{\cdot}5708d\sqrt{h^2 + \tfrac{1}{4}d^2}.$$

3. Measure with callipers the outside diameter of the piece of lead piping given you. Suspend it by a strong thread from the spring balance and obtain its weight. Now let it hang in a bucket of water and determine the apparent diminution in weight produced by the water. Record these observations in your book, and from them calculate

(i) the volume of the lead pipe;
(ii) the weight of 1 c. cm. of lead;
(iii) the internal diameter of the pipe; and
(iv) the thickness of the wall of the pipe.

Check results (iii) and (iv) by direct measurement with the callipers.

4. Draw in your book a square which will have as nearly as possible the same area as the surface of the sphere supplied to you.

State what measurements you make, and indicate clearly how you make the square from them.

5. Determine the volume of the given mass by weighing in air and in water.

Fill the bucket as full as it will hold, lower the mass into the water, collect the overflow in the cylindrical vessel supplied, measure the diameter of the vessel and the depth of the water.

Hence obtain the volume of the mass, and compare with your former result.

6. Measure the length of the given steel chain, and count the number of links.

Suspend from the spring balance the dry tin vessel and find its weight. Put the chain into the vessel, and again read the balance. Then, without taking the chain out, fill up the vessel with water to the mark on the scale, and take another reading. Lastly, remove the chain, fill the vessel up to the mark, and weigh. From these weighings and measurements find the weight and volume of a metre length of chain and the average weight of a link.

7. Hang the given tin can from the spring balance supplied to you, and note the reading of the balance. Put the given lead shot into the can while it is still suspended, counting carefully the number of shot, and again note the reading. Now set the can on the table and fill it with water up to the mark which you will find inside. Suspend the can from the spring balance again, and read the weight. Take off the can and empty the contents into the vessel supplied. See that no lead shot is left clinging to the sides. Fill the can up to the mark with water only, and again determine the weight.

From these readings determine the volume of the whole of the lead shot and the average volume of one.

8. Obtain by measurement the total area of the two faces of the given perforated metal plate.

By means of the spring balance supplied find its weight, and, making use of the principle of Archimedes, its volume.

From these results calculate the thickness of the plate, and check the thickness by means of the callipers supplied.

9. Hang up the strong spiral spring from the beam, and suspend on it weights 0·5, 1, 1·5, 2, 2·5 kg. in succession, measuring in each case the extension produced in the spring. Tabulate your readings and results, and draw a graph showing the relation between weight (y) and extension (x).

Find the extension produced by the unknown mass when it hangs (i) in the air, (ii) completely immersed in water; and by means of your graph obtain the weight and volume of the mass and the weight of a c.c. of the material.

10. Draw that part of the graph of the curve whose equation is $y=\frac{1}{4}x^2$ which lies between $x=0$ and $x=4$, and determine by weighing the ratio of the area contained between the curve, the axis of y, and the line $y=4$, to the area of the square whose sides are the axes and the lines $x=4$, $y=4$.

Take 1 inch as unit of length.

11. You are given a rectangular block having two points P and Q marked on two parallel faces. (See the figure.)

Trace in your book the face CDEF. Turn the block round CF until face CFVW rests on your paper, and trace out this face. Now turn the block back round CF to its original position. Then turn it round DE until face DEYX rests on your paper, and trace out this face. Now mark accurately the position of the points P and Q on your diagram; draw the straight line PQ cutting CF and DE in A and B respectively, and measure its length.

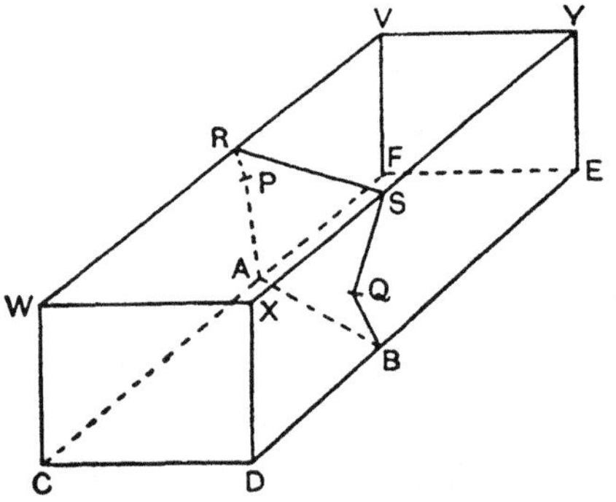

Fig. for Qn. 11.

Now trace out on your book face VWXY and construct on WV and XY respectively the rectangles CFVW and DEYX in the same manner as before. Again marking P and Q accurately, join PQ cutting VW and XY in R and S respectively, and measure the length of PRSQ.

State whether you find PABQ or PRSQ to be the shorter, and draw the shorter path along the surface of the given rectangular block.

12. You are given a rectangular block having points A B C marked on three parallel edges. Draw accurately in your book the section of the block made by a plane passing through the three points, and find its area.

Mark a point D on the block where the plane passes through the fourth parallel edge, and record its distances from the ends of the edge.

13. You are provided with certain solids. In the case of each give its name or sufficient description for identification, count the number F of its faces, the number C of corners, and the number E of edges, and fill up a table like the following:

Name of Solid	F	C	E	F+C−E

What inference seems probable from your table?

Imagine the vertex cut off the square pyramid, one face being thus added. By how many are the corners increased? and the edges? Give the value of F + C − E for this new solid. Also suppose a different corner of the square pyramid cut away, and see what effect this has on the number F + C − E.

Prove the general truth of the inference you draw from the table.

14. Draw on squared paper a segment of a circle ADB whose base AB is 16 cm. and whose height CD is also 16 cm. Estimate its area by counting the squares.

Draw any chord AQ cutting CD at P. If CP is x cm. and the area of the segment ADQ is y sq. cm., tabulate corresponding values of x and y for six cases.

Draw a graph coordinating these values and from it find x when AQ bisects the segment ADB.

Construct this line in your figure and mark this position of Q by the letter E.

15. Let A B C D be four corners of the upper surface of your table taken in order. Suppose that the vertical lines through these points meet the floor at E F G H respectively. Determine the angles EAF EAG EAH.

Draw the triangle AHF on a scale of one-sixth and measure its angles and area. Check these results by calculation. What is the volume in cubic decimetres of the figure AEFH?

16. Place two blocks in a clamp and suspend a simple pendulum between them.

Make careful observations of its length and of the time of a single swing.

Repeat for two other lengths differing from the first length and from each other as greatly as is convenient.

Draw a graph coordinating the length and the square of the number of seconds in the time, and from it deduce the length when the time of the single swing is 1·3 seconds.

17. Suspend the compound pendulum by the hole farthest from the loaded end, and making a suitable number of observations, determine the time of a single oscillation.

Calculate the length of the simple pendulum whose time of swing is the same. The time of swing of a simple pendulum of length k centimetres is $\pi\sqrt{(k/g)}$ seconds.

Check your result by suspending a simple pendulum of this length, obtaining a definite point of suspension by fixing two blocks in a clamp. Measure carefully the time of its swing, and see whether the two pendulums will swing together.

PART VII.

TYPICAL QUESTION PAPERS BY VARIOUS EXAMINING BODIES.

Section 55. London University Matriculation Examination.

ARITHMETIC AND ALGEBRA. (3 HOURS.)

1. The length, width, and depth of a box are 8·79, 4·63, and 5·26 inches respectively, to the nearest hundredth of an inch. Show, without actually multiplying out, how you would find the limits between which the volume must lie, and find the volume correct to as many significant figures as possible.

2. A manufacturer buys a machine for use in his factory, and estimates that its value decreases in each year by 10 per cent. of its value at the beginning of that year. If he sells it after three years for half its original cost, what percentage of the estimated value at the time of sale does he lose?

3. English Consols stand at $77\frac{7}{16}$ and pay $2\frac{1}{2}$ per cent., income tax at 1s. 2d. in the pound being deducted from the interest. French Rentes stand at $94\frac{1}{2}$ and pay 3 per cent. free of income tax. Find which is the better investment and, to the nearest penny, by how much the yields differ per £100 of capital invested.

4. A sphere of a certain material costs $4{\cdot}73x^3$ shillings, where x is its diameter in inches. Find the cost in francs of a sphere of diameter x centimetres, giving the multiplier to two significant figures.

£1 = 25 francs. 1 metre = 39·37 inches.

5 Give any justification you can for the statements:

$$3-(-2)=3+2, \qquad 3-(5-7)=3-5+7.$$

Find the values of x between which the expressions $4x+7$ and $3-2x$ are both positive. If a graphical method be used, the values required must be calculated subsequently.

6. Simplify the expression:

$$(2x^2+3x+1)^2-(2x^2+3x+1)(x^2+7x-6),$$

arranging the result in descending powers of x, and verify your work by calculating the values of the expression and your result when $x=2$.

7. Draw the graph of the expression $y=x^3-12x+6$ for values of x between ± 4.

Find from your graph, as nearly as you can, for what values of y there is more than one positive value of x.

8. Solve the equations:

(i) $ax+by=bx-ay=a^2+b^2$;

(ii) $\dfrac{2x+3}{x-1}-\dfrac{2-3x}{1+x}=\dfrac{25}{4}$.

9. Without quoting any formula, find the nth term and the sum to n terms of the arithmetical progression whose first and fifth terms are 3 and 19 respectively.

Find which term of the series is most nearly equal to 1000, and by how much it differs from 1000.

10. A man borrows £A at $100a$ per cent. per annum simple interest for two years, and lends it for the same time at $100x$ per cent. per annum compound interest, paid yearly. Prove that he makes a profit of £$A(x^2+2x-2a)$, and deduce, by equating this expression to zero, or otherwise, that he will lose unless x is greater than $\sqrt{2a+1}-1$.

GEOMETRY. (3 Hours.)

1. Draw carefully a triangle with AB 3 inches long, BC 6 inches, and CA 5 inches. Find points P, Q in the side BC such that (i) P is equidistant from the sides CA, AB; (ii) Q is one inch nearer to the side CA than to the side AB. State the steps of your construction, but do not prove their correctness.

Measure the distances of P and Q from the sides CA, AB.

2. Define parallel straight lines, and show that if two pairs of parallel straight lines cross each other, the opposite sides and angles of the enclosed figure are equal.

ABCDE is a convex pentagon with the side BC equal and parallel to AE, and the three sides AB, CD, DE all equal to one another. If the angle at A is 70°, find each of the other angles of the pentagon.

3. ABE is any triangle, DC drawn parallel to the base BE cuts AB, AE in D, C respectively. Show that the triangles ABC, ADE have equal areas.

Hence show how to draw a triangle of the same area as a given triangle having one of its angles the same as one of the angles of the given triangle, and having one of its sides containing that angle of a given length.

4. ABC is a triangle right angled at A. AD is drawn perpendicular to BC and produced to E so that DE = BC. The rectangle DEFC is completed, and the square ACGH, on the side of AC remote from B, described. Show by a geometrical construction that the area of the square is the same as that of the rectangle.

Show that if three right-angled isosceles triangles are described on the three sides of a right-angled triangle as hypotenuses, the largest of them is equal to the sum of the other two.

5. State and prove geometrical theorems equivalent to the algebraical identities:

(i) $a(a+b) = a^2 + ab$;

(ii) $(a+b)^2 = a^2 + 2ab + b^2$.

6. If ABC is a triangle with the angle B acute, show that AC^2 is less than $AB^2 + BC^2$.

Show how to draw accurately a rectangle whose area shall be the excess of $AB^2 + BC^2$ over AC^2.

7. Show how to complete practically a circle of which an arc is given.

AB, CD are two given finite straight lines in a plane. Show how to draw two separate concentric circles, one having AB for a chord, the other having CD for a chord. When is this impossible?

8. Draw a circle of 2 inches radius. Draw a straight line whose distance from the centre of the circle is $2\frac{1}{2}$ inches. Find a point P in the line 3 inches from the centre of the circle. Then from P draw a tangent to the circle. State the steps of your construction and prove their correctness.

Also find another point Q in the straight line, such that the length of the tangent drawn from Q to the circle may be double that of the tangent from P to the circle.

9. Show how to inscribe a regular hexagon in a given circle. State the steps of your construction and prove their correctness.

Show also that the perimeter of the hexagon is three-quarters of that of the square circumscribed to the circle.

10. The bisector of the angle A of a triangle ABC meets BC in D. At D the angle CDE is made equal to the angle BAC, DE cutting CA at E. Show that DE = BD.

Section 56. London University Senior School Examination.

ARITHMETIC. (2 Hours.)

1. Find the least whole number which contains each of the fractions $\frac{12}{35}$ and $\frac{45}{14}$ an integral number of times.

2. Multiply £327. 13s. $4\frac{1}{2}$d. by 538.

3. A book is sold to the public at 10s. 6d. and the publisher allows the bookseller 30 per cent. off the published price. How much profit does the bookseller make on the sale of 500 copies?

4. One kilogram is equal to 2·2046 lb. Find the weight nearest to a ton which contains an exact number of kilograms.

5. A cubic centimetre of copper weighs 8·93 grams, and a cubic centimetre of zinc weighs 7·12 grams. Find the weight of a cubic centimetre of an alloy containing 33 parts by weight of copper and 17 of zinc.

6. Find, to the nearest penny, the amount of £450 for 3 years at 4 per cent. compound interest.

7. If a dozen yards of ribbon cost 8s. 6d., find the cost in francs of 10 metres of ribbon.

1 metre = 1·1 yards. £1 = 25 francs.

8. Find the difference between the lengths of fences surrounding a rectangular field 63 yards long and 18 yards broad, and a square field enclosing the same area.

9. Find, to three significant figures, the value of

$$\frac{1}{\sqrt{7\cdot583}-\sqrt{2\cdot512}}$$

10. Show that the difference between a number and the sum of the digits composing the number is always divisible by 9; and explain why this is the case.

ELEMENTARY MATHEMATICS: ALGEBRA. (2 HOURS.)

1. If $x+y=1$, prove that

$$(x^2+y)^2=(y^2+x)(1-xy).$$

2. (i) Find the numerical value which must be given to a in order that

$$x^2-2x+a$$

may be divisible by $x-15$, and write down the other factor.

(ii) Find the factors of

$$ap^2+3a-a^2-3p^2.$$

3. On a certain day a watch indicates one minute past nine when a clock indicates 9 a.m., and one minute to three when the clock indicates 3 p.m. The clock is right at 9 a.m. and gains 5 minutes per day. When does the watch show the correct time, and how much does it gain or lose per day?

4. Solve

$$(a+b)x+(a-b)y=a^2+b^2,$$
$$(a^2-ab+b^2)x+(a^2+ab+b^2)y=a^3.$$

5. Calculate the value of the expression

$$x^2+5x-8$$

when $x=2\cdot5$; and if p be this value, find, to three significant figures, the values of x which make the value of the expression equal to $p+0\cdot1$.

6. Draw, with the same axes, the graphs of

(i) $y=2x+3$,

(ii) $y=x+x^2$,

and write down the values of x at the points of intersection of the graphs.

7. Find a formula for the sum of n terms of an arithmetical progression whose first term is a and whose common difference is d.

A man's yearly income is £600. He invests $\frac{1}{x}$ of it at the end of each year and allows it to accumulate at 4 per cent. simple interest. If the sum standing to his credit immediately after making the twentieth annual investment is £6624, find the value of x.

ELEMENTARY MATHEMATICS: GEOMETRY. (2 HOURS.)

In every case where a construction is made a complete explanation of the work must be given.

1. Prove that if a triangle has two equal angles it has two equal sides.

In the quadrilateral ABCD, the angles at A and C are equal and the side AD is equal to the side DC. Prove that the side AB is equal to the side BC.

2. Prove that a diagonal of a parallelogram divides it into two parts of equal area.

Two straight lines AP, CR (not produced) intersect at O, and the triangles AOC, POR are equal in area. If the parallelograms COPS, AORT are completed, prove that the line ST passes through O.

3. If a straight line be divided into two parts, prove geometrically that the square on the whole line exceeds the sum of the squares on the two parts by twice the rectangle contained by these parts.

Construct the position of the dividing point if this excess is equal to the square on half the line.

4. Prove that, if A, B, C, D are four points on a circle, the angles ABC, ADC, are either equal or supplementary.

P, Q, R, S are four points in this order on a circle, and the arcs PQ, RS are together equal to the arcs PS, QR. Prove that PR is perpendicular to QS.

5. The point A is one inch from the centre C of a circle of radius 2·3 in., and P is a point on the circumference of this

circle such that the angle ACP is 35°. Construct a circle passing through A and cutting the given circle at right angles at P, and measure its radius. Each step of your construction must be explained and justified.

6. Circles are drawn through two fixed points A, B, and tangents are drawn to them from a fixed point C, which is collinear with A and B. Prove that the locus of their points of contact is a circle whose centre is at C.

If those circles only be drawn which have their centres on one side of AB, show clearly in your diagram where the corresponding points of contact lie.

7. Show how to construct the inscribed and escribed circles of a triangle.

If the given triangle is equilateral, prove that the centres of the escribed circles form a triangle of four times the area, and that the circumscribed circle of the given triangle is the inscribed circle of this triangle.

Section 57. Board of Education. Preliminary Examination for Teacher's Certificate.

ARITHMETIC. ($1\frac{3}{4}$ HOURS.)

All the questions may be attempted, but Candidates are recommended not to attempt more questions than they can do well.

Candidates are warned that their answers must be clear and easily intelligible.

1. Find, to the nearest centime, the simple interest on 32,127 francs for 4 years at $3\frac{1}{2}$ per cent. (1 franc = 100 centimes.)

2. You are given that $271352 = 13 \times 20873\frac{3}{13}$. State, to five significant figures, the quotient when ·0271352 is divided by ·013.

3. A number of square tiles are arranged in one large square. How many tiles are used in making the large square if the total number along its two diagonals is 37?

4. A cricketer played 37 innings in Australia during the winter and 41 innings in England during the summer. His average number of runs in an innings was 18·25 in Australia

and 20·12 in England. Calculate, to two places of decimals, his average for the year.

5. Find the smallest whole number by which 12096 must be multiplied that the product may be a perfect square.

6. From a cask of spirits one-hundredth part is withdrawn and replaced by water. From the mixture one-hundredth part is drawn and replaced by water, and a similar operation is again performed. Find the percentage of spirit left.

7. A pint mug weighs 7 ounces when empty and 2 lb. 3 oz. when full of treacle. By how much does the weight of a gallon of treacle exceed the weight of a gallon of water?

8. A rectangular table is 5 feet 9 inches by 4 feet 3 inches. A rectangular cloth, 8 feet 6 inches by 3 feet 10 inches, is spread unfolded on the table so that the longer sides of cloth and table are parallel. What fraction of the area of the table top remains uncovered by the cloth, and what fraction of the cloth hangs perpendicularly over the edges of the table?

9. Twenty-five children are making cardboard trays. A square of side 8 inches is used for each tray. Find the total amount of cardboard used, if each child makes one tray.

The cardboard was bought in sheets 18 inches long and 10 inches wide. How many sheets would be required, and how much cardboard would be wasted?

10. The number of school children in a town increases by 4 per cent. in a year. How many children will there be at the end of two years if there are now 36,250 children at school?

11. Work the following questions as shortly as you can:

(1) Find the cost of 167 articles at 1s. 8d. each.

(2) Simplify $\sqrt{112} \times \sqrt{175}$.

(3) Which is the greater:

$$\frac{3\frac{1}{2}}{7\frac{1}{8}+\frac{1}{11}} \text{ or } \frac{3\frac{2}{5}}{7\frac{1}{8}+\frac{1}{10}}?$$

(4) A bicyclist travels at the rate of 12 miles an hour. How far will he travel in 5 hours $52\frac{1}{2}$ minutes?

12. Draw a rough diagram to illustrate the usual steps in the addition $\frac{1}{3}+\frac{1}{4}$, showing clearly that the sum is $\frac{7}{12}$.

MATHEMATICS I. ($1\frac{1}{2}$ HOURS.)

All questions may be attempted, but Candidates are recommended to begin with the questions which they can answer best.

Squared paper will be supplied on application to the Presiding Officer.

1. A rectangular grass plot, which is 23 feet long and 18 feet wide, is bordered by a gravel walk 4 feet wide. Find the cost of the gravel at 4s. 6d. per cubic yard if the walk is gravelled to a depth of 3 inches.

2. A railway passes in succession through three towns A, B, C. The fare from A to B is 1s. 6d., from A to C 1s. 11d., and from B to C 6d. For all journeys up to and including 20 miles the railway company charges at the rate of 1d. per mile; whilst for every mile thereafter it charges at a uniform lower rate. What is this rate?

3. A watch chain consists of 65 circular links. The outside and inside diameters of each link are 7 and 4·8 mm. respectively. Calculate the extreme length of the chain. What would be the length if one link were cut off?

4. Find the value of a in the expression $8x^2 - 3ax + 10$

(1) when $x - 2$ is a factor;

(2) when $x - 2$ divided into the expression leaves a remainder b.

5. The price of coal in a town is 5s. per ton more than at the pit's mouth, and 1 cwt. less can be bought in the town for 5s. than at the pit's mouth. Find the two prices for a ton of coal.

6. Factorize: $6x^2 + 47x - 95$, $a^2 + a - b - 2b^2 + ab$.

Simplify:

$$(p+2)^3 - (p+2)^2(3p-3) + (3p+6)(p-1)^2 - (p-1)^3.$$

7. (*a*) Solve the equations:

$$\left.\begin{array}{l}(x-1)^2 + 3(y-1)^2 = 13\\ (x-1)^2 - 2(y-1)^2 + 7 = 0\end{array}\right\}.$$

(*b*) Solve the equation:

$$\frac{3{\cdot}58}{x(6-x)} - \frac{1}{(x-6)} = \frac{2\frac{1}{3}}{x}.$$

(*c*) Is it possible for two numbers to have 80 for their sum and 2000 for their product? Give reasons for your answers.

8. A grocer sells a certain kind of tea at k pence per lb. To increase his profits he mixes $(z-1)$ lb. of this tea with 1 lb. of tea which cost him z pence per lb. less and sells the mixture at k pence per lb. as before. By how much does he increase his profit on each pound of tea sold?

9. A traveller walks a certain distance. If he had gone half-a-mile an hour faster he would have walked the distance in $\frac{4}{5}$ of the time; if he had gone half-a-mile an hour slower he would have been $2\frac{1}{2}$ hours longer on the road. Find the distance travelled.

10. Draw on squared paper a graph of the curve $y=x^2$ for values of x between 6 and -6.

Take five small divisions for your unit of x and one small division for your unit of y.

Use your graph to solve the equation $3x^2-12x=35{\cdot}54$.

MATHEMATICS II. ($1\frac{1}{2}$ Hours.)

Candidates are not limited to any particular number of questions, nor are they expected to answer them all.

Squared paper will be supplied on application to the Presiding Officer.

1. A triangle is such that the square on one side is equal to the sum of the squares on the other two sides of the triangle. Prove that one angle of the triangle is a right angle.

The sides of two triangles are 9, 12, 15, and 12, 16, 20 centimetres in length respectively. Show that the triangles can be placed side by side so as to form together a right-angled triangle, and find the area of the triangle formed in this way.

2. Show how to describe a triangle on a given base when the sum of the sides and one angle at the base are given.

Construct an isosceles triangle of perimeter 14 centimetres and altitude 4·5 centimetres. Justify your construction, and calculate from your figure the area of the triangle.

3. The end B of a straight line AB is so near the edge of a sheet of paper that the line cannot be produced in the direction AB. Give a construction by which a perpendicular can be drawn to AB at B without the use of a protractor. Justify your construction.

4. A straight line is drawn through the centre of a circle perpendicular to a chord. Prove that it bisects the chord.

The angles A, B, C, D of a quadrilateral are 60°, 70°, 120°, 110° respectively. Prove that the straight lines which bisect the sides at right angles meet in a point O.

Find the length OA if $BC = CD = 1\frac{3}{4}$ inches.

5. A, B, and C are three points in a straight line, and P is any other point; the straight lines drawn through A, B, and C at right angles to PA, PB, and PC respectively form a triangle DEF. Prove that P, D, E, and F lie on a circle.

6. The sides AB, BC, CA of a triangle ABC are 5, 6·5, 5·5 centimetres in length respectively. Describe a circle to touch CA produced, CB produced, and AB. Measure its radius.

If J be the centre of this circle and KJL be a straight line parallel to AB meeting CA produced in K and CB produced in L, prove that the triangles KAJ, LBJ are isosceles and that $CK + CL - KL = 12$ centimetres.

7. A, B, C are three points in order on a straight line. Through B any straight line is drawn and on it a point P is taken in such a way that the angle PBC is equal to twice the angle PAC. Find the locus of P.

8. A cylindrical bucket filled with moist sand measures internally 12 inches in diameter and 16 inches in height. The bucket is emptied and a conical mound is built with the sand on a base whose diameter is 18 inches. Find the volume and height of the mound and the area of its curved surface.

Section 58. Board of Education. Teacher's Certificate Examination.

MATHEMATICS—PART I. (2 HOURS.)

Candidates are not limited to any particular number of questions, nor are they expected to answer them all.

Squared paper will be supplied on application to the Invigilator.

1. The L.C.M. of two numbers is 27,643 and their H.C.F. is 359. Find the product of the two numbers, giving your reasons for the process you use.

Find the possible values of each number.

2. Find as shortly as you can the product of 61·83981 and 25·27967 correct to the first place of decimals only.

Give your reasons for the contractions used in your method of working this example.

3. A towel 6 feet in length hangs over a roller 2·8 inches in diameter. One end of the towel touches the floor, and the other end is 1 foot 6 inches above the floor. Find the height above the floor of the centre of the roller.

4. The length and breadth of a room have been measured as accurately as possible, and the measurements are known to be not more than 5 per cent. larger or smaller than the true measurements. Find the greatest possible percentage of error in the calculated area of the room.

5. A road has two rows of cottages, facing one another, on either side of it. There are 100 cottages in all, those with even numbers being on one side of the road and those with odd numbers on the other. The door of No. 1 is exactly opposite that of No. 2 and 36 feet from it, while it is 20 feet away from the door of No. 3. A postman has letters to deliver at each cottage, and takes the cottages in the order 1, 2, 3 ... 100, afterwards walking back to No. 2. Find what distance he would have saved if he had taken all the cottages of odd number first and those of even number next, finishing at No. 2.

6. The rateable value of a house is 85 per cent. of the rent. The rates are 5s. in the pound on the rateable value and, in addition, the water rate is $\frac{1}{40}$ of the rent. These two rates together with the rent come to £50. 14s. 9d. What is the rental of the house?

7. A cylindrical tin of coffee measuring internally $3\frac{3}{4}$ inches in height and $2\frac{1}{4}$ inches in diameter is sold at 5d., and one $4\frac{3}{4}$ inches in height and $2\frac{3}{4}$ inches in diameter at 9d. Compare the price of the coffee in the larger and smaller tins.

8. Each of $(3m - 20)$ boys contributed 3d. to a fund and each of $(2m - 9)$ girls contributed 2d. How many pence were collected?

Show that m must be a whole number and cannot be less than 7.

Show that the total sum will contain 1s. 1d. an exact number of times.

Find the number of girls if the total sum is £1. 15s. 9d.

9. (i) Write down the coefficient of x^6 in the product

$$(ax^5 + bx^3 + c)(x^3 + 3x + 1).$$

(ii) Solve the equation :

$$\frac{x+1}{x+2} - \frac{x-3}{x-4} = 0.$$

(iii) $$\frac{1}{8}\left\{4(x - \cdot 5) - \frac{3}{x - \cdot 5}\right\} - \frac{1}{2}(x - \cdot 5) = \frac{11}{24}.$$

(iv) Show that $(215)^4 - (157)^4$ is divisible by 58.

10. If a boy walks from his home to school at the rate of $2\frac{1}{2}$ miles per hour he arrives 10 minutes late. If he runs at 4 miles per hour he arrives 15 minutes before time.
How far did he live from the school ?

11. A class room would be a square whose sides are 18 feet long if it were not for a large fireplace which stands in one corner of the room. The front of the fireplace is 8 feet long and makes an angle of 45° with the wall of the room. Draw a plan of the class room on a scale of 5 feet to the inch, and find the area of the floor space in any way you please.

Prove that the area of the floor space of any square room with a fireplace of a similar kind is equal to the area of a rectangle whose sides are respectively greater and less than the sides of the square by half the length of the front of the fireplace.

12. An Australian sheep-farmer in riding round the boundary of his estate travels first 9 miles to the north-east, then 6 miles to south, then 8 miles to the south-east, then 20 miles to the west and then straight back to the starting point. Find, in any way you like, the total length of his ride, and the area of his property.

13. The lengths AB, AC of two adjacent sides of a triangle are given and the perpendicular distance of A from the base BC. Show that two triangles can be constructed satisfying these conditions.

Find in any way you please the difference between the lengths of their respective bases when AB = 8·1 centimetres, AC = 6 centimetres, and the distance from A to BC = 3·6 centimetres.

14. You are required to cut from a circular disc of metal the largest possible triangle having two of its angles equal to 50° and

80°. The diameter of the disc is 10 centimetres. Draw the necessary figure as accurately as you can. Explain your construction, and be careful to show sufficient construction lines to make your explanation intelligible.

MATHEMATICS—PART II. (3 HOURS.)

Candidates are not limited to any particular number of questions, nor are they expected to attempt them all. Squared paper will be supplied on application to the Invigilator. Tables of logarithms are supplied.

1. The roots of the equation $ax^2+bx+c=0$ are p and q. Find the roots of the equation $cx^2+bx+a=0$, and form an equation whose roots are $2p-1$ and $2q-1$.

2. The rate of interest on Consols is $2\frac{1}{2}$ per cent. Construct a graph to show the interest that can be obtained by investing £100 as the price of Consols falls from par to 50. Find the interest when Consols are at 77.

3. Two numbers $5x7$ and $6x3$ have the same digit x in the tens place. If their product is 363,721, find the value of x.

4. (i) Solve the equations:

$$\frac{(x-4)(x+2)}{(x+4)(x-2)}=\frac{(x+4)(x-2)}{(x-4)(x+2)},$$

$$\left.\begin{aligned}\frac{x}{3}+\frac{3}{x}&=\frac{y}{3}+\frac{3}{y}\\ x+y&=10\end{aligned}\right\}.$$

(ii) Show that $\dfrac{a+d}{1-ad}=\dfrac{b+c}{1-bc}$ if $\dfrac{a-b}{1+ab}-\dfrac{c-d}{1+cd}=0$.

(iii) Factorize: $15x^2-88x-12$.

5. In a certain district there are two evening schools. In the first, 20 per cent. of the students are adults and 50 per cent. boys; the remainder being girls. The second school contains boys and girls only. If the two schools were amalgamated there would be 8 per cent. adults and 56 per cent. boys. Find the percentages of the boys and the girls in the second school.

6. (i) Make use of your tables to discover which is the greater, $\sqrt[3]{\frac{3}{2}}$ or $\sqrt[6]{\frac{5}{2}}$.

(ii) Find approximately a number such that if it be decreased by 10 its logarithm to base 10 is decreased by ·5.

7. If a, b, c be three quantities in arithmetic progression, of which a is the least and d is the common difference, show that

$$a(b^2 - c^2) + b(c^2 - a^2) + c(a^2 - b^2) = 2d^3.$$

8. n players, A, B, C, D, etc., enter for a chess tournament, in which each player has to play one game with every other player. How many games in all must be played?

A is the best player whom neither B nor C can possibly beat. What is the greatest number of games that B and C can win between them?

9. The coefficients in the expansion of $(1+x)^n$ are C_0, C_1, C_2, C_3, etc. Prove that

(i) $r(C_{r-1} + C_r) = (n+1)C_{r-1}$;

(ii) $C_0 - C_1x^2 + C_2x^4 \ldots = (1+x)^n(C_0 - C_1x + C_2x^2 - C_3x^3 + \ldots)$.

10. Two weights A and B resting on the sloping sides of a roof are connected by a cord which passes over the ridge of the roof. Find the locus of the middle point of AB as the weights move up and down in a vertical plane.

11. A railway compartment is 7 feet from window to window and 5 feet from back to front. A passenger facing the engine notices that the sun when due south-east is shining diagonally across the compartment from the corner nearest the engine on his left. Find the angle which the direction in which he is travelling makes with a line drawn due south, and state whether he is travelling to the east or west of south.

12. A, B, C are three given points not in the same straight line. Prove that any number of equilateral triangles can be drawn having their three sides passing respectively through the three given points.

13. A straight line AB is bisected at C. Prove that the perpendicular from C on any straight line which does not intersect AB is equal to half the sum of the perpendiculars from A and B.

ABCD is a parallelogram and XAY any straight line which

passes through A and at the same time falls outside the angle DAB. If b, c, d are the distances of the points B, C, D from XAY, prove that $c = b + d$.

How would this statement require to be modified if the straight line XAY fell within the angle DAB?

14. Two circles touch externally. Prove that the common tangent at the point of contact bisects each of the external common tangents.

Two circles touch externally at A, and are touched by a straight line at B and C; show that the angle BAC is a right angle.

15. A person standing 12 yards from a river bank observes a post on the further bank directly opposite to him. He walks 30 yards parallel to the bank and then walks directly towards the post, reaching the bank after walking 15 yards. Calculate the breadth of the river, assuming the banks to be straight.

16. ABC is a triangle right angled at B whose sides AB, BC are 5 inches and 6 inches long. P and Q are points in AB, BC distant 2 inches from A and C respectively. If PQ meets AC produced in R, find the length of QR.

17. The middle points of the three edges of a cube which meet at A are B, C, and D. Prove that the triangle BCD is equilateral and calculate its area, supposing the length of an edge to be 1 foot.

If the portion ABCD be sawn off the cube, calculate the surface area of the remaining portion of the cube.

18. On September 10 the altitude of the sun above the horizon is 40° at noon. Find

(i) the length of the shadow of a post 10 feet high;

(ii) the area covered by the shadow of a wall which runs due north-east and south-west and is 25 feet long and 10 feet high.

[Draw your figures to the scale of 5 feet to an inch.]

Section 59. Cambridge University Senior Local Examination.

ARITHMETIC. (2 HOURS.)

The WORKING *of the answers is to be sent up. Answers without working will not count. The use of algebraical symbols and processes is permissible.*

1. Find the greatest common measure of 1768, 4862, and 11271.

2. (i) Simplify $\dfrac{4\frac{1}{4}-2\frac{1}{2}}{1\frac{1}{3}-1\frac{1}{4}}-\dfrac{4\frac{1}{4}\text{ of }2\frac{1}{2}}{1\frac{1}{3}\text{ of }1\frac{1}{4}}$.

(ii) Find $\dfrac{13\cdot5}{9\cdot57}\times9\cdot68$ correct to two places of decimals.

3. What will it cost to carpet a room 35 ft. 4 in. long and 27 ft. wide with carpet 27 in. wide at 3s. 4d. per yard?

4. A grocer buys sugar at £1. 7s. 1d. per cwt. and sells it at $3\frac{1}{4}$d. per lb. Calculate his gain per cent.

5. Find to the nearest penny the simple interest on £1870. 10s. for eight months at $3\frac{1}{4}$ per cent.

6. A square field contains 15 acres. Calculate the length of one of its sides in chains, correct to one place of decimals. (A chain is 22 yards.)

7. Taking £1 as worth 25·25 francs, and 97 dollars as worth 505 francs, find the value of 5000 dollars in English money.

8. One clock gains 4 minutes in 12 hours, and another loses 2 minutes in 12 hours. They are set right at noon on Monday. Find the time indicated by each clock, when the one has gained $16\frac{1}{3}$ minutes on the other.

9. A man finds that if he invests his money in 4 per cent. stock at 125 his income will be greater by £9. 7s. 6d. than if he invested it in $2\frac{3}{4}$ per cent. stock at 88. What sum had he to invest? [Brokerage is to be neglected.]

10. A man is due at a certain place at a certain time. If he walks at the rate of 3 miles an hour, he will be 15 minutes late; if he walks at the rate of 4 miles an hour, he will be 15 minutes early. Find the distance he has to walk.

GEOMETRY. (2 Hours.)

The answers to questions marked A *and* B *are to be arranged and sent up to the Examiner in two separate bundles.*

N.B.—*All construction lines should be clearly shown. Attention is called to the alternative questions* B ix, B x, B xi *at the end of the paper.*

A.

A 1. In a triangle ABC, if $\angle B = \angle C$, prove that AB = AC.

PQRS is a quadrilateral in which $\angle P = \angle Q$ and $\angle R = \angle S$. Prove that PS = QR.

A 2. Construct a quadrilateral ABCD in which AB is parallel to CD, having given that AB = 11·2 cm., CD = 5·3 cm., $\angle A = 78°$, $\angle B = 55°$.

State the steps of your construction, and find the area of the quadrilateral.

A 3. Prove that the opposite sides and angles of a parallelogram are equal, and that the diagonals bisect each other.

PQRS is a parallelogram. Prove that the internal bisectors of the angles P and Q are at right angles.

A 4. Prove that the opposite angles of any quadrilateral inscribed in a circle are supplementary.

Two circles cut at Q and Y, and through Q and Y are drawn straight lines PQR, XYZ, which cut one circle at P and X, and the other circle at R and Z. Prove that PX is parallel to RZ.

B.

B 5. If a straight line touch a circle, and from the point of contact a chord be drawn making an acute angle with the tangent, prove that this angle is equal to the angle in the alternate segment.

B 6. If in two triangles ABC and DEF, $\angle A = \angle D$, and

$$AB : AC = DE : DF,$$

prove that the triangles are similar.

In a triangle PQR, PQ = PR = 2 inches, and QR = 1 inch. In the side PQ a point S is taken such that QS = $\frac{1}{2}$ inch. Prove, *otherwise than by measurement*, that the triangle QRS is isosceles.

B 7. P is any point on the circumscribing circle of the triangle ABC. L, M, N are the feet of the perpendiculars from P on the sides of the triangle, produced if necessary. Prove that LMN is a straight line.

Given a quadrilateral, find a point such that the feet of the perpendiculars drawn from it to the four sides, produced if necessary, are collinear.

B 8. Prove that, if a straight line is perpendicular to each of two straight lines at their point of intersection, it is also perpendicular to the plane in which they lie.

From a point P two perpendiculars PM, PN are drawn, one to each of two intersecting planes; prove that the line of intersection of these planes is perpendicular to the plane PMN.

N.B.—*One or more of the following questions* B ix, B x, B xi *may be taken instead of an equal number of the questions* B 6, B 7, B 8, *but lower marks will be assigned to them.*

B ix. Prove that, if in the triangles ABC, DEF, AB = DE, $\angle A = \angle D$, and $\angle C = \angle F$, then the triangles are congruent.

In a triangle XYZ, YM is drawn perpendicular to XZ, and ZN perpendicular to XY; YM cuts ZN in P. Prove that, if PY = PZ, then XY = XZ.

B x. Illustrate and explain by means of a figure the geometrical theorem corresponding to the algebraical identity

$$a^2 - b^2 = (a + b)(a - b).$$

PQR is a triangle, the angle R being obtuse. PN is drawn perpendicular to QR produced. If $PQ^2 = PR^2 + 3QR^2$, prove that R is the middle point of QN.

B xi. Draw two circles with radii 1·5″ and ·9″ and with their centres 3·8″ apart. Construct a common tangent, and measure the length between the points of contact. State the steps of your construction.

ALGEBRA. (3 HOURS.)

Squared paper and tables of logarithms, etc., can be obtained from the Presiding Examiner.

N.B.—*Attention is called to the alternative questions* A, B, C, D *at the end of the paper.*

1. Simplify the expressions:

(a) $(x^2 - 2x - 5)^2 - (x^2 + 4x + 1)^2$:

(b) $$\frac{x(3+x)^2 - (1+3x)^2}{(1-x)^2}.$$

2. Verify that $x - 1$ is a factor of

$$12x^3 - 19x^2 - 3x + 10,$$

and find the other factors.

3. Solve the equations:

(1) $(x+5)^3 = x^3 + 35$;

(2) $$\frac{a}{x-b} + \frac{b}{x-a} = \frac{a+b}{x - \frac{1}{2}(a+b)};$$

(3) $$x(1+y) = b(a-1),$$
$$y(1+x) = a(b-1).$$

4. In a certain town eggs are being sold at $2x$ pence a dozen, and in another town they are sold at x eggs for a shilling. By buying six dozen eggs in the latter town and selling them in the former a profit of 1s. is made: find the buying and selling prices of the six dozen eggs.

5. An alloy is made of 850 grammes of copper, 100 of tin and 50 of zinc; find how much copper and how much tin should be melted with the alloy to produce a mixture of which two-thirds is copper, and in which there is 650 grammes more copper than tin.

6. Write down and prove the formulae for the sum and product of the roots α, β of the quadratic $ax^2 + 2bx + c = 0$, in terms of the coefficients a, b, c.

Find the quadratic whose roots are $\alpha^2 - \alpha\beta$, $\beta^2 - \alpha\beta$.

7. Plot the following values of x and y in a graph:

x	0	1	2	3	4
y	1·5	3·9	4·5	5·7	6·7

There is reason to suppose that one pair of the given values is wrong. Determine which it probably is, and verify that the others satisfy an equation of the form $y = ax^2 + bx + c$, finding the values of the coefficients a, b, c.

8. Find the sum of the first n terms of an arithmetical progression whose first term is a and whose common difference is d.

Prove that, if the sum of the first n terms of a series is equal to $bn^2 + cn$, the series is an arithmetical progression, and find the first term and common difference.

9. Establish the formula for the number of combinations of n things taken r at a time.

How many different sets of five letters can be chosen from the word *algebraic*?

10. Write down the first four terms, and the ratio of the $(r+1)$th to the rth term, in the binomial expansion of $(1+\frac{1}{20})^{16}$.

Calculate, to the nearest shilling, the compound interest on £100 for 16 years at five per cent.

11. Write down the series for e^x in powers of x, and verify by direct multiplication that the product of this series by the series for e^{-x} is equal to unity, as far as the terms in x^4.

Apply the series to calculate e^x for $x = \frac{1}{4}$ to four decimal places, and compare the result so found with that given by the table provided. [The approximate value of e is 2·718.]

12. If the coefficient c is such that $x = a$ is a root of the equation

$$x^3 - 3x + 1 = cx(x-1),$$

verify that $1 - \frac{1}{a}$ is another root, and find the third root in terms of a.

N.B.—*Any of the following questions may be taken instead of any of the questions* 9, 10, 11, 12, *but considerably lower marks will be assigned to them.*

A. Find two factors of the type $2x^2 + ax + b$ of the expression

$$4x^4 + 11x^2 + 9.$$

B. Solve the equations:

(1) $\frac{x}{a}+\frac{y}{b}=2, \quad \frac{3x+y}{3a+b}=\frac{x+3y}{a+3b};$

(2) $\frac{12}{x-1}+\frac{24}{x-2}+35=0.$

Verify your solutions by substitution.

C. A man pays income tax at 1s. in the £ on unearned income and at 9d. in the £ on earned income: his earned income is £200 more than his unearned, and the total income tax paid is £29. 7s. 6d. Find his total income.

D. Plot from the tables a graph of

$$y=10\log_{10}x$$

from $x=1$ to $x=7$, and find approximately where the graph cuts $y=x$.

Section 60. Oxford University Senior Local Examination.

ARITHMETIC. (1¾ HOURS.)

All necessary work must be shown. No credit will be given for answers without sufficient work.

1. What is the smallest sum of money that must be added to £29. 2s. 6d. to make an exact multiple of £2. 1s. 6d.?

2. What decimal of £7 is $\frac{3}{4}$ of £1. 3s. 11d.?

3. Find the side of a square lawn whose area is 2025 sq. yd. How many times must a man go *completely* round it so as to have walked at least a mile?

4. An empty tank 40 centimetres long, 30 centimetres wide, and 25 centimetres high, was filled with water from a tank a metre long and 60 centimetres wide. If the depth of the water in the larger tank is now 35 centimetres, what was it at first?

5. A man puts money into a $3\frac{1}{2}$ per cent. stock at 96. Find, correct to two decimal places, the rate per cent. which he makes on his money.

6. If £1. 6s. 11d. is divided between 8 boys and 5 girls so that a boy has half as much again as a girl, what does each receive?

7. A man, having contracted to construct a railway in one year, employs 1000 men for $4\frac{1}{2}$ months and then finds he has done $\frac{3}{10}$ of the work. How many more men must he employ for the remaining time in order to complete his contract as agreed?

8. A man walking along a road at $3\frac{1}{2}$ miles an hour is overtaken by a motor-car. From the moment it passes him to its disappearance round a corner he takes 27 steps, and walking on reaches the corner with 135 steps more. Assuming that he walks uniformly, find the speed of the car.

9. During one year the population of a town increased 5 per cent., and during the next year it diminished 5 per cent. If at the end of the second year the population was 49,875, what was it at the beginning of the first?

GEOMETRY. ($1\frac{3}{4}$ HOURS.)

No credit will be given for any attempt at a question in Practical Geometry if any of the construction lines are erased. The steps of the construction must be described in words.

1. If two triangles have two angles of the one equal to two angles of the other, each to each, and any side of the one equal to the corresponding side of the other, show that the triangles are equal in all respects.

Under what conditions may two triangles be shown to be congruent?

If ABC, DEF are two triangles having $AB = DE$ and $AC = DF$, and $\angle ABC = \angle DEF$, prove that the angles ACB and DFE are either equal or supplementary.

2. What is the locus of a point which is equidistant from two intersecting straight lines? Prove the correctness of your answer.

Draw two straight lines intersecting at an angle of 60°, and find a point which is at a distance of 5 cm. from one of the lines and 7 cm. from the other.

3. Prove that the angle at the centre of a circle is double of the angle at the circumference on the same arc.

Construct a triangle of base 2 in. and vertical angle 60° so that the bisector of the vertical angle may cut the base at a point $1\frac{1}{2}$ in. from one extremity.

4. Describe a circle about a triangle whose sides are 3·2 in., 4 in., 5 in. respectively, and measure its radius in millimetres.

Prove geometrically that the radius of the circumscribed circle of an equilateral triangle is equal to the diameter of the inscribed circle.

5. Prove that, if two chords of a circle intersect *outside* the circle, the rectangle contained by the parts of the one is equal to the rectangle contained by the parts of the other.

6. Prove that, if a straight line is drawn parallel to one of the sides of a triangle, it divides the other two sides proportionally.

Prove that three parallel straight lines cut any two intersecting straight lines proportionally.

ALGEBRA. ($1\frac{3}{4}$ HOURS.)

All necessary work must be shown. No credit will be given for answers without sufficient work.

Each candidate will be supplied with one piece of squared paper.

1. Prove that $a - x + y$ is a factor of

$$(a-x)^2 + (2y-z)(a-x) + y^2 - yz.$$

2. If $F(x) = 3x^2 - 2x + 1$ and $f(x) = 4x^2 - 3x - 2$, find the value of $3F(3) - 2f(2)$.

3. Simplify:

$$(1)\quad \frac{a-2b}{a^2+2ab-3b^2} - \frac{a-3b}{a^2+ab-2b^2};$$

$$(2)\quad \frac{x^2+2xy+y^2}{x^2+xy+y^2} \times \frac{x^3-y^3}{x^2+xy} \div \left(1 - \frac{y^2}{x^2}\right).$$

4. Solve the equations:

$$(1)\quad \frac{x^2+1}{x} = \frac{a^2+1}{a};$$

$$(2)\quad 2x - y = 5, \quad xy = 12.$$

5. The areas of a side wall, an end wall, and the ceiling of a rectangular room are respectively 270, 184, and 345 square feet. Find the length, breadth, and height of the room.

6. (1) If α and β are the roots of the equation

$$x^2 - px + q = 0,$$

prove that $\alpha + \beta = p$ and $\alpha\beta = q$.

(2) Form the quadratic equation whose roots are the squares of the roots of the equation $x^2 = ax + b$.

7. The first term in an arithmetical progression is 169 and the common difference is -13; how many terms can there be in the series if the sum of all the terms amounts to 988? Account for the double answer.

8. How many different sets of 6 boys can be selected out of 10 boys, if one particular boy must be included and another particular boy excluded?

9. (1) Prove that the coefficients of terms equidistant from the beginning and the end in the expansion of $(a+b)^n$ by the Binomial Theorem are equal.

(2) Find the coefficient of x in the expansion of

$$\left(x^2 + \frac{1}{x}\right)^5.$$

10. Draw the graph of $y = x + \frac{16}{x}$ between

$$x = -8 \quad \text{and} \quad x = +8.$$

Find from the graph the values of the two factors of 16 whose sum is the least possible.

No marks will be given for a solution obtained by any other method, nor unless the axes and the units used are clearly indicated.

Section 61. Matriculation Examination of the Northern Universities (Manchester, Liverpool, Sheffield and Leeds).

ARITHMETIC AND ALGEBRA. (2 HOURS.)

1. The earth's circumference, measured along a meridian, is 40 million metres, or 21,600 nautical miles. Assuming that a metre = 39·370 inches, find the number of feet in a nautical mile.

2. What is the present value of an estate bringing in an annual net rent of £275. 15s. per annum, the first payment being due a year hence, and compound interest being reckoned at £3. 6s. 8d. per cent.?

Also, what is the present value if the first payment is to be made three years hence?

3. Find the factors of

$$6x^2 - 5xy - 6y^2, \quad x^4 - x^2y^2 + 16y^4, \quad x^4 - 9x^2 + 30x - 25.$$

4. (i) A certain sum of money would pay A's wages for a days, or B's wages for b days, or C's wages for c days. How long will the money last if A, B, C are employed together?

(ii) A man rows down stream a certain distance and returns. His speed of rowing relative to the water is m miles an hour, and the current is n miles an hour. Find the ratio of the time he takes to what it would have been if there had been no current.

5. State and prove the rule for forming the square of an expression of the form $a + b + c + \dots$

Find the square root of $(x^2+1)^2 - 4x(x^2-1)$.

6. At an election there were 25 electors, each of whom voted for two out of the three candidates A, B, C. A received 18, B 17, and C 15 votes. How many electors voted for each pair of candidates?

7. (i) Prove that the expressions $3x^2 - 5x - 1$ and $2x^2 - 3x - 4$ cannot be equal for any real value of x. For what value of x will the difference be least?

(ii) Solve the equations:

$$\left.\begin{aligned} xy + 2(x-y) &= 4, \\ x + y - 3xy &= 5. \end{aligned}\right\}$$

8. The distances which a man travels in successive days are in arithmetical progression. At the end of the fifth day he has gone 100 miles, and at the end of the seventh 150 miles. How long will he take to complete 300 miles in all?

9. (i) Find the ratio of the sum of the squares of the first n terms of the series $a + ax + ax^2 + \dots$ to the sum of the first n terms.

(ii) Prove that

$$1 - \frac{x}{1-x} + \left(\frac{x}{1-x}\right)^2 - \dots = 1 - x$$

for all negative values of x. For what positive values of x does the statement hold?

GEOMETRY. (2 HOURS.)

1. State and prove a construction for drawing a line which shall be perpendicular to a given line from a given point without the line.

P, Q are given points on opposite sides of a given line KL. Find a point R in KL so that KL shall bisect the angle PRQ.

2. Define a *parallelogram*, and prove that (1) each diagonal bisects the figure, (2) the diagonals bisect one another.

Show how to trisect a given parallelogram by lines drawn through a corner.

3. ABC is a triangle, right angled at A, and AD is perpendicular to the hypotenuse BC. Prove that the square on AB is equal to the rectangle whose sides are equal to BC, BD.

If the sides AB, AC are 2·8, 9·6 centimetres respectively, calculate the lengths of BD, DC, AD.

4. Divide a given straight line into two parts so that the square on one part may be equal to the rectangle contained by the whole line and the other part.

Prove that, if the line be so divided, the sum of the squares on the whole line and on one part is equal to three times the square on the other part.

5. Investigate the locus of the mid points of (1) parallel, (2) equal, (3) concurrent chords of a circle.

6. If PQ and RS are two chords of a circle intersecting at a point O within the circle, prove that the rectangles PO.OQ and RO.OS are equal.

7. ABC is a triangle, AD bisects the angle BAC and meets BC in D. Prove that BD : DC = BA : AC.

If I is the centre of the circle inscribed in the triangle, prove that AI : ID = BA + AC : BC.

8. Prove that the areas of similar triangles are in the same ratio as the squares on corresponding sides.

ABC is a triangle and BE, CF are drawn perpendicular to the opposite sides. Prove that the triangle AEF is similar to ABC; and if the angle at A be 60°, compare the areas of these triangles.

Section 62. Central Welsh Board. Senior Certificate.

ARITHMETIC. (3 HOURS.)

1. Divide $(12\frac{19}{27} - 3\frac{3}{8})$ by $(6\frac{5}{8} + 2\frac{13}{18} - 4\frac{7}{144})$, and express the quotient in decimals correct to three places.

2. Find, correct to three places of decimals, (1) the square, (2) the cube, (3) the square root of 2·71828.

3. A merchant bought a certain quantity of coal and sold one-half of it at a price which was 10 per cent. above the cost price, the remainder being sold at 19s. 2d. per ton. If, on the whole transaction, he gained $12\frac{1}{2}$ per cent., find what was the cost price per ton.

4. The water supply of a district containing 750,000 inhabitants is drawn from a lake whose area is 10,000 acres. Allowing 26 gallons per day for each person, find how many days' supply are represented by a depth of 1 foot of water in the lake, taking a gallon to contain 277 cubic inches.

5. Find the weight to the nearest lb. of a cubic foot of iron which weighs 7·8 grammes per c. cm., taking a metre to be 39·37 inches and a kilogramme to be 2·204 lb.

6. A circular plate of metal of uniform thickness, and 2 feet in diameter, is pierced by 20 circular rivet holes, all of the same diameter. If, by the piercing of the holes, the weight of the plate is reduced by $1\frac{1}{4}$ per cent., find the diameter of each hole.

7. A contractor undertook to carry out a certain work in 100 days reckoned from the date of the beginning of the work. He employed 160 workmen, and after 30 days one-third of the work was done. Owing to unforeseen difficulties work was suspended for 30 days, and it was found that 2 men would be required to do as much work in a day as was at first done in a day by 1 man. How many additional workmen must be employed in order to finish the work within the stipulated time?

8. A sum of £1000 is lent on condition that it shall be repaid, together with compound interest at 3 per cent. per annum, in two equal instalments at the ends of the second and third years respectively. What is the amount of each instalment?

ELEMENTARY MATHEMATICS: GEOMETRY. (3 HOURS.)

1. Show how to construct a quadrilateral ABCD whose opposite sides AB and CD are parallel, given the lengths of the parallel sides and the lengths of the diagonals AC and BD.

Construct the quadrilateral when the lengths of AB, CD, AC, BD are respectively 3 inches, 2 inches, 3·4 inches, and 4·1 inches, and measure the lengths of the sides BC and DA.

2. Prove that the straight line joining the middle points of two sides of a triangle is parallel to the third side and equal in length to half that side.

Prove that the medians of a triangle meet in a point.

3. How many conditions must be satisfied in order to secure that two triangles shall be equal in all respects? State, without proof, three different sets of conditions for the equality of two triangles.

O is a fixed point and P is any point on a fixed straight line which does not pass through O. If Q is the vertex of an equilateral triangle OPQ described upon OP, find the locus of Q.

4. Prove that the square on one side of a triangle is greater or less than the sum of the squares on the other two sides by twice the rectangle contained by either of these sides and the projection of the other on it, according as the angle contained by the two sides is obtuse or acute.

Prove that twice the sum of the squares on the medians of a right-angled triangle is equal to three times the square on the hypotenuse.

5. P is a given point within a given circle. Prove that the rectangle contained by the segments of any chord of the circle drawn through P is constant.

If the radius of the circle is 1·8 inches and the distance of P from the centre is 1 inch, construct geometrically, without extracting a square root, a chord through P such that one segment of the chord shall be double the other. Measure the length of the chord.

6. Prove that every parallelogram inscribed in a circle must be a rectangle.

A, B, C, D, E, F are the angular points taken in order of a six-sided figure ABCDEF inscribed in a circle. If AB is parallel to DE, and BC parallel to EF, prove that CD is parallel to FA.

7. Give and prove a construction for describing a circle which shall pass through two given points and touch a given straight line. Consider the number of solutions.

8. Prove that the bisector of the vertical angle of a triangle divides the base of a triangle in the ratio of the sides.

Construct a triangle, given that the base is 9 cm. long, the vertical angle is 49°, and that the other two sides are in the ratio of 2 to 1.

Construct the circle inscribed in the triangle, and measure its radius.

ELEMENTARY MATHEMATICS: ALGEBRA. (3 Hours.)

1. Resolve into four factors:

$$(x^2 - 7x)^2 + 4(x^2 - 7x) - 96.$$

Find the remainder when $ax^2 + bx + c$ is divided by $x - h$.

2. (1) If $x^2 + 2x = 1$, prove that

$$x^4 - 6x^2 + 1 = 0;$$

(2) If $x^2 + y^2 = 1$, prove that

$$\frac{x+y-1}{x+y+1} + \frac{x+y+1}{x+y-1} = 2\left(1 + \frac{1}{xy}\right).$$

3. Solve the equations:

(1) $$\frac{2x-3}{3x+4} + \frac{3x+4}{2x-3} = 5\tfrac{1}{6};$$

(2) $$\frac{x-a}{a-b} + \frac{y+3a}{a+b} = 3, \quad x + y = a + b.$$

4. Draw the graph of the equation: $y = x^2 - 6x + 3$.

Determine from your drawing between what values x must lie in order that $x^2 - 6x + 7$ may be less than 2.

5. Find two positive fractions such that their sum is unity and the sum of their squares is $\frac{5}{8}$. Give your answers correct to two places of decimals.

6. The sum of 11 terms of an A.P. is 17, and the sum of 19 terms is 13. Find the first term, the common difference, and the sum of n terms.

7. If $a : b = b : c$, prove that

$$\frac{a+b}{a-b} + \frac{c+b}{c-b} = 0.$$

If $\frac{a^2+c^2}{b^2+d^2} = \frac{ac}{bd}$, prove that $\frac{a}{c}$ is equal to $\frac{b}{d}$ or $\frac{d}{b}$.

8. By selling articles at the rate of a articles for b pence a dealer makes a profit of c per cent. What would be his percentage of profit if he sold them at the rate of b articles for a pence?

Section 63. Scotch Education Department. Leaving Certificate Examination.

MATHEMATICS. LOWER GRADE: FIRST PAPER. (2 HOURS.)

Before attempting to answer any question, candidates should read the whole of it very carefully, since time is often lost through misapprehension as to what is really required.

All the figures should be accurately drawn, and, where geometrical constructions are employed, the construction lines should be shown. Proofs of geometrical constructions need not be given unless specially asked.

The value attached to each question is shown in brackets after the question. In addition 10 *marks are allowed for neatness and good style.*

SECTION I. *All the questions in this Section should, if possible, be answered.*

1. Prove that two triangles which have two sides and the contained angle of the one equal to the corresponding parts of the other are equal in all respects.

In the triangles LMN and PQR the sides LM, LN and the angle MLN are respectively equal to the sides PQ, PR and the angle QPR; O is a point in MN, and S is a point in QR such that MO is equal to QS: prove that LO is equal to PS. (18)

2. Prove that parallelograms on the same base and between the same parallels are equal in area.

Construct a parallelogram with an area of 14 square centimetres and with two adjacent sides of 5 centimetres and 6 centimetres, explaining your method. (20)

3. Prove that, if two chords of a circle are equal, they are equally distant from the centre.

The radius of a circle is 6·5 centimetres. Two chords are respectively 5 cm. and 7·8 cm. in length. Calculate their distances from the centre, testing your result by a carefully drawn figure. (20)

SECTION II. *One and only one question out of each of the pairs of alternatives should be answered.*

Either—**4***a*. ABC is a right-angled triangle, C being the right angle; AC is produced through C to F and BC through C to G, so that CF = CB and CG = CA. If X is the mid point of FG, prove that XC is at right angles to AB. (16)

Or—**4***b*. Two straight lines AB, AC meet at A, and from D, a fixed point in AC, DE is drawn parallel to the bisector of BAC. If P be any point in DE from which perpendiculars are drawn to AB and AC, prove that the difference between the lengths of these perpendiculars is a constant quantity. (16)

Either—**5***a*. ABC is a triangle and points D, E are taken in AB, AC such that AD = $\frac{1}{2}$AC, AE = $\frac{1}{2}$AB; prove that a circle can be described through the points B, C, E, D. (16)

Or—**5***b*. A circle is inscribed in a triangle PQR and touches QR at L, RP at M, PQ at N. If the angles Q, R are respectively 50°, 62°, calculate the size of each of the angles of the triangle LMN. (16)

MATHEMATICS. LOWER GRADE: SECOND PAPER. (2 HOURS.)

Before attempting to answer any question, candidates should read the whole of it very carefully, since time is often lost through misapprehension as to what is really required.

Square-ruled paper is provided for the graphical work.

The value attached to each question is shown in brackets after the question. In addition 10 marks are allowed for neatness and good style.

SECTION I. *All the questions in this Section should, if possible, be answered.*

1. Show that

$$(ax + by)^2 + (bx - ay)^2 = (a^2 + b^2)(x^2 + y^2).$$

Resolve into factors:

(1) $15x^2 - 16x - 15$;

(2) $x^2 - (a - b)x - ab$. (14)

2. Divide $6 - 17x - 14x^2 + 32x^3 + 5x^4 - 6x^5$ by $2 - 5x - 3x^2$; and verify your answer when $x = 1$. (14)

3. Solve the equations:

(1) $\frac{1}{2}(x - \frac{1}{3}) - \frac{1}{4}(x - \frac{1}{5}) = \frac{1}{6}$;

(2) $ax - by = c,\ ax + by = d$;

(3) $6x^2 = x + 1$. (15)

4. Solve the following problems algebraically:

(1) £100 was divided between A and B, so that after each had spent £6, A had three times as much money left as B: find their shares.

(2) In x cricket innings a boy made on the average $9\frac{1}{2}$ runs; in his next two innings he made 17 and 52 runs respectively, and now his average is $14\frac{1}{2}$ runs: find x, assuming that the boy was out at the end of each innings. (15)

SECTION II. *One and only one question out of each of the pairs of alternatives should be answered.*

Either—**5a.** If $ax^2 + bx$ is equal to 4, when $x = 2$, and is equal to $7\frac{1}{2}$, when $x = 3$, what is its value when $x = 5$? (15)

Or—**5b.** Simplify:

(1) $\frac{2a - x}{2a + x} - \frac{4a^2 + x^2}{4a^2 - x^2} + \frac{4ax^3}{16a^4 - x^4}$;

(2) $\frac{1}{x^2 - 3x + 2} - \frac{4}{x^2 + 2x - 3} + \frac{5}{x^2 + x - 6}$. (15)

Either—**6a.** Draw the graphs of

(1) $xy = 4$, (2) $3x + 4y = 16$,

from $x = 1$ to $x = 4\frac{1}{2}$, taking one inch as your unit.

Find from your diagram the values of x and y which satisfy both equations, and verify your results by substituting these values in the equations. (17)

Or—**6b.** ABC is a triangle, in which AB is 20 feet long. From C, CM is drawn at right angles to AB, meeting AB produced in M.

If BM is 6 feet and CM 15 feet, draw the triangle to scale on your square-ruled paper, and calculate

(1) the area of the triangle,

(2) the lengths of AC, BC correct to one decimal place.

If from A a perpendicular AN is drawn to CB produced, calculate the length of AN to one decimal place. (17)

MATHEMATICS. LOWER GRADE: THIRD PAPER. (2 HOURS.)

Before attempting to answer any question, candidates should read the whole of it very carefully, since time is often lost through misapprehension as to what is really required.

All the working must be shown in its proper position in the answer, and the different steps of the calculation should be shortly indicated in words.

The value attached to each question is shown in brackets after the question. In addition 6 *marks are allowed for neatness and good style.*

1. Passengers by a certain line of steamers are allowed 20 cubic feet of luggage. A passenger wishes to take two boxes, each 3 feet long and $2\frac{1}{2}$ broad, and of the same height. What is the greatest height he can have them made so as not to exceed the limit allowed? (7)

2. A father leaves £5600 to be divided between his sons, aged 30, 34, and 41, so that their shares may be in the same ratio as their ages. Find the share of each. (7)

3. A tank is 2·45 metres long, 1·46 metres broad, and ·56 metre deep. Find its volume in cubic metres, and also how many litres it contains, a litre being equal in volume to a cubic decimetre. (9)

4. From January to July of last year a Scottish railway company sold 821,670 first-class tickets and 18,150,737 third-class tickets, receiving £60,966. 0s. 8d. for the first-class and £614,993. 6s. for the third-class. Find the average costs of a first-class and a third-class ticket, correct to the nearest farthing. (9)

5. A building is insured for an annual premium equal to $\frac{9}{80}$ per cent. of the value of the building. If this premium is £5. 2s. $4\frac{1}{2}$d., find the value of the building. (9)

6. The following table gives the times at which a train passed the different stations (A, B, etc.), and the distance in miles of each station from the starting point. Taking a horizontal inch on the square ruled paper to represent 20 minutes, and a vertical inch to represent 10 miles, draw a curve to show the relation between the time spent and the distance travelled; and find from your curve between which stations the train went most quickly and between which it went most slowly, explaining how you come to your conclusion:

	A	B	C	D	E	F	G	H	I
Time	9.0	9.16	9.24	9.45	10.0	10.10	10.24	10.48	10.54
Distance	0	5	8	17	25	30	39	58	62

(13)

MATHEMATICS. HIGHER GRADE: FIRST PAPER. (2 HOURS.)

Before attempting to answer any question, candidates should read the whole of it very carefully, since time is often lost through misapprehension as to what is really required.

The value attached to each question is shown in brackets after the question. In addition 10 marks are allowed for neatness and good style.

SECTION I. *All the questions in this Section should, if possible, be answered.*

1. Draw a geometrical diagram illustrating the algebraical identity

$$(x+y)(x-y) = x^2 - y^2;$$

explaining how your figure shows that the identity has been verified.

ABC is a triangle in which AB = AC, and D is a point in BC produced. Prove that $BD.CD = AD^2 - AB^2$. (18)

2. If a quadrilateral is inscribed in a circle, prove that the sum of each pair of opposite angles is two right angles.

A variable circle passes through Q, R, the ends of the base of the triangle PQR, and cuts PQ, PR in S, T. Show that ST is fixed in direction; and employ this property to draw the tangent at P to the circle circumscribing PQR, stating and justifying your construction. (18)

3. If a tangent and a secant are drawn to a circle from an external point, prove that the square of the tangent is equal to the rectangle contained by the segments of the secant intercepted between the point and the circle.

OC is a radius of a given circle, whose centre is O; A and B are points on OC and OC produced such that

$$OA.OB = OC^2.$$

If a circle passes through A and B, and cuts the given circle in E and F, prove that OE and OF are tangents to this circle. (18)

SECTION II. *One and only one question out of each pair of alternatives should be answered.*

Either—**4***a*. Prove that the internal and external bisectors of the vertical angle of a triangle divide the base into segments which are in the ratio of the sides of the triangle.

ABC is a triangle whose sides BC, CA, AB are respectively 16, 9, 13 inches in length; the bisectors of A meet BC and BC produced in D and E. Calculate, from a rough figure, the lengths of BD, BE. (18)

Or—**4***b*. If two triangles have one angle of the one equal to one angle of the other, and the sides about these angles proportional, prove that the triangles are similar.

A is a fixed point and P is a variable point, both on the circumference of a fixed circle. Q is a point in AP such that $AQ = \frac{2}{5}AP$. Prove that as P moves round the given circle, Q traces out another circle, and find the centre and radius of this circle. (18)

Either—**5***a*. Prove that a straight line, which is perpendicular to each of two intersecting straight lines at their point of intersection, is perpendicular to the plane in which they lie.

Find the locus of a point which is equidistant from two given parallel planes, and also equidistant from two given points. (18)

Or—**5***b*. State the volume of a tetrahedron in terms of the area of its base and the altitude of its vertex.

OA, OB, OC are concurrent edges of a cube and are each one inch in length. Find the volume of the tetrahedron OABC as a fraction of a cubic inch.

If the corners of the cube opposite O, A, B, C are O′, A′, B′, C′, show that it is possible to form a regular tetrahedron by removing from the cube the tetrahedron OABC, and the equal tetrahedra whose vertices are A′, B′, C′. Find the volume of this regular tetrahedron and the length of its edge. (18)

MATHEMATICS. HIGHER GRADE: SECOND PAPER. (2 HOURS.)

Before attempting to answer any question, candidates should read the whole of it very carefully, since time is often lost through misapprehension as to what is really required.

Square-ruled paper and four-place logarithmic tables are provided.

The value attached to each question is shown in brackets after the question. In addition 10 *marks are allowed for neatness and good style.*

SECTION I. *All the questions in this Section should, if possible, be answered.*

1. A man pays income-tax at the rate of 1s. 2d. in £1 on one-third of his income, and at the rate of 9d. in £1 on the rest. If his net income is £1146. 13s. 4d., find how much income-tax has been deducted. (13)

2. Find the factors of:

(1) $18x^2 - 9xy - 20y^2$;

(2) $a^4 + a^3b - ab^3 - b^4$;

(3) $(pq + rs)^2 + (pr - qs)^2$. (13)

3. Solve the following equations, verifying your solutions:

(1) $x^2 + bc = bx + c^2$;

(2) $\sqrt{3x+2} + \sqrt{3x-1} = \sqrt{6x+5}$;

(3) $\frac{1}{5}(x^2 + y^2) = xy - y^2 = 1$. (14)

4. Taking an inch as unit, draw the graphs of

$$y = \tfrac{1}{5}x^2, \quad y = \frac{4}{x+2}$$

from $x = -1$ to $x = 4$.

Write down the coordinates of the point of intersection as determined by your figure.

Show that the value of x at the point of intersection is a root of the equation:

$$x^3 + 2x^2 = 20.$$ (14)

SECTION II. *One and only one question out of each pair of alternatives should be answered.*

Either—**5***a*. Three motor-cars leave Edinburgh at 9, 9.30, and 10 a.m., travelling along the same road at uniform rates. The second travels 3 miles an hour faster than the first, and the third 5 miles an hour faster than the second. At a certain time the first and second are the same distance from Edinburgh and the third is one mile further from Edinburgh than they are. Find this time and the rates at which they have travelled. (18)

Or—**5***b*. ACB is a semicircle of which the diameter AB is 5 inches in length, and D is the projection of C on AB; find the length of AD when C is so chosen that the sum of the lengths of AD and CD is 6 inches.

Show that there two positions of D which satisfy this condition, and find the length of CD in each case. (18)

Either—**6***a*. If $\frac{a}{b}$ is a fraction, in which a and b are both positive quantities, and a is less than b, prove that its value becomes more nearly equal to unity when the same positive quantity is added to both numerator and denominator.

OX, OY are the positive directions of the axes of coordinates and OR is the bisector of the angle YOX. P is a point within the angle ROY whose coordinates are x, y. From P, PQ is drawn parallel to OR, and the coordinates of Q are x', y'. Show that the fraction $\frac{y'}{x'}$ is got from $\frac{y}{x}$ by adding the same quantity to both numerator and denominator, and prove from your diagram that its value is nearer to unity than that of $\frac{y}{x}$ is. (18)

Or—**6***b*. Find an expression for the sum of n terms of a geometrical progression, of which a is the first term and r the common ratio.

OP, OQ are two straight lines; from A, a point in OP, AB is drawn perpendicular to OQ; from B, BC is drawn perpendicular to OP; from C, CD is drawn perpendicular to OQ; and so on. Show that the lines AB, BC, CD ... form a geometrieal progression; and find their common ratio when OA = 13, AB = 5. (18)

Section 64. Intermediate Education Board for Ireland. Senior Grade.

ARITHMETIC. (1 HOUR.)

Logarithmic Tables may be obtained from the Superintendent.

1. A father when dying left three-fifths of the value of his property to his son, and two-fifths to his daughter. The son afterwards gave three-eighths of his share to his sister; what ratio did the son's portion then bear to the daughter's portion?

2. A man invested £4032 in two equal parts, in 3 per cent. stock which he purchased at 84, and in $3\frac{1}{2}$ per cent. stock which he purchased at 96. Find (*a*) the difference between his annual dividends from the two stocks, (*b*) his total annual dividend from both stocks. [Brokerage may be neglected.]

3. A man bought a horse for £50, and sold it immediately after, receiving a bill for £60, payable at the end of six months. He cashed the bill on the same day at a discount of 4 per cent. per annum: find his gain or loss per cent. on his outlay.

4. Find to the nearest tenth the number of Avoirdupois pounds in a kilogram, being given that 36 cubic feet of water weighs 2250 lb., 1 cubic metre of water weighs 1000 kilograms and 1 metre = 3·28 feet.

5. A man has £10,000 invested in a business which pays 10 per cent. per annum on capital. If he draws £800 a year of his income from this source and adds the remainder to his capital, how much money will he have in the business at the end of 4 years?

6. The value of a property increases at a uniform rate of $6\frac{1}{2}$ per cent. yearly. If the present value is £1500, what will the property be worth at the end of 7 years?

7. The height of a room is three-fourths of the breadth, and one-half of the length. At the rate of $1\frac{1}{2}$d. per square foot it costs £5 to decorate the walls, after deducting one-sixth of the wall space for door, windows, and fireplace. Find the dimensions of the room.

8. Prove and exemplify the rule for dividing a fraction by a fraction.

PASS GEOMETRY. (2 Hours.)

Theoretical proofs are to be given except for Questions 6 and 7.

1. Two of the angles of a triangle are equal. Prove that the sides opposite to them are also equal.

2. Show how to divide a given straight line so that the rectangle under the whole line and one part may be equal to the square on the other.

3. AB and CD are two chords of a circle such that AB is greater than CD. Prove that AB is nearer to the centre than CD.

4. Equal angles are drawn at the centres or at the circumferences of two equal circles. Prove that the arcs on which they stand are equal.

5. Show how to circumscribe to a given circle a triangle equiangular to a given triangle.

6. Draw two circles of radii 5 cm. and 3 cm. with their centres at a distance 7·5 cm. apart. Draw and measure one of the common tangents.

7. Construct a triangle with sides 8 cm., 5·8 cm., and 4·7 cm. Inscribe a triangle equiangular to it in a circle of radius 6·4 cm. Measure the longest side of the inscribed triangle.

8. AB and AC are two chords of a circle such that the angle BAC is one-third of a right angle. Prove that the chord BC is equal to the radius of the circle.

9. M is a point in the base BC of the isosceles triangle ABC. Prove that $AB^2 - AM^2 = BM . MC$.

10. Three straight lines are given which intersect at the points A, B, and C. Show that in addition to the inscribed circle of the triangle ABC three other circles can be drawn to touch the three lines. Show how to construct one of these circles.

HONOURS GEOMETRY. (2 Hours.)

1. In two triangles ABC and PQR, AB : AC : : PQ : PR, and the angles at C and R are equal. Prove that the angles at B and Q are equal or supplementary.

2. Similar figures are similarly described on the three sides of a right-angled triangle. Prove that the figure on the hypotenuse is equal to the sum of the other two.

3. Prove that straight lines drawn across three parallel planes are divided proportionally.

4. The perpendiculars from A and B on the opposite sides of the triangle ABC meet at O. Prove that AO is double the perpendicular from the circumcentre of the triangle on the side BC.

5. Perpendiculars are let fall on the sides of a cyclic quadrilateral from any point on the circumscribing circle. Prove that the rectangle under the perpendiculars on one pair of opposite sides is equal to the rectangle under the perpendiculars on the other pair of opposite sides.

6. Three circles pass through two common points A and B. Two straight lines drawn through A meet the circles again in points P, Q, R, and X, Y, Z respectively. Prove that

$$PQ : QR :: XY : YZ.$$

7. Show how to construct a triangle ABC, being given the positions of the angular point A and the points where the bisectors of the angles B and C meet the opposite sides.

8. A and B are two points on a circle of centre C. M is a point on AB produced. The tangents at A and B meet the perpendicular through M to CM in the points P and Q. Prove that $MP = MQ$.

9. Given two non-intersecting straight lines in space, show how to draw a straight line meeting them both at right angles, and show that this line gives the shortest distance between the two given lines.

10. A, B, C, and D are four non-coplanar points. Points X and Y are taken in AB and CD respectively, so that $AX : XB :: DY : YC$. Points P and Q are taken in BC and DA respectively, so that $BP : PC :: AQ : QD$. Prove that the straight lines XY and PQ intersect.

PASS ALGEBRA. (2 HOURS.)

In solving quadratic equations the formula giving the roots must not be assumed.

1. Divide

$$(b+c)^3+(c+a)^3+(a+b)^3+a^3+b^3+c^3 \quad \text{by} \quad a+b+c.$$

2. Calculate to two places of decimals the values of x which satisfy the equation:

$$6x^2-13x-4=0.$$

3. Find the values of x, y, z, which satisfy simultaneously the equations:

$$5x-3y-2z=11,$$
$$3x-y-3z=-4,$$
$$2x-5y-5z=-9.$$

4. If the ages of a man and of his son are expressed in years, the father's age is half the square of his son's, and five years ago the father's age was three times the square of his son's; find their ages.

5. Prove that $(x^2+4xy+y^2)^2+3(x^2-y^2)^2$ is a perfect square.

6. Express in its simplest form:

$$\frac{7x-6}{6x^2+7x-3}-\frac{3x^2-6x+5}{3x^2-7x+2}+\frac{2x^2-2x+3}{2x^2-x-6}.$$

7. Find x from the equation:

$$\frac{2}{x-3a}-\frac{2}{x-3b}+\frac{3}{a-b}=0.$$

8. Find a fraction such that, if 3 be subtracted from its numerator and at the same time 2 be added to its denominator, a fraction is obtained which is five-ninths of the original, while if 4 be subtracted from the original numerator and at the same time 5 from the original denominator, a quantity is obtained which is ten-ninths of the original.

9. Without assuming any of the laws of indices, prove that

$$(a^4)^3=a^{12}.$$

Assuming the laws of indices, prove that

$$\log_b c=\frac{\log_a c}{\log_a b}.$$

10. Draw graphs representing the equations:

$$y = \frac{(x-1)^2}{3}, \qquad x + 2y = 6,$$

for values of x between -3 and 4, and from your drawing find the values of x and y which simultaneously satisfy these equations.

HONOURS ALGEBRA. (2 HOURS.)

1. Find the values of x, y, z, which simultaneously satisfy the equations:

$$x + ay + \frac{z}{a^2} = a^2,$$

$$x + by + \frac{z}{b^2} = b^2,$$

$$x + cy + \frac{z}{c^2} = c^2.$$

2. A man runs 4 miles and walks 3 miles in a total time of $1\frac{1}{2}$ hours; if he increases his rates of running and walking each by 1 mile per hour, he could run 3 miles and walk 2 miles in a total time of 50 minutes; find his rates of running and walking.

3. Given $ax + by + cz = 1$ and $x + y + z = 0$, prove that

$$(a-b)(a-c)x^2 + (b-c)(b-a)y^2 + (c-a)(c-b)z^2 = 1.$$

4. Define $a^{\frac{m}{n}}$, where m, n are positive integers.

Without assuming any of the laws of indices, prove that, if a is a positive quantity, the real positive value of $a^{\frac{7}{4}}$ is equal to the real positive value of $a^{\frac{21}{12}}$.

5. If $x = a(b^3 - c^3)$, $y = b(c^3 - a^3)$, $z = c(a^3 - b^3)$, prove that

$$\frac{x^3 + y^3 + z^3}{xyz} = \frac{a^3 + b^3 + c^3}{abc}.$$

6. Investigate the number of combinations of n dissimilar things taken r together.

In how many different orders can n persons enter a room (1) so that, of two given individuals, one, A, shall immediately precede the other, B, (2) so that A shall precede B, but not necessarily immediately.

7. Assuming the proof of the Binomial Theorem when the exponent is a positive integer which has any particular value, prove directly that it is true when that value is increased by unity.

Obtain the coefficient of x^r in the expansion of

$$\frac{16x-7}{3+13x-10x^2}$$

in ascending powers of x.

8. Assuming the series for $\log_e(1+x)$ in ascending powers of x, deduce that for $\log_e\left(\frac{1+x}{1-x}\right)$.

Calculate $\log_e \frac{5}{4}$ and $\log_e \frac{128}{125}$ each to 5 places of decimals, and hence calculate $\log_{10} 2$ to 4 places of decimals.

ANSWERS AND HINTS.

THE following pages contain answers to numerical and other questions, and hints towards solution in suitable cases.

In many cases a numerical result is given in two forms, the second form in brackets. The first form is the reasonable answer to give, and the bracketed one is shown as an aid to judging the correctness of the method employed by the student.

The answers to drawing questions have in many cases been calculated and given for the master's convenience to greater accuracy than is possible by drawing.

PART I.

Section 1.

1. £6. 9s. 11½d. **2.** £1. 2s. 0½d. **3.** £150.

4. Second man does for 6·94d. the work for which the first takes an hour and charges 7d.

5. Gain is £11. 4s. 8d. **6.** 0·170.

7. (i) 0·01d.; (ii) £4. 9s. 11d. **8.** Value is 1½d. less than 50s.

9. Profit £738. 15s. **10.** Payment £345.

11. Shipper made £240. 12s. 6d.

12. Author gets £550.

13. Kerosene 8·15d. or 8¼d.; Crystal Oil 6·67d. or 6¾d.

14. 227 articles; price may be increased by 3½d.

15. 47 bicycles; reduced by 22s.

16. £3·693. Sum is 7 farthings less than £3·7; and 7 far. = £0·007 to the accuracy required.

17. The multiplier being odd, the correct answer will contain an odd halfpenny.

18. £12. 19s. 4d.

19. $\frac{1}{2}$ of £5. 5s. = £2. 12s. 6d.; £2. 12s. 6d. + 9d. = £2. 13s. 3d.

20. £1. 5s. $2\frac{1}{2}$d.

21. 14s. 10d.

22. Profit £9. 14s. 8d.

23. Gain 10s. 8d.

24. Wheat 6s. 8d.; barley 6s. 10d.; oats 6s. 5d.

25. To nearest penny cost is £8. 12s. 7d.

26. Value of sugar 17s. 3d.

27. Potatoes 6s. 8d.

28. Coal £950. 17s. 8d.

29. Costing £157. 2s. 5d. to nearest penny.

30. Copper £709. 10s. 7d. to nearest penny.

31. Apples £19. 6s. $11\frac{1}{2}$d.

32. Castings £204. 16s. 3d. to nearest penny.

33. Tea £355. 14s. 5d. to nearest penny.

34. Troy ounce worth £3·894.

35. 123·2745 grains.

36. £17. 17s.

37. 262 trucks to nearest integer.

38. 1200 lb., 1700 lb., 1500 lb. These numbers should be given to two significant figures only, as 6·7 is given only to that degree of approximation.

39. First offer best; profit 3s. 10d.

40. 1320 yards.

41. 38,400 chairs.

42. 35 cm.; 23 m.

43. 5 mm. and 0·2 in. represent 54 and 55 miles respectively Thames is represented approximately by 2 cm. or $\frac{3}{4}$ in.

44. 4 desks.

45. 1020 yards.

46. 0·0018 in.

47. 110 ft.

48. 288 ft.

49. $4470\frac{1}{2}$ miles.

50. $12\frac{21}{32}$ in.; $13\frac{1}{8}$ in.

51. £25,410.

52. 56 plots.

53. £7. 5s. $6\frac{1}{4}$d.

54. 14·3 c. ft.

55. The answer 97,354 gallons may be accepted, but the clever student, who knows that the number 277 is an approximation and realises that the quantity of material cannot be accurately measured, will give 97,400 or 97,000 gallons.

56. 107 min.

57. In square feet to the nearest integer: 72 80 71.

58. (i) The product of two odd numbers is odd.

(ii) The cost is an exact number of shillings, because the area is an exact number of square metres.

(iii) The answer is a ratio and not a sum of money.

(iv) The new field is longer than the old and consequently narrower.

59. 3·8 years. **60.** 56 hours.

61. 16 yrs. and 30 yrs. to nearest integer.

62. Saves £223. 7s. 6d. **63.** Width 47 miles.

64. At 3.6 p.m. C pours in his 32nd bucketful and makes the tank overflow by $2\frac{1}{2}$ gallons, A having already made 47 and B 38 journeys. This is an unnatural question in that it assumes the men to keep exact time, to measure exactly, and to spill no drop on the way.

65. 7920 strides; exactly an hour. **66.** 4 hrs. 56 min.

67. $3.27\frac{1}{2}$ p.m. **68.** Monday, 1st May, 1905. **69.** $1\frac{1}{2}$ tons.

70. Gains 1 min. $27\frac{1}{2}$ sec. and shows 9.1 a.m. to nearest minute.

71. Wednesday. **72.** Monday.

73. Sunday, 18th April. **74.** 3,000,000; 300.

75. 0·0210. The last zero is a significant figure, the number written being a closer approximation than 0·0211 or 0·0209.

76. 0·3125. **77.** 4004 men. **78.** 102 and 105 perforations.

79. From 100,000 he can make 1851 sets and there are left 25,960 red, 22,258 green and 46 blue cards. 20, 21, 27 thousand of the respective colours will make exactly 500 sets.

80. £789. 18s. 6d. **81.** 117,138 tons.

82. 15 pence. **83.** 5s. $7\frac{1}{4}$d.

84. 740,000 workmen, to two significant figures. Another reasonable form for the answer is $\frac{3}{4}$ million.

85. 120 acres. **86.** Eldest son £18,551; each other child £6184.

87. (i) 17/63; (ii) £2406. 2s. 9d. to nearest penny.

88. £525,000.

91. (i) Probably no fraction of the box is made till well on in the 7 hours, the earlier stages being occupied in the preparation of the material.

(ii) The sums possessed by Jane and Emma are not possible in English money.

(iii) The time B will take is a function of his energy rather than of the size of his shovel.

92. 37 min. 20 sec. after 3 o'clock, when one has done more than 10 laps and the other more than 11.

93. 1,247,400.

94. $2^3.3.5.199$ and $2^2.3^3.7.199$; H.C.F. $2^2.3.199$; L.C.M. $2^3.3^3.5.7.199$.

95. $2^3.3^6.5^3.7^3$ or 630^3 or 250,047,000; to make it a square divide by 2.5.7 or 70.

96. 1·7 lb. per sq. in. to the nearest tenth.

Section 2.

1. 15·5 sq. cm.

2. The numbers (nearest point) are all 5 mm. from the left side of the rectangle, and their distances from the bottom end are 122, 55 or 56, 37, 19, 7 mm.

5. £3. 1s. 8d. **6.** 2 acres.

7. Greater than an acre by 60 sq. yds.

8. 25 acres, to the nearest integer.

9. Total £15,732; per hectare £4500.

10. 21 ares to nearest integer; 3520 blocks.

11. Periphery 172·9 cm.; area 1853 sq. cm.; ratio 1·201; all to four significant figures.

12. Area 96 sq. ft. ($96\frac{1}{3}$); 14 yards tape (13 yd. 8 in.).

13. 12 sheets used; profit 27·5 %.

14. 9 sq. ft. **15.** (i) 56 tins; (ii) 49 tins.

16. 31 sq. cm.; 10 c.cm. **17.** 175 lb. or 1 cwt. 63 lb.

18. 38 kilos to nearest integer.

19. 9,400,000 gallons to two significant figures.

20. 89 lb. The figure is obviously intended to have the two arms equal and of the same breadth all along. As drawings are rarely perfect, it is well to take several measurements of each dimension and to use the average.

21. 432 litres to nearest integer. **22.** 175 gallons.

23. 1100 bricks to three significant figures (the first zero being a significant figure).

24. 1719 lb. to nearest integer.

25. 216 c. in.; 56 lb. to nearest integer; 1·9 c. in.; 1·3 sq. in.

26. 23 grams; see remark with answer to Qn. 20.

27. 9·9 c. yd. to nearest tenth; most reasonable answer 10 c. yd., since 10 loads or cubic yards would have to be ordered. Cost £2. 2s. 8d. for the exact quantity, £2. 5s. 0d. for 10 loads.

28. £14. 0s. 5d. to nearest penny.

29. 9600 gallons to two significant figures. **30.** 10 pence.

31. 25 c. ft. **32.** 440 tons.

33. The length is 18·4 cm.; the breadth (best measured with the book closed) 12·2 cm. Thickness of 1 cm. takes 156 leaves. One c. cm. weighs 1 gramme (0·97).

34. 20 cm. **35.** 3/70 or 0·043 inch.

36. Plate 3200 lb. to two figures (3238); cubic foot 488 lb.

37. 0·0047 in. **38.** 21,500 gallons. **39.** 21 sq. cm.

40. The platinum is 9 ounces heavier.

41. 1040 sq. in.; 190 inches (189·2).

42. 3,800,000 litres. **43.** £62,831. 4s.

44. 2·7 times. **45.** 53 lb. to nearest integer.

46. 1600 bricks (1616). **47.** 12 lb. (12 lb. 3 oz.).

48. 100 tons (101).

49. 1010 tons; not affected to the specified degree of accuracy.

50. A tenth of an inch (0·092).

51. 58 minutes. **52.** 6 million tons. **53.** 280 tons.

54. (i) $V = kbh$ c. yd. $= 192$ c. yd.

(ii) $A = 2(kb + bh + hk)$ sq. yd. $= 208$ sq. yd. $V = 8v$; $A = 4a$.

55. 255 days.

56. Reservoir 400 acres (406 for the specified six months, somewhat less for exact half year); catchment area 29,000 ac. or 46 sq. ml.

57. 4·9 million c. ft.; 170 million c. ft.; 7·1 in. Rectangle will have area of $6\frac{1}{4}$ sq. in.

Section 3.

1. 110 acres. **2.** Distance 52 km. Area lake 55,000 hectares.

3. 31 sq. cm.; 1·3 hectares.

4. 144 sq. miles (140 appropriate if 0·36 is to be supposed an approximation).

5. £15. **6.** 1 sq. yd. (1·0). **7.** £22 or £23.

8. Data give 256 hectares to the sq. mile (true value to three figures 259).

9. 3·8 sq. m. 27s.

10. Represents 7 sq. miles (7·2). Cost £3000 (3100).

11. 50 kilograms to nearest integer.

12. 320 c. ft. Will hold 10 tons 1 cwt. (but it would have been better to ask for the answer to the nearest ton, no closer approximation being reliable).

13. 1200 sq. m. (1150); 35,000 c. m. (34,500). For each dimension a number of measurements should be taken, and the results should be obtained from their average values.

14. 68,000 c. ft. air space; 270 people.

15. (i) 8·2 in., 4·8 in., 2·7 in.; 110 c. in. (ii) Weight 3·9 lb. (iii) 110,000 blocks; cost £750.

16. Page measures 7·25 in. by 4·8 in. (breadth is best measured with book closed); number of blocks 250,000; number of loads 170 (174); cost £110 (111).

17. 10 cwt. as accurately as data allow. For each dimension several measurements should be taken and the average used.

18. I get average measurements of: inches 35·0, 16·8, 13·4; centimetres 88·7, 42·5, 33·8; but two significant figures are enough to expect. Volume 28 gallons or 130 litres.

19. Dimensions in cm. 35, 20, 10. Volume 7000 c. cm.

20. 14,000 gallons; 100,000 litres to two significant figures.

21. Second side of rectangle measures 21 units for Germany, 1·6 units for Switzerland.

22. Scale of 1/5,780,000 (to three figures) fits exactly for breadth and with a margin for length. Area is represented by 39·4 sq. cm.; and if breadth of map is taken as one dimension, the other is 4·4 cm.

23. Pond covers 0·22 of field. Surface lowered by 13 cm. Cost just over £3.

24. The map requires a length of 178 mm. and breadth of 172 mm. The page has length 184 mm., which is enough, and a breadth of 122 mm., which is not enough.

Section 4.

1. 67s. or 68s.

2. Second dimension 33 or 34 in. Weight of punt 220 lb., about 1040 lb. less than water displaced.

3. 11 sq. cm.; 11,000 c. m. **4.** 5·9 tons; £6. 10s.

5. 230,000 or 240,000 litres; 39s.

6. 32 or 33 men. Ratios 0·004, 0·000,02, 0·000,000,07.

7. 54 or 55 lb.

8. Frontages 110 ft. and 40 or 41 ft.; area 5000 sq. ft.

9. 31 tons; 8 acres. **10.** 10½ acres; 3·8 hectares.

11. 22 sq. cm.; 25s.; each to two figures.

12. 2·0 sq. in.; 8·2 sq. ft.

13. 22,000 cubic yards. The closer approximation 22,500 is less appropriate, as the data are evidently given only to two figures.

14. 12,000 sq. ft.

Section 5.

1. He should cut off 11 rounds, as 10 rounds fall a little short of 18 yards.

2. The vulgar fraction is 1/791. Difference for radius of 28 cm. is about 0·99 sq. cm.

3. 9 or 10 turns. Knot travels same distance as any point of blind, viz. 45 inches.

4. The marked spokes are first at right angles after a distance of 27 feet (27·5), and next after 5 seconds (4·7) from the beginning of observation.

5. (i) One medal 20 sq. cm. (ii) Uncovered board 640 sq. cm.

6. 2800 c. mm.

7. 47 or 48 gallons; 24 inches, represented on the figure by 1·2 inches.

8. 36 cwt. or 1·8 tons. **9.** 90 cubes.

10. 550 lb. (552 lb.) or 5 cwt. (4·9 cwt.). **11.** 98 metres.

12. Length 31 yd. (31·4); vol. 9 c. in. (8·88). **13.** 19 lb.

14. About 300 (or between 270 and 320). **15.** 57 or 58 lb.

16. 3·1 tons. **17.** £7. 8s. **18.** 33 kilos (33·3).

19. 29 lb. (29·4). Weight increases to 30·5 lb., which is by 3 %.

20. Surface of one tube 226 sq. in.; of all 785 sq. ft. Whole weight 14 cwt. (13·7).

21. 0·092 in. (0·0918). **22.** 4200 c. cm. **23.** 16 cwt.

24. £29. 2s.; £7. 6s. **25.** A trifle over 1 gram (1·02).

26. 262 cubic feet. **27.** 54 c. mm.

28. One fill of 5 feet and 18 fills of 2 feet, *i.e.* 2100 gallons.

Section 6.

1. (i) 2·3 tons. (ii) Yes; it will lift 5½ tons safely.

2. 102,000 c. ft.; 5700 tons. **3.** 11·5 gallons; 6s. 2½d.

4. 4·5 kilos (4·47); 5s. 3d. (5s. 2½d.).

5. Between 276 and 277 kilos; £4. 12s. to nearest shilling.

6. 1·1 sq. in.; 4·1 or 4·2 c. in.

7. Cavity of 160 c. in. (exactly). **8.** 815 c. in., weighing 56 lb.

9. Scale roughly 4000 miles to an inch, more accurately 3900 miles to an inch. Area of tropics 59 or 58 million square miles and 5·2 or 5·1 times British Empire, according as the less or more accurate scale is used; 60 million sq. miles and 5 times are quite appropriate answers.

10. 5 pints (5·15).

11. More accurately 1 acre is represented by $209/x$ inches, the numerator being given to the nearest integer. Result obtained by the rule is wrong by 4 %.

12. Capacity of tyre 210 c. in., of pump 4·4 c. in. Pressure after 40 strokes 1·8 atmospheres or 27·5 pounds per square inch.

Section 8.

1. 3007 yards to the nearest integer. The boundary requires special consideration.

2. 12,960 yards.

3. 80 or 81 threads each way in a square inch (80·5 to three significant figures).

4. 1536 meshes. 197 feet, and with knots 209 feet.

5. £4. 5s. and £5. 1s.

6. 3586 lb. Among the data 1·7 is evidently an approximation, so that our result is valueless beyond the second significant figure.

7. The weight must be such that with the weight of the balloon and gas it is equal to the weight of an equal volume of air. Weight is 300 or 400 kilos (366).

PART II.

Section 9.

1. £3. 1s. 6d. **2.** 125 rupees. **3.** £7. 13s. 0½d.; £3·235.

4. £920. **5.** 6s. 10½d.; 4s. 9⅛d.; 3s. 9⅝d. **6.** 5 farthings.

7. Dearer in France by 2 % of the English price (1·76).

8. £641. 14s. 11d. **9.** 75 cents (73).

10. 51 kilos greater by 0·2 lb. as nearly as data allow. (A more accurate relation between kilo and lb. shows 51 kilos greater by 0·4 lb.)

11. 252 ten-franc pieces, to the nearest integer (252·4).

12. The body of the table is:

1904	£21,094,994	£21,352,994.
1905	22,125,640	20,890,501.

13. 9·14 francs; 1·75 dollars; 7·42 marks.

14. Costs 25 shillings more in France.

15. Whole cost £532,000; per mile £8000.

16. 12 pence per pound. **17.** 20 bushels per acre (19·6).

18. Denmark, 0·6 % is wheatland yielding 15·6 bush. per ac.; Great Britain, 3·0 % is wheatland yielding 30·8 bush. per ac.

19. 4s. 3d.

20. There are 274 bottles, and a quantity insufficient for a bottle which we must suppose wasted. Profit per bottle is 6·7 pence.

21. £481.

22. The data being given to three figures only, the multiplier will not be reliable to more than three figures (if to so many) and may be stated as 1·11. But it is not unreasonable to state the multiplier as 1·113 because, although the 3 may be wrong, the result of multiplying by 1·113 will be nearer the truth than by using 1·11.

23. 12·5 grammes bluestone and 7·8 of lime, to nearest tenth.

24. 772 or 773 cm. per sec. according to the data selected from the front cover. The difference shows that the data will not give the result to the required degree of accuracy.

25. £3. 1s. **26.** 22 kilos. **27.** 37·5 %.

28. 56 lb. to the nearest integer.

29. (i) 5·5 centimes per km.; (ii) 86 centimes per 50 kg. per 100 km.

30. 576,000 litres per hectare; 200 tons per acre.

31. (i) The weight in grammes of a leaden bullet is 6·0 times the cube of its diameter in centimetres (5·95).

(ii) They cost 21·25 francs per 1000.

32. The respective quantities are, to one decimal place: for A 1·8 gm., 10·0 gm., 100 c. cm.; and for B 0·9 gm., 0·7 gm., 100 c. cm.

Section 10.

1. London 60·617 per ac., 64 times density of England. Westmoreland 0·127 per ac., 0·13 of density of England. England 0·945.

2. United Kingdom 357, India 158.

3. Belgium 614, denser than the United Kingdom; China 260, less dense.

4. 74 million. **5.** 735, denser than the United Kingdom.

6. City has 40 persons per acre, County 61; City 7 persons to a house, County 8; City 6 houses to acre, County 8.

7. Cardigan 13 %, Flint 2 %.

8. 4,670,000 to three significant figures.

9. On 10 acres in a year 16 people are born and 9 die, giving an increase of 7; each to the nearest integer.

10. Percentage who are electors 18 17 16 in the three countries; number represented by a member 67,000 62,000 43,000 respectively; all to two significant figures.

11. In each year 15 % of the population attended school. 2,432,000 boys attended in 1903, and 2,485,000 in 1904.

12. The respective percentages are: 54, 26, 5, 10, 5.

13. £3,154,000 in 1903, and £2,895,000 in 1904; decrease 8 %.

14. From United States 27·5 %: from British Possessions 32·3 %, Canada and India sending 97·8 % of this.

15. 31·28 bush. per ac. £196,608.

16. Greater in 1904 by £2592.

17. Sugar, 83 lb. and 7s. 4d.; tea, 6 lb. and 4s. 7d. (These are wholesale prices, and the retail prices are greater by the cost of distribution.)

18. £2. 15s., £3. 3s., £2. 17s. in the respective years. Falling off in quantity 9 %, in value 19 %.

19. 30 gal.; 6s.

20. Beer 15 %, wine 52 %, spirits 37 %. Total consumption of beer increased by about 340,000 gallons.

21. 390 gallons. **22.** 200 million gallons (181).

23. The three increases are 9 % 8 % 9 %.

24. 99 % to nearest integer.

25. Respective percentages are 89, 13, 0, 9, 40.

26. Decrease for flax 43 %, for potatoes 35 %.

27. 40 ounces; £4. 0s. **28.** Decrease £2. 2s.

29. The respective numbers are:

595	489	82 %
219	270	123 %
186	200	108 %
164	144	88 %

30. Foreign, 34·5 %; Colonies, 41·3 %; Total, 36·7 %.

31. The four percentages are: 38, 8, 12, 5. The last percentage is almost exactly 5·5.

32. £13 per head in each year, to the nearest pound.

33. For the three years: £8. 9s., £8. 2s., £8. 7s.

34. In millions of pounds sterling Canada's exports of farm products were 14½ and Britain's imports 72.

35. £805. 11s. 8d.

36. 0·7, 6·4, 9·2, 10·8, 12 pence in the pound respectively.

37. Higher income pays a virtual rate 2·3d. in the pound higher.

38. £2. 5s. 1d. a year; 10·4d. a week.

39. (i) 3·3, 2·1; (ii) 8·7, 2·3; (iii) 61, 0·13.

40. (i) ·72, ·72; (ii) £4. 15s., £30. 3s.

41. $10\frac{1}{2}$d. **42.** Percentages: 25, 22, 22. **43.** 4 %.

44. Net balance, £95,158,000. Percentages, 87, 11, 2. Revenue per head, £3. 12s., £3. 9s., £2. 4s.

Section 11.

1. Cambridge were heavier, averaging 12 stone as against 11 st. 13 lb. for Oxford.

2. 0·28 inch. **3.** 2·14 inches; 7·1 or 7·2 sq. in.

4. Odell is better on each separate innings, but Thompson on the match as a whole; the cost of a wicket being as follows:

	1st innings.	2nd innings.	Whole match.
Odell,	21	9	13·8
Thompson,	22·5	9·8	13

5. 36s. $0\frac{3}{4}$d.

6. £1. 7s. 4d. to the nearest penny; or in half-crowns, to correspond with the data, £1. 7s. 6d.

7. 65 letters per head. **8.** Birth-rate 29·4.

9. $6\frac{1}{4}$d., $2\frac{1}{4}$d., 1d., $2\frac{1}{4}$d. **10.** 35·82.

11. £513 exactly. **12.** 64° F.

Section 12.

1. £1. 13s. 4d. The work is often simplified by starting from a convenient point; in this case, reckon first what happens with goods the cost price of which is £100.

2. 9000.

3. 217,000 in 1911; and 137,000 in 1871.

4. 82 %. The data are obviously approximations, so that 81·6 is too exact a value to give as an answer, and the most appropriate answer is 80.

5. 10 % is profit (9·9). **6.** (i) 7650 cwt.; (ii) 3 % (3·4).

7. Taxation per head increased by 8 % (7·6).

8. Loses $5\frac{1}{2}$ %.

9. Expenses were 64 % of receipts (63·8).

10. 66 % to men, 34 % to women.

11. (i) Percentages: 6·2 increase, 16 decrease, 39 increase, 1·4 increase. (ii) Total values in hundreds of thousands of pounds: 13, 19, 24, 2·7.

12. Ought to charge £11. 15s. Actual profit, 50 %.

13. 54·3 %. **14.** 92 % (91·5).

15. Shrinkage, 1·6 %. **16.** 2·4 bushels.

17. 17,000 oz. (17,136). **18.** 25 % increase (25·0).

19. Decreased by 4 % (exact, in the sense that calculation leads to exactly 4; but, of course, the 8 per cent. increase is a rough approximation).

20. Exactly 300 candidates.

Section 13.

1. (i) 39,743,000,000. (ii) 0·025. (iii) 3·83.

2. The three rules give 0·7857 0·7855 0·7854; the last agreeing with the tables, and the first differing most from them.

3. For most of us it is shortest to take five figures of each factor and multiply out. While some of the partial products thus used may be discarded, it is quicker to put them down than to spend time deciding which to discard; and we thus avoid the chance of an error in the decision as to discarding. The use of contracted methods leads to serious waste of time when the exact result hangs nearly evenly in the balance between two values and it is necessary to obtain several more figures of the result to decide between the two values. The product is 1,340,000,000.

4. (i) £7010. 11s. 8d.; 1s. 8d. is 1/12 of £1.

(ii) 72; the expression becomes $1\frac{1}{8} \times 64$ because $1\frac{1}{8} = 1{\cdot}125$.

(iii) £2. 11s.; we calculate 1/3 of £7·69569.

(iv) £5. 14s.; 38 lb. at $14\frac{1}{2}$d. is the same as 76 lb. at $7\frac{1}{4}$d., and the required result is therefore 76 lb. at $1\frac{3}{4}$ shillings.

5. Five more significant figures for the exact product.
Product to three places is 0·311.
Remainder 0·25. [The term "remainder" is unusual, except in connection with integral divisor and dividend; presumably it means the fractional part of the quotient multiplied by the divisor.]

6. 55 c. mm. (or to three figures 55·1).

7. The lines being measured to two significant figures (7·8 cm. and 6·2 cm.), the area is best given to the same approximation, as 48 sq. cm. If each measure may be half a millimetre wrong, the true area lies between 48·36 ± 0·7 sq. cm. (to two decimal places). The greatest possible area thus differs from 48 sq. cm. by 1·06 sq. cm.

8. The length lies between 17·35 and 17·45 m., the breadth between 7·75 and 7·85 m. The area from the given dimensions is 136 sq. m. to the nearest integer or 140 to two figures. But the true area may be (to three figures) anything between 137 and 134 sq. m., so that only *one* figure of our result is reliable. The percentage error of 136 sq. m. may be a little more than 1.

9. 3·5 gallons. Result may be wrong by 6 %.

10. The greatest possible errors are, to two figures: (i) 1·1 sq. cm.; (ii) 9·6 c. cm. If we are limited to reliable figures, we must state the area as 30 sq. cm. and the volume as 200 c. cm.; but if from these results we wish to proceed to further results (such as to find the weight of the cube when given its density) it is desirable to retain more figures, perhaps giving the area and volume as 31·4 sq. cm. and 176 c. cm., and to bear in mind that we have retained unreliable figures.

11. 0·0154 gramme, 14 % more than in catalogue.

12. Percentage errors: (i) 0·085 too small; (ii) 0·182 too great.

13. Errors in length and breadth, 15 and $12\frac{1}{2}$ inches; in area, 30 sq. ft.

14. 3·1 acres (3·13) or 3 ac. 20 poles.

15. Possible error in excess, 8·2 %; in defect, 7·8 %. (V and L being true volume and length, greatest possible area is $\dfrac{1{\cdot}05V}{0{\cdot}97L}$ and least $\dfrac{0{\cdot}95V}{1{\cdot}03L}$).

16. Percentage of milk between 96·7 and 72·5.

17. Speed between 25·7 and 27·3 miles an hour.

18. 6170 5940 4720 millions of dollars for the three years, each to three figures. With the given limits to the accuracy of the data, the number 6170 may be too great or too small by 30.

19. The percentage from Ceylon is between 33 and 35. In pounds it is between 71 and 76 million, so that neither 70 million nor 80 million is reliable. We are therefore forced to state it as 100 million; which is reliable to one figure, being nearer to the true quantity than is either 200 million or zero.

20. For wheat two figures are reliable, £7,000,000 (the first zero being a significant figure). For barley two figures and for oats three figures are reliable; values £8,700,000 and £12,800,000.

21. Cost per mile for the five roads, given to the nearest pound, £309 £146 £175 £150 £202. The true value of the third number (£175) may be greater by £1; if we limited ourselves to reliable figures we should have to enter it as £200.

22. The number 1·571 is accurate to four figures, its true value being 11/7.

23. £1865. The accuracy of the number £1·2155062 is excessive; even on a loan of £10,000 the last figure makes a difference of less than a halfpenny.

24. 5 litres (5·0). The formula of the text gives for a sphere (by putting $h=2r$) $4\pi r^3/3$ or $4{\cdot}1888r^3$; so that the rule in question which gives $4{\cdot}2r^3$ is reliable only to two figures.

25. Substituting 0·54 increases the product by £0·1844; discarding 7s. 6d. reduces the product by £0·2021. The net result is a reduction of less than £0·02; this could only affect a result hanging in the balance between two integers, which the clerk's result of £99 does not.

26. 34 minutes. The greater dimension of the rectangle being less than 1 km., the area is much less than 1 sq. km.

27. (i) One factor being a multiple of 9, the product must be so also.

(ii) Each of the given numbers being thought of as an improper fraction, we see that one numerator contains the factor 3, and also one denominator; so that the product cannot contain 3 as a factor. Or again, both denominators of the improper fractions are odd and both numerators even, so that the product must (if an integer) be a multiple of 4.

(iii) The result must be a pure ratio, not a sum of money.

(iv) The length of the required field being greater than that of the given one, its breadth must be less.

(v) There must be in the mixture more of the cheaper tea than of the dearer, for the price of the mixture is nearer to 1s. 7d. than to 2s. 6d.

(vi) A metre is more than a yard, so that 1 c m. is greater than 1 c. yd. or 27 c. ft.

(vii) The average for boys and girls together must lie between the average for boys and the average for girls.

28. More accurately area is 1·57 acres.

29. The number of tons is $74 \times 24 \times 12 \div 43$. 24×12 is 288, almost 300; and $74 \div 43$ is more than $1\frac{1}{2}$; so that the number is well over 300. Or 74×12 is 888; and $24 \div 43$ being more than 1/2, the number is more than 444.

30. The length is ample; the breadth not quite enough, breadth of figure being less than 1·1 in.

31. Distance by water is 506 miles, and $30 \times 18\frac{1}{2}$ is greater than 506, so that distance is greater by land.

32. Number of gallons supplied is $3\frac{1}{7} \times 1760 \times \frac{1}{12} \times 6{\cdot}23 \times 10$, which is more than 17,600 since $3 \times 6 > 12$; so that 2-inch pipe is large enough.

Section 14.

1. Total 1,500,000 quarters (1,490,000). Decimals in least and greatest years 0·05 and 0·16.

2. Lowest price 53s. 6d. at end of November 1904; highest 74s. 9d. on 1 November 1905. Times at which price changed from rising to falling: 1 May 1904, in October 1904, 1 March 1905, 1 November 1905. Times at which it changed from falling to rising: in January 1904, in September 1904, at end of November 1904, 1 August 1905. Price was rising fastest in September and October 1905.

3. Corner square 5 cm. in the side gives the greatest volume, which is 2000 c. cm.

4. In 1875, 85 % of imports were for home consumption; 81 % of exports were British produce. In the period the exports of British produce increased in the ratio 2·7; the re-exports of imported produce in almost exactly the same ratio, the ratio being possibly a little less (2·67), but so little less that on the data we cannot be sure it is less.

5. The population in thousands was 32,830 or 32,840 on 1 Jan. 1902 and 33,200 on 1 Jan. 1903. Birth rates at beginning and end of year a trifle more and a trifle less than 29·2 per 1000; say 29·2 for the year. Number of children born in the year 960,000 (959,000). Death rates at beginning and end 18·2 and 18·0; total number of deaths 590,000 (594,000). Births exceed deaths by 370,000 (365,000). Population increased by 360,000 or 370,000; so that to the accuracy to which we are able to work, the births and deaths account for the change of population.

6. For 30 payments the instalment is about £5. 4s.; for 40 it is about £4. 10s. About 21 payments of £640 are necessary.

7. The fourth run is the fastest, the speeds in miles per hour being respectively 55 54 53 57.

8. Trains pass at 5.28 p.m.

9. Should pass C at 10.50; speed 37 miles an hour.

10. At 8.50 p.m. (The nearest 10 minutes is enough to expect from a graphical method; the nearest minute is 8.51 p.m.)

Section 15.

1. 3007.

2. Approximate value 29·6. Its square is less than 879 by 0·3 %.

3. 0·31331.

4. 4·8 pence and 9·6 pence.

5. (i) 0·20; (ii) 0·0031; (iii) 0·50; the final zeros in (i) and (iii) being significant figures. Short methods could easily be devised for these questions, since 0·0625 is 1/16, and 22 is a factor of both 198 and 7854. But we are asked, not for short methods, but for contracted methods; that is, we must not carry out a division too far, and must not carry out useless multiplication. The wisest course is to carry the multiplication and division a figure or two beyond what we judge absolutely necessary, the time spent on these extra figures being less than the time it would take to judge accurately how many figures are necessary.

6. Additional length 10 inches (10·4). New diameter 0·78 inch.

7. Cost £1800 (1770, or to nearest integer 1775); 30 gallons (28); 4 chains to the inch (4·6).

8. 11·2 cm. **9.** 0·96 second. **10.** 6·28 cm.

11. Time from the data is 8·3 minutes. With given reliability of the data, time may be anything between 8·37 and 8·32, so that to one decimal place it may be 8·4 and not 8·3. To the nearest integer, 8 min. is certainly correct.

12. 261,000 and 0·0260.

13. 17,900,000 in 1851 and 8,900,000 in 1801 (17,928,000 and 8,901,000).

14. Base and height of triangle give area 8·8 sq. cm. Formula gives 35, *i.e. four* times the area. But rougher work will serve to answer the question. Thus, if the sides are taken to the nearest cm. as 7 6 3, area is 9 sq. cm., while formula gives $\sqrt{16 \times 80}$ and $\sqrt{16 \times 81}$ would be exactly 4 times 9.

15. The values 988 and 0·000207 are reliable, but the fourth significant figure in each case would be unreliable. The values of *a b c* are assumed to be stated as exactly as is possible with four significant figures.

16. 1·42 inches.

17. 110,000 feet (115,000); 0·2 inch (0·19).

18. 45 feet. **19.** 8·75 cm. **20.** 13 inches (12·6).

21. 280 gallons to two figures; 3 ft. 3 in. to nearest inch.

22. 85 gallons per minute, to nearest integer.

23. Given drains carry off 33 230 690 1500 cubic feet per minute respectively, each to two significant figures. Diameters required for given flow 8½ 16 20½ inches respectively.

24. 170 c. ft. per min., to two figures.

25. Suppose D the foot of a perpendicular from A on a horizontal plane through BC. The pyramid ABCD (whose volume is one-third of its height multiplied by its base) is to be rearranged as a prism on the same base. A is thus lowered by 2/3 of 2·3 m. or say 150 cm., while BC is raised by 1/3 of 2·3 m., say by 80 cm.

Section 16.

1. 54 miles an hour. **2.** 29 feet per second.

3. 168 kilometres an hour. Increase of speed 4 %.

4. Train is faster, steamer's speed being 38 miles an hour.

5. The former is faster by 1·1 miles an hour.

6. Speed must be 59 kilometres an hour. At speed of 57 kilometres an hour, train would be 18 minutes late.

7. Scheduled speeds 52 54 52; actual speeds 48 61 52½; all to the nearest integer, except the last entry, in which the half is exact.

8. 4 minutes (exactly). **9.** 1 km. per hr. is 27·8 cm. per sec.

10. (i) True, for $0·07 \times 15 > 1$ and $0·07 \times 14 < 1$.

(ii) 30 in. = 76 cm., to nearest cm.

(iii) 15 m. per hr. = 22 ft. per sec.

(iv) True to nearest thousandth of a £.

(v) 35·6 ft. per lb. is 148 lb. per mile.

11. (i) Speed is half a mile (or 2640 ft.) in 60 sec., *i.e.* 44 ft. per sec.

(ii) Since $16\frac{1}{2} = 15 \times 1·1$,. result is £30. 5s. 7d., to nearest penny.

(iii) Express the difference of the two squares in product form $13·79 \times 1·93$ and use logarithms. Result 83·6, to three figures.

(iv) 1·5241. To be sure of accuracy in the last figure, each partial product must be taken to 6 decimal places.

12. 65·1 miles an hour.

13. While the man does 100 yards at 12 miles an hour, the car does 75 yd., so that he could beat it by 5 yd.

14. A is $2\frac{1}{4}$ minutes late. The passing takes place 15/4 minutes late, so that A was at the usual passing place 15/4 − 6/4 minutes late.

15. Down train is 0·3 mile (528 yd.) beyond usual passing place when it meets up train. When down train is at usual passing place, up train is 3/4 mile away; they approach one another by 3/4 mile in 0·6 min., in which time down train covers 0·3 mile.

16. The slow train must lie by at its 2nd station (9th station from point of departure of fast train). It reaches this station 37 minutes from our time-origin, and would not reach the next till 54 minutes. The fast train passes the slow at 56 minutes, and passes the preceding station at 50 minutes. A graphical solution is appropriate.

17. The fast train passes 10 slow trains. It overhauls a slow by 1 mile in going 11/6 miles. The first slow overtaken had a start of 2/3 mile and was overtaken in 22/18 miles. The second had a further start of 5/3 miles and was overtaken in 55/18 miles more; and so on. Let us take 55/18 miles as unit; the whole journey is $9\frac{8}{11}$ units; and the trains are overtaken at distances 0·4 1·4 2·4 ... 9·4, so that 10 are overtaken.

A graphical solution is appropriate.

18. Time direct 37·5 sec., by path 35 sec. For distances 76 77 78 79 the direct route is slower by 16 17 18 19 sixths of a second respectively. Reduction of distance by 1 yd. reduces difference by 1/6 sec., and times are equal when AD is 60 yd.

19. The number may be 4 or 5 according to circumstances. A car meets others at intervals of 5/12 mile.

20. I travel 800 yd. between meeting two cars. Interval between overtaking two is 25 minutes (in which I travel 5 miles).

21. Buses overtake the man at intervals of $2\frac{2}{3}$ miles. There must therefore be one, and cannot be more than two.

22. 4·2 sq. ft.

23. 40 ft. of pipe contain 14·0 c. ft., to nearest tenth. Running all 24 hours it will take 3 days 6 hours.

24. 5 hours, to the nearest tenth.

Section 17.

1. £59. 1s. 1d. **2.** £21,132,193. 2s. 4d.

3. £58. 14s. 6d. Four-figure tables of logarithms do not give the nearest penny; more accurate tables must be used, or their help dispensed with. It simplifies work without tables to notice that 3285 is a multiple of 365; but it is inadvisable to watch for such relations, as there is more time lost in the search for them than is saved by their use.

4. £2. 7s. **5.** £4. 1s. 4d., to nearest penny.

6. To one decimal place the rate per cent. is 3·1 in each case.

7. (i) The multiplication of a whole number of pence by 240 gives a whole number of pounds.

(ii) 1/10 lb. would make three more packets.

(iii) The interest is less than 5 %, which is £211.

(iv) 3 % gives to the nearest tenth £153, less than the given result even before deduction of income tax.

(v) The area being an even number of square feet, there will be no odd 6d. in the cost.

8. The amounts in 17 31 45 years are respectively £150 £215 £303 (or 302). It takes 38 or 39 years to amount to £260.

9. One per cent. below bank rate is better by 12s.

10. £808 has an error of less than 10s.; the interest being (more exactly) a few pence more than £807. 10s.

11. (i) Amount is $p(1+r/100)^n$ pounds. (ii) £15,000.

12. Compound interest exceeds simple by 15s. 3d.

13. (i) £306. 3s. 8d., (ii) 29·21 francs, (iii) \$292·07; to the nearest penny, centime, and cent respectively.

14. £361. 13s. 4d.

15. £13. 12s. 5d., to nearest penny (besides the returned £60).

16. 3 per cent. per annum.

17. £461. 6s. 8d., to the nearest penny.

18. It will double itself and a little more. (To one decimal place the multiplier is 2·1.)

19. Balance £147. 10s.

20. Rent £64. This will enable him to pay interest, ground rent, and rates, and have his 8 % left.

21. £174. 7s. 7d., to nearest penny.

22. £6453. 13s. The present value of £1 to be paid in a year and £1 to be paid in two years is $£(1{\cdot}05+1)/1{\cdot}05^2$; so that two payments of $£12{,}000 \times 1{\cdot}05^2 \div 2{\cdot}05$ will liquidate the debt.

23. £12. 10s. **24.** $3\frac{3}{4}$ % exactly.

25. A £833. 9s. 4d. and B £496. 10s. 8d.

26. The respective shares are: (i) £948. 10s., £799, £649. 10s.; (ii) £2450. 13s. 4d., £1725. 6s. 8d., £1000.

27. 207 days for A, 64 for B, 34 for C. A works 6 times and B 13/7 times as many days as C; so that together the three work 62/7 times as many days as C, or C works 305 × 7/62 days.

28. 0·67 (2/3 is exact). **29.** 9·7 % ($9\frac{2}{3}$ is exact).

30. Minerals fell by 75·5 % (exactly) of their former value. (Express the value of each undertaking, at the earlier and at the later date, as a fraction of the total value of all.)

31. A is out of pocket because of the decision of the Court by £244 (to the nearest pound) more than he would otherwise have been.

32. 5·5 % on ordinary shares. **33.** £1383. 6s. 8d.

34. (i) £75. (ii) 3·0 %, to the nearest tenth.

35. He can pay 14s. 11d. in the £, to the nearest penny.

36. Equivalent price £$97\frac{1}{4}$, to the nearest sixteenth of a £. £100 sterling yields £3. 1s. 9d.

37. Net income £423. 15s., being 3·2 % (to one decimal) on the sum invested.

38. 3·5 %, to one decimal place (exactly 220/63).

39. 3·0 %, to one decimal (exactly 560/189).

40. The chief difficulty here is what to do with the brokerage. It might be treated as a deduction from interest or as an addition to the purchase price. To treat it in the latter way is to ignore the difference between money which is invested and may be recovered by realising the investment and money which is spent and gone for ever. The brokerage is spent and not invested, and should be treated as a deduction from the first year's interest, as in SOLUTION I following. SOLUTION II shows the other method of treatment, and is given because the method is customary. SOLUTION I.: He must invest in the 6 per cents. £4588, to the nearest pound; the produce of the £6000 (with brokerage deducted) is less than at 4 % by £60 × 96 ÷ 79; £1 in the 6 per cents. produces £1308 ÷ 82300 more than at 4 %; so that

$$£(82300 \div 1308) \times 60 \times 96 \div 79$$

in the 6 per cents. will make good the shortage on the 3 per cents. SOLUTION II.: If brokerage is treated as raising the purchase prices to 99 and 103⅛, he must invest £3200 in the 6 per cents.

41. Paid in duty £787. 10s. Heir's income £540.

42. He held £2500 stock, and income was £87. 10s. On £100 stock he lost £24, so that amount held was £600 × 100 ÷ 24.

43. £11,900. By selling £100 stock he increases his income by £29 ÷ 17; so that £20,300 × 17 ÷ 29 will make the required difference.

PART III.

Section 18.

3. AE = 1·1 inches (1·12).

4. 5100 yards (5080) or 2·9 miles (2·885).

5. ∠P is greater than ∠A when P is inside circle of centre B and radius BA; while ∠B is less than ∠A when P is on the same side as A of the perpendicular bisector of AB. The area required is the smaller area enclosed by these loci.

6. Follows from $CB + BF > CF$ and $CA + AF > CF$.

7. If the three bisectors meet in O, the triangles OBC and OCD are isosceles. (The three bisectors will meet in a point only when BC = CD.)

9. Two adjacent sides are together greater than the diagonal which forms a triangle with them. The four equations formed by writing down this relation, in all possible ways, give, when added together, the required property.

10. (i) AL is always 5·6 cm. long. (ii) If angle ABC is also given, draw BK making this angle with BC and 19 cm. long; join CK and make angle KCA equal to CKB.

11. Longer edge of face measures 4·2 inches (4·24), diagonal 5·2 inches (5·20).

Section 19.

1. 88 feet (87 ft. 11 in.). Shadows are least at midday when the sun's altitude is greatest.

2. 8·5 cubic feet (8·52).

3. 1100 kilograms, to three significant figures.

4. 33 cwt.

5. Diameters 16·0 and 10·6 cm. Area 130 sq. cm. (133).

6. The five values, each given in the two ways, are: 5·860, 5·86; 28·698, 28·7; 8·540, 8·54; 1·156, 1·16; 0·035, 0·0346.

7. 1·59. **8.** 5·8 (5·83). **9.** 3·7 (3·693).

10. 17 gallons (17·2). **11.** 2·2 kilos (2·21).

12. It will not break with 2 tons. The breaking weight in cwt. is $15 \times 22 \times 4{\cdot}5^2 \div 128$; $4{\cdot}5^2$ is greater than 20 while 15×22 or 330 is more than twice 128, so that the whole expression is > 40. OR: the expression is

$$(4{\cdot}5/4)^2 \times (15 \times 22 \div 8);$$

the first factor is greater than 1 and the second than 40.

13. 10 inches (10·4). **14.** 98 tons (98·15).

15. (i) 16. (ii) 56 (accurately 55·875 or $55\frac{7}{8}$).

16. 100 c. yd. (102 or 103).

17. The first formula gives 99 gallons, the more accurate one 103, or 4 gallons more.

18. Cavity 123 cubic inches (but 122 almost as close); possible limits 141 and 104 c. in. The iron has a volume of $72 \times 1728 \div 480$ c. in., which, subtracted from volume of sphere, gives volume of cavity.

19. 106·6 c. cm.

20. Diameter 0·615 inch. A ball of weight 1/12 lb. has diameter 0·729 in., and the weight is proportional to the cube of the diameter, so that a ball of weight 1/20 lb. will have diameter of $0{\cdot}729 \times (12/20)^{1/3}$ inch.

21. The results for the three cases are, the number of days and number of pounds being given to the nearest whole number: (i) 10, 7, 5 days. (ii) Cost of coal £250, £562 (or £563), £1000. (iii) Other expenses £2000, £1333, £1000. (iv) Total cost £2250, £1896, £2000.

22.

t	0·5	1	1·5	2	2·5	3	3·5
Horizontal distance, metres	5·5	11	16·5	22	27·5	33	38·5
Vertical depth, metres	1·2	4·9	11·0	19·6	30·6	44·1	60·0

23. The true formula is $\pi\sqrt{k/g}$. It gives $3{\cdot}14\sqrt{100/981}$ sec. as the interval between ticks of a metre pendulum, and by writing 3 for 3·14 and 900 for 981, we see that, roughly, the interval is 1 sec. More accurately we note that $3{\cdot}14^2$ being $= 9{\cdot}86$, the formula gives $\sqrt{986/981}$ sec., or, to a pretty close approximation, 1 second.

24. 305 feet, to the nearest integer.

25. The faster train consumes 8 cwt. more.

26. (i) $a10000 + b1000 + c100 + d10 + e$. (ii) 10942.

Section 20.

1. The length of AO may be 2·8 in. (2·80) or 1·1 in. (1·07).

2. AP = 69 m. AQ = 98 m.

3. △s ACD and BCE are isosceles. **4.** The angle is 45°.

5. The line must make 40° with the parallel to AB through C.

6. The triangle, if accurately drawn, will be equilateral, because each angle is 60°.

7. There is no doubt about the position of R, the second intersection of the two circles which fix it lying beyond the boundary AB. There are two possible positions for S, and either should be accepted.

8. (i) Distance is 1·1 inch (1·13). (ii) Further revolution through 16°.

Section 21.

1. Successive values of area in sq. m. 144 200 216 224 224 216. Greatest area is 225 sq. m., when there are 15 hurdles in each side.

2. Profit is greatest at 18d., and is £600.

3. Least number to pay is 2800 copies. To get the graphs within the limit of your paper, notice that nothing less than 1000 copies or £60 need be shown.

4. The 87th rod is 11·2 feet long, the 175th is the 25th from the other end, and is 52·19 ft. long. The figure represents an area of 2300 sq. yds.

By drawing against each rod a rectangle of the length of the rod and one yard broad, and drawing such a rectangle against one of the pillars also, we can cover up the side of the bridge except for approximately triangular pieces along the chain, some of these pieces being in excess and some in defect. The positive triangular pieces have a breadth of 1 yard and a total length equal to that of a pillar; the negative triangular pieces have the same property; so that the sums of positive and of negative are approximately equal. The number of yards of steel rods plus the length of a pillar, being the number of yards in the sum of the lengths of the rectangles, is also the number of square yards in the total area of the rectangles, or in the area of the side of the bridge, and is therefore 2300. The error in taking this as the total length of steel rods is the length of a pillar, 85 feet, which is small compared with the whole length.

5. The number who obtain less than 25 marks is about 150 (perhaps a few less); the number who obtain less than 45 is about 500. So that the remainder who neither fail nor obtain the first class are about 350 in number, or perhaps a few more.

6. There are about 680 men of height 64 in., about 1260 of height 69 in., and about 250 of height 72 in.

7. The point travels 31 or 32 feet in 4 sec., and about 61 ft. in $5\frac{1}{2}$ sec.

8. The successive values of b^2+4bh for the table are 73 64 63·4 68 76·4 88. The best value for b is 4·6, calculation showing that this gives a smaller area than 4·5 or 4·7; and the corresponding value of h is 2·3 to the nearest tenth.

[Notice that we say "the height of the box is 2·3 inches" and "h is equal to 2·3." Since h is defined in the question to be a mere number, namely the number of inches in the height, we must not say that h is "2·3 inches."]

9. Successive values of $r^2/7+20/r$ are 20·14 10·57 7·95 7·29; we must therefore go further, to greater values of r, in order to get the least value of the expression. The best value of r is 4·1 (but the graph cannot be expected to give anything better than 4·0 or 4·2); and the least possible surface area 160 square feet.

10. A millimetre may be taken to represent a penny, or a half-penny if the paper is large enough. Selling prices are 10d. or 11d., 2s. 1d., 3s. 6d., 5s. 11d. or 6s.; but errors of a penny or more must be expected on the scale suggested for the graph.

11. The weight is 18·4 lb., to the nearest tenth; but the prescribed scale will not give the decimal accurately.

12. The Fahrenheit reading is 68°, but 67 or 69 is enough to expect on the scale recommended.

13. First train travels 45 miles an hour (44·9); it starts from Oxford at 7.23 p.m.

14. Distance apart at 10.48 is 166 or 167 miles; time at 200 miles distance is 11.10 a.m. In each case an error of a few units should be expected and accepted.

15. The revised marks are 49 and 63. A deviation of a mark or two should be allowed.

16. It is 11 or 12 days within the year; but the graph is not one that allows of accuracy, and a large deviation is to be expected.

17. The town is 47 miles from A. A graphical solution should give the distance within a mile or two.

18. The successive values of RD + DP + DQ are, in miles: 20·9 20·0 19·1 18·2 17·4 16·7 16·0 15·5 15·2 15·3 16·0; best value for DR is 8·3 miles. All these numbers are given more accurately than can well be obtained by drawing.

19. When OX is greatest, OQ is 18 feet, to the nearest foot.

Section 22.

2. The perpendiculars from B and C on AO being parallel, one of them will lie on the opposite side of BC from O, and therefore outside the triangle. The image of O in it (which is either D or E) will lie still further outside the triangle.

3. (i) Suppose the two lines which lie up and down the page to converge either upwards or downwards; then each difference of two angles is equal to the angle between these converging lines. Suppose the two lines up and down the page to be parallel; then each difference vanishes. (ii) The four values are 109° 126° 75° 92°.

4. Six equilateral triangles just fill the space round a point; so do four squares, or three regular hexagons; and a pavement could be made in each of these cases. But three regular pentagons would not fill up the space round a point and four would overlap; and such figures cannot make a pavement.

5. The angles at the base must be made equal to $67\frac{1}{2}$ degrees.

6. The vertical angle is 26°, each base angle 77°.

7. The original data show that $P-R=20°$, while Q may have any value less than 116°. The further data give $P=64°$ $Q=72°$ $R=44°$; which may be the angles of a triangle.

8. The angle BAC is obtained by bisecting a right angle. Then ABD is made equal to 60°, the angle of an equilateral triangle; and ABE is made equal to 15°, obtained from 60° by bisecting twice.

9. (i) In degrees $\angle ACD = 180 - 2x$ and $\angle BCD = x - y$.
(ii) Lay down CB equal to 10 cm., and CD equal to 7 cm., at an angle of 20°. A lies on BD produced, at a distance of 7 cm. from C.

10. $2\angle AEB = \angle C + \angle D$, since the angle-sum property of the triangle ABE and of the quadrilateral ABCD show that each member of the equation is equal to $360° - \angle A - \angle B$.

11. The angle $ABC = 2\angle EBF = \angle EFH = 60°$, and so with $\angle ACB$.

12. Suppose the right angle of ABC to be at B, and D to be the middle point of the hypotenuse AC; suppose O such a point on AC that $\angle ABO = \angle CAB$, then also $\angle CBO = \angle OCB$. Hence $OA = OB = OC$, so that O coincides with D. *Or*, Consider the figure obtained by folding the two acute vertices into coincidence with the right vertex.

13. The three exterior angles are together equal to 4 right angles, so that $\angle CAB$ is right and the triangle is right angled.

14. Superpose the triangles so that EF lies on BC. ADC is then an isosceles triangle.

15. $\angle BOD$, the exterior angle of $\triangle OAB$, is equal to the angles OAB and OBA. $\angle DOC$ in the same way $= \angle OAC + \angle OCA$. The triangles named, having two pairs of angles equal, have also the third pair equal; that is, the halves of the angles ABC and ACB are equal.

16. Draw the perpendicular BK from B on AC. The bisector of ∠ABK cuts AC in the foot of the perpendicular from P.

17. Suppose the sides AB and DC of the quadrilateral ABCD to meet in P beyond B and C, while BC and AD meet in Q beyond C and D; and that the angles B and D of the quadrilateral are equal. Then

$$\angle A + \angle B + \angle Q = 180^\circ = \angle A + \angle D + \angle P.$$

18. Each angle of the new quadrilateral forms, along with the halves of two angles of the original quadrilateral, the three angles of a triangle.

19. The order of magnitude of the sides of △CDE is the same as of the opposite angles. The angles are $78\frac{3}{4}^\circ$ $56\frac{1}{4}^\circ$ 45°.

20. (i) Draw any line joining the two given lines. If it is possible to lay a second line equal in length to the first between the given lines in such a position that the intercepts on the given lines are equal, the given lines are parallel. (ii) Imagine that you start from the middle of a side and walk round the hexagon. You turn through 60° at each corner, so that after three corners your direction is parallel to your original direction.

21. With the degree as unit, the exterior angle at F is 360/7, so that the two interior angles G and E of the triangle FGE each measure 180 7. The angle GED is 720/7 or 103, since, together with ∠FEG and an exterior angle of the heptagon, it makes 180. The angle BGE is 540/7 or 77, since, along with an exterior angle, the angle FGE, and another equal angle, it makes 180.

22. First Solution: Consider the figure to be hinged at B C D E, and think of the forms assumed as DE and CB turn about E and B from one extreme position in which EDC is a straight line to the other in which BCD is straight. One form assumed is the regular pentagon, and in that form CD and BE are parallel; and in the whole motion these two lines are only once parallel. Therefore, when they are parallel, the pentagon is regular.

Second Solution: Join BE; and draw DF parallel to CB to meet BE in F. Show in succession that triangle ABE, triangle DEF, parallelogram DFBC are congruent to the corresponding figures in a regular pentagon.

23. 2160°.

24. (i) Divide the polygon into triangles by drawing lines from the vertices to an inside point O. The angles of these n triangles are those of the polygon together with 360° at O, so that the angles of the polygon are $180n-360$ degrees.

(ii) The polygon may be concave at E or at F or at both E and F. To make our proposition true for this case, we must consider that two lines, say FD and FE, form two angles, and we must take for the angle F of the polygon the angle through which a line will turn about F in moving from lying along FD to lying along FE, keeping all the time within the polygon.

25. Suppose the triangle ABC to have angle A greater than angle B; suppose the bisector of angle C and the perpendicular from C to meet AB in O and P. Imagine CB to shorten till B comes to B′, where CB′ = CA. During this process AB turns about A, and CP turns through the same angle about C and finally lies along CO. The base angles of the isosceles triangle are each $(A+B)/2$, so that OCP, the angle turned through by OP, is $(A-B)/2$, and in the numerical case 6°.

26. Suppose the bisectors of $\angle B$ and $\angle C$ to be BPQ and CPR, and the external bisector of $\angle A$ to be QAR. The angle-sum property for triangle PBC gives (in degrees)

$$\angle QPR = 180 - B/2 - C/2 = 90 + A/2.$$

Triangle ABQ gives

$$\angle PQR = \angle RAB - \angle ABQ = 90 - A/2 - B/2 = C/2,$$

and so for $\angle PRQ$.

27. BX and CY being the interior and exterior bisectors of B and C, then $\angle CBX = B/2$ and $\angle BCY = 90^\circ + C/2$, so that their sum is equal to $180^\circ - A/2$, less than two right angles. Hence BX and CY meet towards X and Y; that is, on the same side of BC as A.

28. While angle BAO increases from 58° to 180°, angle CAO diminishes from 58° to 0°. Angle CAO can be equal to BAO, but not greater.

Section 23.

1. Areas: $a \times b$ square inches and $c \times d \div 4840$ acres.

2. (i) $a + b/10 + c/100$ metres;
(ii) $(p + q/10 + r/100)(s + t/10 + u/100)$ square metres.

3. (i) 4160 litres; 55 drums. (ii) $a(100b + c)(100d + e)/10$ litres; number of drums $400a(100b + c)(100d + e) \div (\pi ffk)$.

4. The total number of stones being $31680/b$, we still need $31680/b - c$; numerical value 584.

5. $a + c$ lb. were bought for $ab + cd$ pence and sold for $240g$ pence; so that the average buying and selling prices were $(ab + cd)/(a + c)$ and $240g/(a + c)$ pence per pound. The gain was $240g - ab - cd$ pence on an outlay of $ab + cd$ pence, at the rate of $100(240g - ab - cd)/(ab + cd)$ pence on an outlay of 100 pence. Numerical values $11\frac{3}{4}$d. per lb., 15d. per lb., and (to the nearest integer) 28 per cent.

6. The wall having a volume of $36a \times 12b \times c$ cubic inches, and a brick 64 c. in., the number of bricks is

$$36 \times 12 \times a \times b \times c \div 64 \quad \text{or} \quad 27abc/4.$$

Numerical value 45000 (45360).

7. (i) The graph gives 0·4 and 2·6 as the two values of x.
(ii) The area is $3x - x^2$ square feet, and the graph shows for what values of x the area will be 1 sq. ft. Two adjacent sides must have lengths of 0·4 and 2·6 ft.

8. The area is greatest when $x = 6$, and is 33 sq. in. when $x = 4{\cdot}3$ or $7{\cdot}7$.

9. (i) $p = ab/4840$; (ii) $p = 25$; (iii) $q = c^2/4840$.

10. If the height is h feet, then

$$x^2 + 4hx = 36 \quad \text{and} \quad h = (36 - x^2)/4x.$$

The volume is hx^2 or $(36 - x^2)x/4$. For largest tank $x = 3{\cdot}5$, but graph will hardly give the decimal accurately.

11. Cost in shillings is $(3225 + 4x - 2xx)/6$. When $x = 10$ the cost is greatest and is £28 or £29.

12. Walls cost £$3(a + b)d/40$, floor costs £$ab/30$. Numerical results £7. 16s. and £5. 10s.

13. (i) £$(c + b - a)/2$ in first bank and £$(c + a - b)/2$ in second.
(ii) £$(2c + 2b - a)/3$ in first and £$(c + a - 2b)/3$ in second.

14. Length of inner edges $2p+2q-12$ inches. Area enclosed $(p-3)(q-3)$ sq. in.; numerical value 45 sq. in.

15. The volume is $xyz/1728$ c. ft. and the cost $xyzab/1728$ shillings. Numerical case, cost is 0·9 of a penny.

16. (i) The new average is 16 years and $(5n+32)/(n+4)$ months, or $16+(5n+32)/\{12(n+4)\}$ years, or $(197n+800)/(n+4)$ months; with many other possible forms. (ii) $n=8$.

17. The sum of the ages in years being

$$98k/12+(96+x)m/12+89n/12,$$

the average is this divided by $k+m+n$. Average may also be stated as 7 years and $\{14k+(12+x)m+5n\}\div(k+m+n)$ months or as 8 years and $\{2k+xm-7n\}/(k+m+n)$ months.

18. Respective heights above ground floor, in inches, $a+3b$ $a+4b$ $a+9b$ $a+(n-\ )b$; numerical values, in feet, 2 3 6 12. Length of carpet $a+(n-1)b+(n-1)m+24$ inches; numerical value $9\frac{7}{8}$ yards.

19. (i) M the number of sides is $360/(180-n)$. (ii) The value of n expressed in terms of M is $180-360/\text{M}$. Give M any values, say 3, 4, 5,..., and calculate the corresponding values of n, viz. 60 90 108 ....

20. If number thought of is a, the "answer" is $2a+3$, say b; so that a is $(b-3)/2$. The boy must therefore subtract 3 and then divide by 2.

21. (i) To five places 1 farthing is equal to £0·00104. Taking this as absolutely correct, we have n farthings equal to £$1·04n/1000$; and $0·04n$ is equal to 1 to the nearest integer so long as it lies between 0·5 and 1·5, *i.e.* so long as n lies between 12·5 and 37·5. But with the true value of 1 farthing, the difference between n farthings and £$(n+1)/1000$ is £$(n/24-1)/1000$, which is zero to the nearest thousandth so long as n lies between (but not equal to) 12 and 36.

(ii) £0·132.

22. First tramp keeps $(a+b)/3$ loaves, and therefore gives $(2a-b)/3$. Second gives $(2b-a)/3$ loaves. Third pays 3 pence a loaf, $2a-b$ pence to the first and $2b-a$ to the second.

23. They should pay £$50ax/(ax+by)$ and £$50by/(ax+by)$.

24. (i) The alloy contains $xz/100$ ounces of tin worth $xza/1600$ shillings, etc. Whole is worth $(xza + yzb)/1600$ shillings. (ii) $x + y = 100$.

25. The wife gets £$(x + y/240)(1 - m/10 - n/12)$.

26. For proof, suppose a and $a + 6$ to be the numbers. We have to prove the expressions $(a + 6)^3 - a^3 + 108$ and $9\{(a + 6)^2 + a^2\}$ to be equal. Each of them simplifies to $18a^2 + 108a + 324$. As a may denote any number we like to name, we have proved the property for all numbers. On the candidate's plan the property must be separately proved for every pair of numbers. To find it true for a few particular cases makes its general truth likely, but does not make it certain.

27. (i) £134. 11s. 3d.
(ii) $\{x$(£156. 11s. 6d.) $+ y$(£112. 4s. 4d.)$\} \div (x + y)$.

28. (i) Second error 720 times. (ii) $\{(b - c)a + 33\}/(e - d)$ times.

29. Average buying and selling prices, in shillings, $(ap + bq)/(a + b)$ and $(ap + bq)/(a + b) + 5$. Numerical values 50s. 4d. and 55s. 4d.

30. State A's money as $a - 1$ pounds $b + 19$ shillings and $c + 12$ pence, and we can subtract without "carrying." Difference is $a - c - 1$ pounds 19 shillings $12 + c - a$ pence.

31. a men and b women reap $1/x$ of the field in a day, and c men and b women reap $1/y$ in a day. By subtracting, we see that $a - c$ men will reap $1/x - 1/y$ or $(y - x)/xy$ in a day. Hence 1 man reaps $(y - x)/\{xy(a - c)\}$ in a day, and takes $xy(a - c)/(y - x)$ days for the whole field.

32. A and B do $1/z$ of the work in a day; similar expressions for B and C and for C and A. From the three expressions we see by addition that two men of A's power, two of B's power and two of C's will do $1/z + 1/x + 1/y$, or $(xy + yz + zx)/xyz$ of the work in a day. The three men A B C will do half that, and will take $2xyz/(xy + yz + zx)$ days for the whole. Another equally good form for the number of days is $2/(1/z + 1/x + 1/y)$.

33. The difference is $239(a - c)$ pence, and will buy 239 loaves.

34. The excess weight of $p/2{\cdot}2 - 30$ kilos is carried $8m/5$ kilometres. To the payment add the registration charge, and convert to shillings, and we have

$$\{(p/2{\cdot}2 - 30)8m/50 + 10\} \div 125 \quad \text{shillings;}$$

an expression which can also be put in simpler forms. Numerical value 3s. 9d.

35. (i) Gain is $(b - a)/a$ pounds on one pound, or $100(b - a)/a$ on a hundred. (ii) $100(b - a)/(a + x)$ per cent. The denominator being increased, the percentage is reduced. (iii) Numerical values 17 % 15 %.

36. $100x(a + b + x)/ab$ per cent. 4 per cent.

37. The change is greatest when the correction is additive; true area is then $ab(1 + x/100)^2$, and error $ab(2x/100 + x^2/10000)$. Numerical value 1·68 sq. in.

38. (i) Price of an egg in pence may be expressed as $x/12$ or $12/y$; so that $xy = 144$. (ii) 20d. or 21d. per dozen.

39. In a minute he goes $1056x$ inches, and takes $1056x/y$ steps· 136 steps.

40. (i) $z = xy \div 60(y - x)$. (ii) z hr. = 1 hr. 41 min.

41. (i) Between $45a \div 22(x + y)$ and $45a \div 22(x - y)$ miles an hour.

(ii) Between 26·1 and 25·4 miles an hour.

42. 2 hrs. 40 min.

43. For a mile they take $60/x$ and $60/y$ minutes, and A gains $60(x - y)/xy$ min. So that A gains z minutes in

$$xyz \div 60(x - y) \text{ miles or } 88xyz \div 3(x - y) \text{ yards,}$$

which is the distance run.

44. In a minute they do a/p and b/q laps respectively, or together $(aq + bp)pq$ laps. Together they do a lap in $pq/(aq + bp)$ minutes, and their respective fractions of the lap are $aq/(aq + bp)$ and $bp/(aq + bp)$.

45. (i) If the digits are a and b the numbers are $10a + b$ and $10b + a$; so that the sum is $11(a + b)$ and the difference $9(a - b)$. (ii) a and b are to have such values that $10a + b = 6(a + b)$ or $4a = 5b$; the only digits for which this can be true are $a = 5$ and $b = 4$, and the number is 54.

46. Selling price of a pound is $12p$ pence, so that buying and purification cost $12p \cdot 100/(100+q)$ pence. Purification therefore costs $1200p/(100+q)-p$ pence, which can also be written $(1100-q)p/(100+q)$ pence.

47. (i) $100(b-a)/a \qquad 100(c-b)/b.$ (ii) $(a+b+c)/3.$

48. $\{ax+by+cz+w(p-a-b-c)\}/p$ per cent.

49. $pq/(p+q)$ lb.

50. A does 100 yd. while B does $100-x$ yd. and C does $(100-x)(100-y)/100$ yd., so that A beats C by

$$x+y-xy/100 \text{ yd.}$$

Numerical value $9\frac{3}{4}$yd.

51. In metres per kilometre the falls are $(x-y)/a \quad y/(r-a) \quad x/r.$ For the fractional form, each of these must be divided by 1000.

52. Time for a mile down and back is

$$1/(a+b)+1/(a-b) \text{ or } 2a/(a^2-b^2) \text{ hours;}$$

time for two miles on the lake $2/a$ hours. The former is greater by $2a\{1/(a^2-b^2)-1/a^2\}$ or $2b^2 \div a(a^2-b^2)$ hours.

Section 24.

1. Segment of bisector measures 2·1 inches (2·0 is almost as near).

3. Triangles can be made from (i), (ii), (iv), any number of them from (iv). No triangle can be made from (iii), because the sum of the angles differs from 180°. In (v) we begin with the side AC; the two loci whose intersection would give B do not meet, and no triangle is possible.

4. Two distinct triangles. A or C may be 95°, but when we try to make a triangle with B = 95°, the two loci for A do not meet.

5. In the second paragraph the triangles must be congruent if AC > AB, but if AB > AC two forms are possible with the same dimensions. The two triangles in the third paragraph are congruent.

6. Prove that triangles PYZ QZX RXY are congruent.

7. The second triangle is isosceles, each side being half the corresponding side of the original triangle.

8. ABB′ and ACC′ are isosceles right-angled triangles, so that their acute angles contain 45°. The angle between BB′ and CC′ is 90°, since it is the third angle of a triangle whose other two angles measure 45°.

10. BC lies between 2·6 and 3·9 inches.

11. In the second paragraph let angle C be acute. Find R′ in CB, so that AR′ = AC. If we construct on AR′ a triangle congruent to PQR, Q is given by the intersection of R′B and a circle. One intersection is inadmissible, being on R′B produced backwards, and the other is at B. Therefore triangle PQR coincides with ABR′, and the stated properties follow.

12. The image E of A in BP gives a second corner of the triangle; EC meets BP in the third corner, the vertex, which is 2 inches from D.

13. A is 37° or 38°, D is 31°.

14. ∠ABE is 22·5° and ∠AEB is 112·5°.

15. Produce AC to E so that AE = AB. Then △ADE is congruent to △ADB, and greater than △ADC.

16. D and E lie on the bisectors of angles C and B respectively.

17. The angle HBF is a straight angle, for the three angles composing it are equal to the three angles of the triangle ABC.

18. The distance run is twice a diagonal.

19. AD must bisect the base.

20. Triangles ABC and DCB are congruent.

21. Denote the mid points of AX and BY by O and Q, and the intersection of AY with OQ by R. Join XR, and BR. Consideration of the folded figure shows ∠ARO = ∠XRO and ∠BRQ = ∠YRQ. Also ∠ARO = ∠YRQ; so that the vertically opposite angles XRO and BRQ being equal, BRX is a straight line.

22. (i) Denote the mid points of AD and BC by O and Q. Suppose the figure folded so that A falls on D. Then B will fall on C, and the crease passes through O and Q. The angles AOQ and DOQ are right, since they coincide in the folding and are together a straight angle; and so for the angles at Q. (ii) If BM is the perpendicular from

B on AD, we construct the right-angled triangle AMB, and find BM or OQ to be 1 foot long.

23. In the figure AD BD CD bisect the angles of the triangle: AF and BG are perpendicular to AB; and so on. The points within the angle EAF have A for the nearest point; every point within the area FADBG has some point of AB as the nearest, and so on.

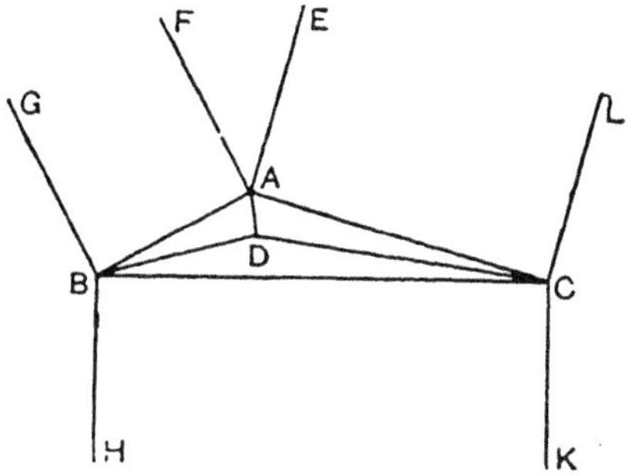

Fig. for Qn. 23.

24. The successive intervals are in hundredths of an inch: 13 53 27 40 40 27 53 13. Drawing will not give them to this accuracy.

26. O is on the locus of points equidistant from A and A′, so that OA = OA′. Likewise OB = OB′. Thus triangles AOB and A′OB′ are congruent.

27. (i) Draw PA; its degree of rotation measures the rotation of the transparent paper. (ii) When X, the new position of A, is given, we find Y, the new position of B, by making ∠BPY = ∠APX. (iii) Rotation about the mid point of BC brings B to C and C to B.

28. (i) follows Question 26. (ii) Subtracting the equal angles BCA and YZX, we have ∠SCA = ∠SZX, and triangles SCA and SZX are congruent. (iii) The drawing pin is to be at S.

29. The triangles OAB OA′B′ are congruent, as in Qn. 28; and the addition of a suitable angle to the equal angles AOB A′OB′ gives ∠AOA′ = ∠BOB′.

30. OA = OB = OC, so that triangle OCA is isosceles.

31. Triangles EAB and CAF are congruent.

32. The second property becomes the first when the line through C is perpendicular to AB.

33. If the perpendiculars are called EM and GN, the triangles ECM and GCN are congruent.

34. D being the image of B in AP, BP + PC is equal to DP + PC, and DAC is a straight line equal to BA + AC.

35. Take D the image of B in AQ, and proceed as in Qn. 34.

36. The intersection of two perpendicular bisectors is equidistant from the three vertices, and therefore lies on the third bisector.

37. The intersection of the bisectors of two angles is equidistant from the three sides, and therefore lies on the bisector of the third angle.

Section 25.

1. (i) k may be expressed as $V \div 0{\cdot}785d^2$ or as $1{\cdot}27V \div d^2$. (ii) 0·77 c. in. (iii) 1,300,000 in. or 21 miles.

2. (i) 93 gallons. (ii) x may be expressed as $p/11{\cdot}4$ or as $0{\cdot}088p$. (iii) 9 inches.

3. (i) 0·7 pint. (ii) $h = 132p/(a^2 + ab + b^2)$. (iii) 5·8 inches.

4. (i) $c = 6{\cdot}7$. (ii) $a = c(f - b)/f$. (iii) 0·75 mm.

5. (i) 1030 lb. to three significant figures. (ii) $d + 2t$ inches. (iii) $3{\cdot}14kd$ and $3{\cdot}14k(d + 2t)$ sq. in. (iv) £63. 13s.

6. (i) $c = \frac{3}{4}a^2b$.
(ii) With the same symbols as before, $b = \frac{4}{3}c/a^2$.
(iii) 0·8 inch (5/6 inch).

7. (i) $3600p/q$ miles an hour. (ii) 30 miles an hour (29·5). (iii) $105600r/s$ yards in a minute.

8. (i) 490 lb. (ii) $P = W/4 + 17{\cdot}5$. (iii) 117·5 lb.

9. (i) $H = LG^2/(3d)^5$. (ii) Head of 40 feet to the nearest foot.

Section 26.

2. There are two forms for the parallelogram. In one form the other side and diagonal measure 37 and 103 mm.; in the other they measure 62 and 125 mm.

5. (ii) Denote the intersection of AB and CD by O; complete the rectangle AOCK and AK is the perpendicular at A.

6. (i) Each side has 8 half panes and each end 6; there are 8 rows, each containing 5 whole panes, and 7 containing 6 panes; total, 82 panes and 28 half panes. (ii) Length of side 3·6 in.; angles 113° and 67°.

7. The triangle BEF is isosceles and has base angles of 60°; it is thus equilateral and has EF = BE = CD. Also EF is parallel to CD, since ∠BEF = 60° = ∠BAC. (This method is simpler than that suggested by the first paragraph, to use equality of triangles EAC and FAC to show FE ∥ AC.)

8. The diagonals, which bisect one another, must both be diameters of the circle, and the rhombus would have to be a square. The answer is Yes or No, according as the definition used includes the square as a rhombus or excludes it.

9. F is the intersection of the bisector of angle ACB with AB.

10. OO′PP′ is a parallelogram and O′P′ = OP; and so on. The construction amounts to sliding the circle bodily to a new position.

12. Through the middle point of CP draw a parallel to BC. Its intersection with AB is a point on the required line.

13. Let the perpendiculars be called BP CA DR. Drop a perpendicular BS from B on CQ, and show that CS = DR.

14. (i) Euclid's is the best proof:—D being the middle point of AB and E the middle point of AC, triangles DBC and EBC are equal, since each is half of ABC; the heights of their vertices D and E above BC are therefore equal. (ii) Draw PQ parallel to CD to meet AB in Q. Take a point R on AB, such that OR = OQ. PR is the required line.

15. (i) CF and DE are parallel because CD is equal and parallel to EF. Hence AX = XY and BY = YX (ii) $n-1$ equal lengths must be laid off along each of the perpendiculars.

16. Divide each side of an inch square into quarter inches. The scale is made by suitable joining of the points of division.

17. Bisect the external angle between AX and AY, and from the bisector cut off AR equal to the given length. A parallel through R to one of the given lines intersects the other in one end of the required line.

18. Suppose ABCD the parallelogram, A and C the smaller angles. Comparison of triangles ABC and BAD gives the base AC greater than the base BD.

19. PQ = RS because triangles PBQ and RDS are congruent; similarly PS = RQ.

20. (i) The quadrilateral whose vertices are the mid points of the sides of a given quadrilateral is a parallelogram, for two opposite sides of it are parallel to a diagonal of the given quadrilateral and equal to half the diagonal. The lines joining the mid points of opposite sides of a quadrilateral bisect one another because they are diagonals of a parallelogram. (ii) If these two lines bisect one another at right angles, the parallelogram is a rhombus; and each diagonal of the original figure is twice a side of the rhombus.

21. (i) Suppose the bisector of angle A meets BC (not produced) in P. Complete the rhombus ABPQ. BQ is the bisector of angle B, and meets CD beyond Q and D. (ii) Let the bisector of A meet BC produced at R. Complete the rhombus ABRS. BS is the bisector of angle B and meets CD (not produced).

22. If there were two such lines AEB and CED, A and C lying on OX, we should have the triangles EAC and EBD congruent, and the exterior angle of the triangle OCD equal to an interior and opposite angle. *Otherwise*: Suppose AEB one such line, A lying on OX; through B draw a parallel to OX; a line drawn from C on OX through E to D, such that E bisects CD, will have its end D on this parallel and not on OY.

23. (i) The block has 4 diagonals, and its faces have 12 diagonals.
(ii) One face of the block being ABCD and the opposite face EFGH, consider the diagonals DF and BH. The parallel edges BF and DH are equal; also, the triangles BCD and FGH being congruent, the lines BD and FH are equal. Thus BFHD is a parallelogram, and its diagonals bisect one another.

24. (i) E and F are the middle points of AB and CD. Joining E and F to one another and to B and D divides the original parallelogram into four equilateral triangles. (ii) Height of pyramid 2·4 inches (2·5 almost as close). Volume 3 c. in. (3·2).

25. Suppose the two lines in question to meet CA in L and M. Denote the intersections of CA with FE and GK by O and Q. EFGK is a parallelogram, and EF || GK. Hence triangles LOF and AQG are congruent, and LO = AQ. Similarly MO = AQ. So that OL = OM, and L and M coincide.

26. (i) Through P draw a parallel to HKL, and proceed by congruent triangles. (ii) Any two lines may be called HL and PR, and their middle points K and Q; so that the converse is untrue. (iii) Suppose HK = KL and PQ = QR, and KQ parallel to LR. Through H draw HT parallel to KQ to meet PQR in T. Then TQ = QR = PQ; and T and P coincide.

PART IV.

Section 27.

1. (i) The number of pounds may be expressed as
$p/20+(750-p)/8$ or as $375/4-3p/40$.
(ii) $3p/20-225/4$. (iii) £60; £11. 5s. (iv) 100 half-crown and 650 shilling tickets.

2. Nearest integer 21 miles (exact, $20\frac{5}{8}$ miles). [Logically, in a question like this, if an integral answer is desired, it should be the next integer *greater* than the exact answer, but with data so obviously rough this is an excessive refinement. In the present case the *next greater* integer and the *nearest* integer happen to be the same.]

3. 38% class A and 62% class B.

4. 1050 feet per second. **5.** 4·5 gallons to the nearest tenth.

6. The angles of the two partial triangles exceed those of the original triangle by the straight angle where the dividing line meets the base, so that (s being stated in right angles) $s+s=s+2$.

7. At 40 cents per 100 lb.

8. The three distances of the plate from the lens are in inches 4·235 4·068 4·013. x and y being expressed in inches, x may be put in the form $4/(1-4/y)$, which shows that when y is great x is nearly 4.

9. (i) 5·5 mm. (ii) 11 mm. **10.** (i) $k=1/720$ (ii) $4\frac{1}{8}$ tons.

11. (i) 4·5 revolutions. (ii) 5 mm. **12.** AB is 10/9 miles.

13. He must advance the price by 51 cents, to the nearest integer, and sell at 141 cents (1 dollar 41 cents).

14. (i) $T=18+0{\cdot}6t$. (ii) If the whole journey takes 30 min., he walks 4/3 miles.

15. Mean head 11 feet, to the nearest integer.

16. 17 volumes of A to 56 volumes of B.

17. $x=5$ $y=12$ $z=13$.

18. (iii) The sum must be paid with 4 florins and 9 shillings.

19. $21\frac{5}{8}$ miles from home.

20. 2·8 gallons (to nearest tenth) at 10s. $3\frac{3}{4}$d. a gallon, and 10·4 lb. at 3s. $3\frac{1}{2}$d. a pound.

21. Lengths 352 and 440 feet; speeds 70·4 and 88 feet per sec.

22. The angles are, in degrees, $x-y+80$ $y-z+60$ $z-x+40$; and for the special case $x=70$ $y=z=90$.

23. (i) $a=70$ $b=85$; (ii) $z=£19$.

24. Suppose speed of stream in miles an hour to be x, and speed of rowing y. In 18 min. the man rows $0{\cdot}3(y-x)$ miles up stream. He has then to row $0{\cdot}3(y-x)+1{\cdot}5$ miles down stream, and we get the hours taken by dividing this expression by $y+x$. By equating the time he rows after meeting the bottle to the time the bottle takes to reach the starting place, we get

$$0{\cdot}3+\{0{\cdot}3(y-x)+1{\cdot}5\}/(y+x)=1{\cdot}5/x\,;$$

this equation proves to be independent of y, and gives $x=2{\cdot}5$. [We can see that the result must be independent of y by thinking of all motions as relative to the water. Relative to the water, the bottle is at rest, while the man rows away for 18 minutes, and will take 18 min. to come back to the same portion of the water. The starting place on the bank comes up stream (relative to the water) 1·5 miles in 36 minutes, *i.e.* at 2·5 miles an hour.]

25. $a=0{\cdot}700$; $b=0{\cdot}0008$. Price for 30 miles £4. 10s.

26. Stream flows 4·3 miles an hour ($4\frac{1}{3}$). Depth 3·6 feet ($3\frac{5}{8}$).

27. $t=0{\cdot}5$; $d=46$.

28. $x=3{\cdot}2$ (or 35/11) $y=9{\cdot}5$ (or 105/11).

29. (i) $x=10$. (ii) $x=2$ $y=3$.

30. $x=42$ $y=33$.

31. $x=1$ $y=1{\cdot}5$ $z=2{\cdot}5$.

32. $a=1$ $b=2$ $c=1$.

33. $a=0{\cdot}8$.

34. A first is 5 points, a second is 3, and a third is 1.

35. 320 parts of marl (more accurately 316).

36. There were 56 c. cm. of gold and 7 c. cm. of copper, to the nearest integer. Pure gold would have weighed 1216 grammes.

37. 51 children on books.

38. 1 farthing; 1·1 farthings.

39. 700 men. (Use x for number of men in side of one of the squares.)

40. Half-a-crown an hour.

41. 5 lb. of cheaper tea to 4 of dearer.

42. 1 cwt. free.

43. 1 cwt. free.

44. £519.

45. £550.

46. (i) $10(x+2)+2x+5$ or $12x+25$. (ii) Remainder is 1. (iii) The numbers are 34 and 36.

47. 30 miles.

48. 7 boys; 25 apples.

49. He gave £45 and £30.

50. Cow cost £6, sheep £2. 5s., and pig £1. 12s.

51. 25 sovereigns and 50 shillings.

52. Brother's age 14 years, girl's 12.

53. A had 20s. and B 10s.

54. £1856. 5s. and £1987. 10s.

55. The numbers are 53 and 35.

56. The numbers are 68 and 57.

57. 24 ft and 17 ft.

58. 35 % of A and 65 % of B.

59. In a years. This may also be expressed as $b+c+d$ years, and in other ways.

60. The shares are respectively £15,000, £4615. 8s., £3076. 18s., £2307. 14s.

61. Eldest gets £5801. 5s., second and third each £4350. 18s., and fourth £2900. 12s.

62. Lengths 5·6 and 3·6 miles.

63. At 2 o'clock.

64. From P to Q is 20 miles.

65. Speeds 15 and 20 miles an hour; distances 40 and 56 miles.

66. Course is a mile long; A takes 4 min. 57 sec., and B 5 min. 30 sec. Suppose the course k yd. long, A's speed x yd. per sec., B's speed y yd. per sec. The three conditions give $k/x=(k-176)/y$ $k/y=60+(k-160)/x$

$$(k-160)/y=3+k/x.$$

The first and third equations give $y=16/3$; then the first and second give $(k-176)(k-160)=(k-320)k$ or $k=1760$.

Section 28.

3. There is only one point when AB = AC, the bisector of the external angle at A being parallel to BC.

5. The locus is a circle, but at this stage the student cannot prove it.

6. An intersection of the curve with XQ would involve a position of AB with both A and N upon XQ; in which case AB could not pass through Z.

7. The pupils may be told that the curve is called a "parabola."

9. The pupils may be told that the curve is a portion of an ellipse.

11. The locus is a semi-circle, as the pupils will readily see, although they cannot prove it.

12. Greatest breadth 2·6 inches.

13. In a minute E travels 15·4 feet horizontally, and is then 6 feet below AB.

15. Draw OP in such position that ∠AOP = ∠XYZ. Through C the intersection of OP with the curve draw MN parallel to OA. Divide OM at M′ in the ratio 2 : 3. Draw M′C′ parallel to OA to meet the curve in C′. Then ∠AOC′ : ∠C′OC = 2 : 3.
The protractor values are 32°, 64°, 29°, 43° or 44°.

16. AB can turn 43° on each side of AD.

Section 29.

1. Volume of wood in cubic inches:

$$2x(bc + ca + ab) - 4x^2(a + b + c) + 8x^3.$$

2. ABCD $= (a + c)(b + d)$; EFGH $= (a - c)(b - d)$; the difference $= 2(ad + bc)$.

3. The two squares have sides of lengths 6 and 2 inches; the single square measures 4 in. in the side, and each rectangle 6 in. by 2 in. If the sides of the former squares are denoted by a and b, the property is $(a - b)^2 + 2ab = a^2 + b^2$.

4. $ca^2/4840b$ inches of rain fell. Numerical value 0·23 in. to two significant figures.

15. If the three points are not collinear, AP + PB is greater than AB, and what we are asked to show is that $(AP+PB)^2 > AB^2$.

16. In the illustration x is a multiple of a, and y of b, while the relation is true without this limitation.

17. Area 32 or 33 sq. cm.; $b = 8{\cdot}8$ cm.

18. Take the 3-inch side as base AB, and two positions for the vertex are given by the intersection of a circle of radius 5 in. and centre A with a parallel to AB 4 inches away.

19. There are two triangles not identically equal.

20. The sides measure 2·8 and 4·6 inches (2·79 and 4·58).

21. A lies on a parallel to BC, the extreme positions being those in which AB = AC and AB = BC, 1·5 inches apart (1·47). [No restriction to the plane of the paper being stated, it is more correct to state as locus of A all the points on the cylinder described by the segment as it revolves about BC.]

22. Other side 6·2 cm.

23. Suppose ABC the triangle, F the mid point of AB, and FE parallel to BC meeting AC in E. We have two methods of proof. (i) Draw FD parallel to AC to meet BC in D; triangles AFE and BFD are congruent and CDFE is a parallelogram, whence the result. Or (ii) FE being parallel to BC, △EBC is equal to △FBC, which is half the triangle ABC; hence CE = EA.

24. D E F being the mid points of BC CA AB, triangles EBC and FBC are equal, since each is half ABC; hence FE is parallel to BC. Similarly DE || AB and DF || AC. Hence FE = DC = BC/2. (See also answer to Qn. 14 of Sec. 26 on page 389.)

25. P must be inside the triangle DEF of the preceding question.

26. Measurement shows the triangle to be right-angled, area 24 sq. cm.

27. Triangle ODR can be divided into 9 triangles congruent to OBP.

28. DAB being congruent to CBA, CD is parallel to AB; whence triangles ACD BCD are equal.

29. BD = 8 cm. (accurately).

30. (i) BM DN being perpendiculars let fall on AC, and O the intersection of the diagonals, BM = DN; whence triangles BOM DON are congruent. (ii) If angles BAM and DAN are equal, triangles BAM and DAN are congruent, and M N coincide with one another and with O.

32. The join of two consecutive mid points is parallel to a diagonal.

33. (i) Take any line for BD a diagonal of the quadrilateral, and call its middle point O; draw AC through O, making OA and OC unequal, and we have a non-parallelogram satisfying the first condition. (ii) If O bisects BD and in addition AB || CD, triangles OAB OCD are congruent, and the figure is a parallelogram.

34. 121 acres.

35. Angle 67°; area 23 sq. cm.

36. Other side of rectangle measures 9·3 cm. or 3·6 in. [It is a pity to mix up the imperial and metric measures. This question is included only because such questions are customary.]

37. Triangles YAX and PAX are equal; therefore triangles OXY and OAP are equal; that is, OA . OP = OX . OY.

38. On OX mark off OA 4·2 cm. in length, and AD 2·9 cm. On OY, at right angles to OX, mark off OC 3·7 cm. long, and complete the parallelogram OABC. Let DB meet OY in E, and the required number is the length of CE in cm., viz. 5·4 (5·3 is almost as near).

40. (ii) On one side BC of the given triangle ABC mark off BD equal to a given side of the required triangle. CE parallel to AD cuts AB in E, so that △BED has the required area. On a parallel to BD through E take F, in such position that BF is the second given side; and BDF is the required triangle.

41. The diagonal measures 5·6 inches. (For construction see answer to Question 38 of this section.)

42. The area of the rhombus is the same as of a triangle whose base and height are equal to the diagonals. The second diagonal measures 4·1 cm. (4·07).

44. The sides measure 4·1 cm. (4·13).

45. Distance BC to DA is less than AB, which is less than distance AB to CD, which is less than BC.

46. By subtracting the triangles we find area 9·8 sq. in. (accurately 9·776).

47. As P moves from A to O, the diagonal of the inscribed square diminishes from length AC to length BC, and of course the square diminishes with it.

48. Rectangle is greatest when AP = 4·5 in., its side measuring 6·4 in. (6·36).

50. Begin construction with a triangle of sides 9·8 – 6·7, 5·5, 4·2. Area is 34 sq. cm. (35 almost as close). To lengthen parallel sides by x cm. is to add a parallelogram of base x cm. and of the height of the trapezium, and so of area hx, where $h = 4{\cdot}2$ cm.

51. The line bisects the second parallel side also.

52. AD = 8·3 cm.; CD = 11·5 cm.; area 97 sq. cm.

53. $\angle B = 86^\circ$ $\angle C = 74^\circ$ $\angle D = 90^\circ$; areas ABC = 27 sq. cm., ACD = 26 sq. cm., quadrilateral = 53 sq. cm.

54. Lawn has area 1040 sq. ft. (1037).

55. Height of triangle 2·3 in. (2·28 more accurately).

57. Draw CK parallel to DB to meet AB in K. Then AP = $\frac{1}{3}$AK = 1 inch pretty exactly. Area of field 2·4 acres.

58. Area of field 3 acres (more accurately 2·9). BP is 2·6 cm.

59. Area is 266 sq. cm.

60. On OB mark off OE = 4 cm. V is to move so that quadrilateral VAOE = 16 sq. cm., or so that triangle VAE = 2·1 sq. cm. (since triangle AOE = 13·9 sq. cm.). So that locus of V is a straight line.

Section 30.

1. The heights in feet are 100 95 65 45 (or 40) 35 – 20 – 45 – 50 – 45 5 20 30 35 (or 40).

2. Freezing and boiling points of B are 0° and 100°. A reads 32° when B is zero, and B reads – 18° when A is zero. Readings of A and B are the same at – 40°.

5. Value of difference 0·79; so that 8·0425 sq. in., the mean between the extreme values, is within 0·79/2 or 0·395 sq. in. of the exact value. For practical purposes, we say that the area is 8 sq. in. with a possible error of 0·4 sq. in.

6. Values of x are 4·3 and $-0·3$.

7. y has its greatest value, 5, when x is zero; and tends to zero when x becomes large.

8. The equal values of x and y are 0·4 and $-2·1$ (0·36 and $-2·06$).

9. The expression is positive if x is less than 1 or greater than 5. The graph should be drawn for confirmation.

10. $a=9/2$ $b=-4$. The two required values of the expression are -2 and 1/17; which latter may also be expressed as 0·059.

11. $a=-1$ $b=-1·2$ $c=2·6$. The value of y is $-0·6$ when $x=-2$.

12. (i) $d=1/2$. (ii) $a=1·25$, $c=1$.

13. $y+x=(x^2+ab)/(x-a)$. With given values of a and b this becomes $(17x^2-12500)/(17x-550)$.

Section 31.

4. The fraction planed away is 0·2 (more accurately 0·17).

5. The school is 1·3 miles (1·32) from the road.

6. AC = 4·0 inches (4·01). Area = 12·3 sq. inches.

7. Height of pole 11 8 metres. **8.** Area 57 sq. cm.

9. 10 acres to the nearest rood.

11. Third sides measure 5·6 and 3·2 inches (5·61 and 3·19).

12. One lap measures 4200 yards (4160), the three sides measuring 1100 1600 1500 yards (1100 1590 1470).

14. The east-west dimension of the cellar lies between 16·5 and 40·25 feet; no information about north-south dimension. The treasure lies within 16·75 feet of the north wall, and within the same distance of the east wall; between 16·5 and 23·5 feet of the west wall; no information about distance from south wall.

16. Lay down BC. At the proper angle with BC lay down the direction BA, and cut off BD = 3 cm. Complete the isosceles triangle ADC.

$\angle$BAC = 57° $\angle$ACB = 82° AB = 8·9 cm. AC = 5·9 cm.

17. We first make the triangle PQR, and then place PST symmetrically in it. SQ = 7·1 cm., which represents 7·1 feet (7·2 is almost as near).

18. Length of shelf 10·8 or 10·9 inches, height 8·8 inches. Length for n books is $3·9 + 2·3n$ inches.

19. The point is equidistant from D and d, and therefore $12\frac{3}{4}$ inches from ycd. It is equidistant from CD and cd, and therefore also $12\frac{3}{4}$ inches from CD.

Section 32.

1. C has its greatest value 53° (53·1°) when A is 90°. When C = 28°, A is either 36° or 144°.

2. Latest sunrise is about 28 to 30 Dec., earliest sunset 13 or 14 Dec., shortest day about 20 to 23 Dec.

3. The greatest value of AB is 12 cm., the least value 4 cm. The area is greatest when AB = AC.

5. The area is greatest when OQ is 2 inches, and is 1 sq. in. when OQ is 3·4 or 0·6 inches (3·41 or 0·59).

6. The square root is 5·8 (5·83). **7.** Cube root is 4·6 (4·64).

8. Total duty is $100x - 10x^2$ pence, and is greatest when the rate is 5 pence per lb.

9. Total cost of the voyage is £2250 £1896 £2000 for the three speeds named. Least possible cost £1890; for this the speed must be 16 knots (15·88).

10. Cost of journey in £ is $12·5x + 525/x$. Most economical speed is 6·5 miles an hour (6·48), and cost then £160 (162).

11. a = £1; b = 1 shilling. [Note that a and b are sums of money. If $ax + bx^2$ had been given as the cost in shillings, a and b would have been mere numbers.]

12. The cost in £ when the voyage takes n days is $56n + 1400/n$. For 8 days cost is £623. Least possible cost £560, when voyage takes 5 days.

13. For the given values of x the expression has the values 1·3 and 0·3 (1·31 and 0·31).

14. y has its least value 0·2 (0·16) when $x = 3·2$ (3·16).

15. The expression lies between the stated values if x lies between $-3{\cdot}8$ and $-2{\cdot}8$ or between $-1{\cdot}1$ and $+0{\cdot}7$ or between $+3{\cdot}1$ and $+3{\cdot}9$. (These six numbers, more accurately stated, are $-3{\cdot}76$ $-2{\cdot}77$ $-1{\cdot}12$ $+0{\cdot}69$ $+3{\cdot}06$ $+3{\cdot}89$.)

16. The maximum value of y is 2 (when $x=1$), and the minimum 0 (when $x=0$).

17. The expression lies within the given limits, while x lies between $-1{\cdot}8$ and $-1{\cdot}5$ or between $+0{\cdot}9$ and $+1{\cdot}1$. (More accurately these numbers are $-1{\cdot}79$, $-1{\cdot}53$ or $-1{\cdot}54$, $0{\cdot}87$, $1{\cdot}12$.)

18. The missing entries are 0 0·4 3·9 6·9. A triangle of side 2·5 in. has area 2·7 sq. in.; a triangle of area 5 sq. in. has a side of 3·4 in.

19. The values for the table are, in cm., 48 46 45 44 44 44 45 46 48 50 52. The minimum is given by CP = 11 cm., and angles then measure 47°. The join of B to the image of A in CD gives the point P.

20. The graph gives no information about the rain; $x+y$ is equal to 24 for every point on it.

21. $x=17/24=0{\cdot}71$ $y=7/24=0{\cdot}29$.

22. $x=2{\cdot}12$ $y=-0{\cdot}16$.

23. The roots are $-1{\cdot}5$ $-0{\cdot}3$ $+1{\cdot}9$.

24. Both are positive when $x>-0{\cdot}4$.

25. $x=2{\cdot}3$, $y=2{\cdot}6$, and $x=-1{\cdot}3$, $y=-4{\cdot}6$.

26. The straight line and the curve meet without crossing one another.

27. They meet in three points.

28. The second graph is $(x^2+2)/4$, and the roots are $-1{\cdot}1$ $-0{\cdot}3$ $+1{\cdot}6$.

29. The second graph is $y=3/2x$ and the roots are $-2{\cdot}1$ $-0{\cdot}5$ $+2{\cdot}7$. For the first root $-2{\cdot}2$ is almost as good.

30. Original volume 14,400 c. cm. When A is in surface volume is 8100 c. cm., and inclination of AB 31°. Volume when x cm. of AE is under water is $450x+8100$ c. cm., and when y cm. of AB is under water $270y$ c. cm. Inclination of AB when one quarter has run away is 22°, and when half has run away 34°.

31. Length of bottom 62 in.; area sheet iron 14,600 sq. in. or 101 sq. ft.; capacity 420 gallons (418). Data in gallons for graph 0 17 68 150 240 330 420. At 40 gallons height is 15 in., and at 200 35 or 36.

32. The values of y are $-2{\cdot}36$ $-1{\cdot}71$ 0 $2{\cdot}31$. Put $10+w$ for x, and we have $y=2w+0{\cdot}3w^2+0{\cdot}01w^3$ $z=2w$; so that $y-z=0{\cdot}3w^2+0{\cdot}01w^3$, and this is less than $0{\cdot}1$, so long as w lies between $-0{\cdot}5$ and $+0{\cdot}5$.

33. $y=x^2+3x+2$.

34. There are two intersections, one where $x=4$ and $y=6$, the other where $x=0{\cdot}2$ and $y=5{\cdot}4$. ($0{\cdot}1$ is almost as good as $0{\cdot}2$ for the second value of x.)

35. The expression for y is x^2-3x-8, and is least when $x=1{\cdot}5$.

36. The relation is $y=0{\cdot}37x+0{\cdot}43$. For $y=0{\cdot}5$ x is $0{\cdot}2$ (or more exactly $0{\cdot}19$).

37. (i) We denote the base by O and stations 10 20 30 and 40 days' journeys out from O by A B C D. We call one day's provisions for one sled's complement of men and dogs a food-unit. Sled 1 can feed the whole expedition to A and keep food enough to get back to O. Then sled 2 provides 20 food-units to take the expedition to B, and uses the remaining 20 units to get back to O. Sled 3 could get to C and return to O on its own supplies; suppose it to go x days beyond C, and then to turn. It will have done 10 days of the return journey in $40+x$ days, and will then meet sled 1 at a point $20+x$ days out; each sled then having $20-x$ food-units. They return 10 days' journey together, the time then being $50+x$ days from the start, and then ($10+x$ days out) they have $10-x$ food-units each, and meet sled 2 with $30-x$ units. As they require $3(10+x)$ units to take them home, they will just have consumed their food on arrival if $x=10/3$. Sled 3 thus attains a point $33\frac{1}{3}$ days from the base.

(ii) On the plan just discussed, sled 1 did on its second journey a useless piece of travelling by meeting sled 3, while the latter still had food. Let sled 3 go to D, 40 days out, and return to B; its food is exhausted, and it wants help. Sled 1 must therefore delay its second start by 20 days so as to arrive at B simultaneously with sled 3;

its provisions bring them back to A, where sled 2, having waited 20 days at O, meets them with 30 food-units, which will take them home to O.

38. A change in c moves the curve bodily parallel to the y-axis.

Section 33.

1. The angles are 90° 53° 37°.

2. It is the square on the diagonal.

3. The side measures 5·2 inches (5·20).

4. The builder's method is, from a corner of the floor, to measure 3 units along one edge of the floor and 4 along the other. The distance between the points so found should measure 5 units.

5. Draw CM perpendicular to AB. Calculate CM and AM from triangle CMA, and then CB from triangle CMB. CB is 6 miles 4 furlongs.

6. The join measures 4·25 cm. **7.** Area 472 sq. ft.

8. Length of side 15 cm. **9.** Length left over $13\frac{3}{8}$ inches.

10. Show that the line joining the middle point of the hypotenuse of the right-angled triangle to the opposite vertex divides the triangle into two isosceles triangles of the named dimensions.

11. Triangle ACB is congruent to AHE, and DFB to DGE. BC produced cuts the figure into the two squares.

12. GP must be equal to AD. Rotate triangles BPA and FGP about B and F, each through a right angle.

13. EN = 4·7 cm.; EC = 4·8 cm.; area ECD = 5·1 sq. cm. (More accurately the numbers are 4·66 4·78 5·12).

14. Length 7·1 cm. (7·07).

15. The square on each equals $CA^2 + AB^2 + AP^2$.

17. If the edges measure a b c inches, we have $b^2 + c^2 = 8^2$, etc., three linear equations for a^2 b^2 c^2. The values of a b c are 1·3 4·8 6·4.

18. The distance is $\sqrt{2p^2 + q^2}$ miles; numerical value 15·8 miles.

19. (i) $PS^2 = PQ^2 - QS^2 = 4QS^2 - QS^2$.

(ii) $\triangle PSU = \triangle QSU = 2\triangle TSU$, so that $\triangle PST = 3\triangle TSU$.

20. The triangle has a height which is less than its side, and therefore less than twice the height of the rectangle.

21. OPRQ is a rhombus composed of two equilateral triangles. $QR^2 = 49$ sq. cm.; $OR^2 = 147$ sq. cm.

22. (i) $\angle EAB = \angle EAD + \angle DAB = 60° + 30°$.
(ii) $AD^2 = AB^2 - DB^2 = 3AC^2$.

23. Distance AC is 4 miles. (An unnatural question.)

24. (i) Each is equal to $OA^2 + OB^2 + OC^2 + OD^2$, O being the intersection of the diagonals.
(ii) $DA = 7{\cdot}14$ cm. (iii) $BD = 9{\cdot}8$ cm.

25. $OG = \sqrt{12^2 + 16^2 + 15^2}$ ft. $= 25$ ft.

26. $EF = 4{\cdot}24$ in.; area $EBCF = 7{\cdot}5$ sq. in.

27. O being the intersection of the diagonals, we write $OA^2 + OB^2$ for AB^2, etc., and the relation becomes $BO^2 = DO^2$.

28. $AP = 2{\cdot}9$ in. or $0{\cdot}6$ in. (more accurately $2{\cdot}95$ and $0{\cdot}55$). Q is given as the intersection of the locus with a circle of centre A and radius 3 in.

29. (i) AP and BP are equal when $x = 4{\cdot}9$; the perpendicular bisector of AB cuts KL in P. $AP^2 + BP^2$ is equal to AB^2 when $x = 10{\cdot}2$ or $1{\cdot}8$; the semicircle described on AB cuts KL in the two positions for P.

30. Make $(2x)^2$ equal to the given number n. Then draw a right-angled triangle with hypotenuse equal to x^2+1 or $n/4+1$, and one side equal to x^2-1 or $n/4-1$; the remaining side is $2x$ or $\sqrt{n}$.

31. $CD^2 = AD \cdot BD + CB^2 = 2CB^2$, so that CD is the diagonal of the square on BC.

32. Denote the middle point of AB by O, and the foot of the perpendicular from P by M. Then

$$PA^2 - PB^2 = MA^2 - MB^2 = 4OA.OM,$$

so that OM is constant.

33. If AM is the perpendicular on BC, and O the mid point of BC, then $BE = 2OM$. Numerical case $BE = 1{\cdot}6$ cm.

34. $AN = 5$ cm., so that $NQ = 9$ cm., and $PQ = 15$ cm.

35. If a and b denote the length of the radius and the distance of E from O, $BE^2 = a^2 + b^2$ and $CF^2 = (a+b)^2 + (b-a)^2$.

36. The boy would take 67 steps along BC. Along AC the man takes 78 and the boy 104.

37. $AC = 6{\cdot}5$ cm., and $6{\cdot}5^2 = 6{\cdot}0^2 + 2{\cdot}5^2$. Area 18 sq. cm. (17·64).

38. To the nearest tenth of an inch the bannister measures 43·0 inches.

39. $BD = 3$ in.

40. The values are 15·2 and 4·5.

41. (i) The area is $AB(AP + BQ) \div 2\sqrt{2}$.

(ii) $\triangle AUB$ is equal to twice $\triangle AUS$, which is equal to $\triangle PUR$ since $\triangle SAP = \triangle RAP$; and so on.

42. The projection of ED on the dotted line has length 20·5 yd.; the projection of BC on CD 30·1 yd. Area is 1850 sq. yd.

43. The five routes for which AX is 0 1 2 3 4 miles take respectively 2·76 2·64 2·61 2·64 2·72 minutes. The graph indicates $AX = 2$ miles as the quickest route, and calculation confirms that this is better than $AX = 1{\cdot}9$ or 2·1 miles.

44. Spider goes 41 feet, passing corner at height of 5·6 feet.

45. Lengths are (i) 20·8 in., (ii) 26·8 in.

46. The side of the tray is a trapezium, whose breadth is the hypotenuse of a right-angled triangle of sides 1″ and 5/8″, and therefore 1·18″. The lengths of the parallel sides of the trapezium are 4″ and $5\frac{1}{4}$″, and of the slant sides $\sqrt{1{\cdot}18^2 + (5/8)^2}$ inches or 1·33″.

47. Lengths are (i) 30 ft., (ii) 31 ft., (iii) 29 ft. Distance in a straight line 22 ft.

Section 34.

1. It falls 400 ft. in 5 sec., and takes 2·5 sec. to fall 100 ft.

2. Internal diameter 1 inch.

3. Boils at 205°, to the nearest integer.

4. Height is 80 ft. after 1 and after 5 sec., 128 ft. after 2 and after 4 sec.

5. 69 and 87 yards (or more accurately 69·3 and 87·3).

6. 60.

7. Hypotenuse 13 inches.

8. Length 200 feet, breadth 128 ft.

9. Year of birth 1806. The equation to be solved is

$$x^2 + 22 = 1914 - x.$$

10. 129 revolutions per minute.

11. Time of rise 2 sec. The second value (6 sec.) given by the quadratic is the moment of passing the same spot on the descent.

12. (i) 0·7 pint; (ii) diameters 1·9 and 4·2 inches.

13. Denote the line by AB and let the smaller part be next A; its length is 4·14 cm. The other root $-24·14$ means that, if BA is produced beyond A to P, so that AP = 24·14 cm., then $BP^2 = 2AP^2$.

14. $+1·74$ and $-5·74$. **15.** 0·083 and -12.

16. $+9·3$ and $-2·3$.

17. (i) 3 and 4. (ii) $+0·603$ and $-11·603$.

18. $x = 18$. Volume 3·2 c. ft.

19. (i) 3/7 or 0·43. (ii) 2·37 and 0·63.

20. Greater root 1·53. Left member then has value 0·011.

21. Roots $+4·7$ and $-1·4$. For positive root left member becomes 44·05, and right member 44·29.

22. (i) $x = 50/83 = 0·60$, $y = 50/31 = 1·61$. We substitute a and b for $1/x$ and $1/y$, and solve for a and b. (ii) $x = +8·2$ or $-7·1$.

23. (i) The relation is true when $x = -7·5$. (ii) Left member minus right has the value 0; so that the relation is true for all values of x. (iii) Left member minus right has the value -16; no value of x makes relation true. (iv) Left member minus right has the value $5x^2 + 4$. This is positive for all values of x, and no value can make the relation true.

The student is at the present stage unacquainted with the terms "real" and "imaginary," and should not use either term in connection with number (iv).

24. (i) $x = +2$, $y = -3$. (ii) $x = 8$ or -7.

25. (i) $x = +0·04$ or $-2·04$.
(ii) $x = 5/26 = 0·192$, $y = 1/3 = 0·077$.

26. (i) $x = 0·1$, $y = 0·4$. (ii) $x = 1/6 = 0·33$. (iii) $x = +2$ or -7.

27. (i) $x = y = 20$. (ii) $x = 1$. (iii) $x = +13$ or -12.

28. $x=6$ to within a tenth (more exactly 6·02).

29. By solving the equation as a quadratic in $\sqrt{R}$, we find $R=1{\cdot}53$; whence $y=6{\cdot}47$, and depth is 6 ft. 6 in. to nearest inch.

30. $x=\frac{1}{2}$ $y=1\frac{3}{4}$ or $x=-2\frac{1}{2}$ $y=4\frac{3}{4}$.

31. The second route is shorter if x is greater than 6·11 (accurately $6\frac{1}{9}$), the two routes are equal when x has this value, and the first route is shorter if x is less than this value.

32. $x=1{\cdot}56$, $y=2{\cdot}56$; and $x=-2{\cdot}56$, $y=-1{\cdot}56$.

33. (i) The graph is a circle, radius 3, centre at origin.
(ii) $x=2{\cdot}87$, $y=0{\cdot}88$; and $x=-1{\cdot}25$, $y=-2{\cdot}73$.

34. $x=y=4/3=1{\cdot}33$; and $x=7/3=2{\cdot}33$, $y=-2/3=-0{\cdot}67$.

35. $x=0{\cdot}89$, $y=0{\cdot}77$; and $x=-3{\cdot}4$, $y=-7{\cdot}8$.

36. (i) $x=25/7=3{\cdot}57$. (ii) $x=4$, $y=3$, or $x=-0{\cdot}27$, $y=0{\cdot}70$.

37. $x=8{\cdot}5$, $y=0{\cdot}73$ or $x=-1{\cdot}2$, $y=-5{\cdot}3$.

38. (i) $x=5{\cdot}12$. (ii) $x=7$, $y=1$; $x=-6$, $y=-12$.
(iii) $x=7{\cdot}06$ or $-4{\cdot}72$.

39. (i) $x=27$ or 16. (ii) $x=6{\cdot}2$, $y=2{\cdot}8$.

40. (i) $x=0{\cdot}847$ or $-1{\cdot}18$. (ii) $x=3$, $y=6$; $x=-3$, $y=-6$.

41. (i) $x=1$, $y=-2$; $x=-11/9=-1{\cdot}22$, $y=-1/3=-0{\cdot}333$.
(ii) $x=0{\cdot}744$ or $-2{\cdot}04$.

42. (i) $x=2{\cdot}62$ or $0{\cdot}38$; the corresponding values of $x+1/x$ are 3·002 and 3·012.
(ii) $x=8$, $y=7$, $z=5$, or $x=-8$, $y=-7$, $z=-5$.

PART V.

Section 35.

1. AC = 5·94 in. AC is nearest to the centre.
2. Distance is 2·42 in.
4. The centre is inside an acute-angled triangle, on the hypotenuse of a right-angled triangle, and outside an obtuse-angled triangle.
6. C's distance is 5 cm. The sum of the distances of B and C is twice A's distance.
7. Distance from B and C is 5·5 cm. (5·46).
8. Radius 4·2 cm. (4·24).
9. BD = 6 cm. ∠BAC = ∠CAD = 41° (42° is nearly as close). Radius 3 cm. (3·02).
10. Suppose K the centre, KM and KN perpendiculars on the chords. KM = KN, and triangles KMO KNO are congruent. Thus OM = ON, and the result follows.
11. The centre is the intersection of the perpendicular bisectors of the chords. If PC = CQ, this point is at C, and AB is a diameter.
12. Draw radii of a circle at intervals of 15° (by bisecting and re-bisecting an angle of 60°).
13. Triangles ABC AOD are congruent.
14. The perpendicular bisectors of the two lines coincide. Another way to put it is, that the lines joining the extremities are equally inclined to the two parallel lines.
15. Suppose O the centre, CD CE the equal lines drawn from C. Then triangles COD COE are congruent; so that CO, which passes through the centre, is the bisector of ∠DCE. The triangles being congruent, CD and CE are equally distant from the centre O, and the chords are equal.
16. The difference between 360° and the angle between two pins must be a sub-multiple of 360°.

17. (i) r must lie between 2 cm. and 18 cm. (ii) CD is greatest when it is the diameter of the fixed circle; then $r = 12{\cdot}8$ cm. and $\angle CYD = 77°$. (iii) When CD = 5 cm., $r = 17{\cdot}8$ cm. or 3·46 cm.

18. (i) The locus is the straight line which bisects all the chords at right angles. (ii) The three centres must be collinear.

19. Height 9·2 cm. (9·24).

20. It crawls an eighth of the circumference of each circle; the distances are 4·7 cm. and 3·3 cm., the radius of the line of latitude being 4·2 cm. (more exactly, the three numbers are 4·71 3·33 4·24).

21. The lengths are, in cm., 1·7 3·5 5·2 6·8 8·5 10·0 11·5. Chord subtending 48° has length 8·1 cm.

22. There is one axis of symmetry for the isosceles triangle or the semi-circle, two for the rectangle or rhombus, five for the regular pentagon, and six for the regular hexagon.

23. If K is the point in which A crosses OC, then KP is always $3\frac{1}{2}$ in., so that the locus of P is a circle.

24. The extreme position of B is at its greatest possible distance from XY; measurement or general reasoning shows this to be 60° from the mean position. C has its extreme position when A B C are in a straight line; it is 4·9 cm. from its mean position.

Section 36.

2. Length of tangent 3·75 in.

3. String wraps 2/3 of circumference; half the free length is a side of a right-angled triangle whose hypotenuse is 94 cm. and other side 47 cm. Length of string 360 cm. (to nearest integer).

4. Tangent measures 3·6 cm. **6.** Lengths 4·5 in. and 2·4 in.

7. The perpendicular to BC from the fixed point is the radius of the circle.

8. Imagine the secant to rotate about P till it is a tangent, and observe the changes in the angles OPA OQB.

9. Distance between centres 0·74 in. or 3·14 in. Angle 8° or 37°.

11. Height of triangle 3 in.; length of side 3·46 in. Area of square is 0·77 of area of triangle.

12. Suppose the octagon to be made from a square by cutting off the corners, the portion of the side of the square cut off at a corner measuring x cm. Each side of the octagon measures $10 - 2x$ cm., and the theorem of Pythagoras gives the equation $x^2 - 20x + 50 = 0$. Each side of the octagon is 4·142 cm. long and the perimeter 33·14.

13. Largest circle has diameter 7 in. and area 38·5 sq. in. Largest square (found by drawing one of its diagonals from a corner of the rectangle) has side 2·24 in. and area 5·0 sq. in. Area wasted 19·5 sq. in.

14. Radius 3·8 cm.

15. The bisectors (internal and external) meet by threes in four points, each point equidistant from the sides of ABC. The internal bisectors are the perpendiculars in question.

16. The radius of the locus is 5·8 cm.

17. The loci are a circle and a straight line, which intersect in two, one, or no points, according to circumstances.

18. The circle is an escribed circle of the triangle ADE.

19. The lengths are 1·55 in. and 4·42 in.

20. The radius is 2·9 cm.

21. Greatest angle 67°. In-radius 0·8 in.

22. The centre of each semicircle is the intersection of the bisector of an angle with the opposite side.

23. *Either*: The hypotenuse is equal to the sum of the portions of the sides up to the points of contact, and the remaining portion of each side is equal to the radius. *Or*: If c denote the hypotenuse, a b the sides, and r the in-radius, then $(a+b+c)(a+b-c) = a^2+b^2+2ab-c^2 = 2ab$, and $r(a+b+c) = ab$. So that $2r = a+b-c$.

24. Distance is 21 ft. It is half the perimeter, as is readily seen from the equality of two tangents from any point to a circle.

25. Radius is 3·2 cm. (3·17).

26. Any two tangents meeting at 60° will give the distance of P from A, or it can be seen without drawing that it is 2 inches. PC is 1·3 in.

27. Area 4·4 sq. in. In-radius 0·8 in. (0·79), e-radius 1·7 in. (1·725), circum-radius 2·7 in. (2·65).

28. The centres lie on the perpendicular bisector of AB, and one on each of the bisectors of the angles between AB and CD. Radii 1·9 cm. and 7·8 cm.

29. The circle has radius 6·7 cm. Each tangent from A is 9·1 cm. long; and perimeter of ABF is 18·2 cm.

30. One pair of opposite sides is made up of four tangents, which are equal to the four tangents that make up the other pair.

31. The area, being half the product of the perimeter and the radius, is 9·2 sq. cm. (9·23).

32. (ii) Take BC 4 in. long. Draw a circle of centre C and radius $\sqrt{12}$ in. to cut AB in A. Let fall a perpendicular AD on BC, and D is the required point of section. BD is 3·41 or 0·59 in. (iii) Least value of BE^2+EC^2 is got when the circle touches AB; then $BE = EC = 2$ in.

33. (i) The inscribed hexagon consists of six equilateral triangles each of area $r^2\sqrt{3}/4$, r being the radius; and the circumscribed hexagon of six equilateral triangles each of area $r^2/\sqrt{3}$. The ratio is 3/4. OTHERWISE: Suppose OA OB two radii enclosing an angle of 60°, AC BC tangents at their ends, D the intersection of OC and AB. Then $CD = CA/2$ and $CA = CO/2$; so that $OD = \frac{3}{4}OC$, and area OAB is $\frac{3}{4}$ of area OACB.

34. Side of octagon measures 4·1 cm. *Either*: With each corner as centre and semi-diagonal as radius draw circles, which intersect the sides in the vertices of the octagon; prove by calculating lengths of sides. *Or*: Make suitable angles at the centre of the square.

35. Side measures 1·2 in. (1·3 almost as close).

36. Tangent measures 8·8 cm. or 4·1 cm.

37. Tangents measure 4·0 cm. and 2·1 cm. Area of rectangle 8·4 sq. cm.

38. Length of tangent 44 mm. Angle 23°.

39. If tangent at A meets BC in O, $OB = OA = OC$.

40. Length of each tangent 4·90 cm.

41. Length is $\sqrt{\{d^2-(a+b)^2\}}$.

42. If the tangents meet in T, the triangles TPR and TQS are isosceles.

43. Length of strap 36 cm.

44. $AC \cdot BD=(c+a-b)(c+b-a)=c^2-(a-b)^2$.

45. Denote the intersections of the circle with the bisector of $\angle BAC$ by E and F. The circles required are the circum-circles of triangles EBC FBC, and their radii are 1·8 in. and 0·7 in.

46. With the centres of the given circles as centres draw three pairs of circles having radii (i) 2·3 and 2·3 in., (ii) 2·3 and 0·7 in., (iii) 0·7 and 2·3 in. Each pair of circles gives, by their intersections, two centres for the touching circle.

47. There are six touching circles; their centres being given by the intersection of a circle of radius $2{\cdot}5-1$ inches with circles of radii 2 ± 1 inches, and of a circle of radius $2{\cdot}5+1$ inches with a circle of radius $2+1$ inches.

48. Join the intersection of the tangents to the centre of the circle. At one of the points where this line cuts the circumference draw a tangent. The required circle is the inscribed or the escribed circle of the triangle formed by the three tangents.

49. There are two positions for the $\frac{3}{4}''$ circle and for each of these there are two positions for the $1''$ circle.

50. Think of the octagon as made by cutting the corners off a square. Each circle touches one side and the two diagonals of the square.

52. The circles will cut the sides of the triangle at the points of contact of the inscribed circle; their radii are 1·75 in. 1·15 in. 0·65 in.

53. The given circles divide the plane into four regions, one region external to both circles, the others internal to one or both circles. When the touching circle is small there are two positions for it in each region. As the touching circle increases in size, the internal regions become, one after another, too small to contain it; and the number of positions is reduced successively to six, four, and two. When the touching circle has grown large enough to contain the given circles, the number of positions rises to four. At the moment of transition between two of the cases enumerated, the number of positions is odd.

54. (i) The ratio is 2/3 or 0·667. (ii) $r = 8\frac{1}{3}$ in.

Section 37.

1. (i) $x = a + b$. (ii) $x = 4\frac{1}{2}$. (iii) $x = 4$ or -1.

2. (i) $x = 2\frac{1}{2}$. (ii) $x = a - b$. (iii) $x = 2{\cdot}385$ or $-8{\cdot}385$.

3. (i) $x = 7$. (ii) $x = 2a$.

4. (i) $x = 56$. (ii) $x = a + b$. (ii) $x = 33$ $y = 3$.

5. (i) $x = -22/7$. (ii) $x = 1 + b$.

6. (i) $x = 13$. (ii) $x = 58/49 = 1{\cdot}184$, $y = -9/49 = -0{\cdot}184$. (iii) $x = (b^2 + ab - ac)/(c - 2b)$.

7. (i) $x = 1$; this is obvious by inspection. (ii) $x = 1$ or $1/2$; the former is obvious by inspection, and the other obtained by solving in the usual way. (iii) $x = (1 - b)/(1 + a)$, $y = (b - 1)/(1 + a)$.

8. (i) $x = 5$. (ii) $x = 0$ or $-1/2$. (iii) $x = 2b + a$, $y = 2a + b$.

9. (i) $x = 16/7$, $y = 10/33$; the simplest way is to substitute p for $1/(x+1)$ and q for $1/(y-1)$, and first find p and q. (ii) $x = y = 1{\cdot}5$. (iii) $x = 1$, $y = 0$. (iv) $x = 4$ or $-1/2$.

10. (i) $x = 2$ or $-11/6$. (ii) $x = 0{\cdot}0941$ and $y = -0{\cdot}0565$, to three significant figures. (iii) $x = y = ab/2$.

11. (i) $x = 4{\cdot}4$ or $0{\cdot}85$. (ii) $x = p - q$, $y = p + q$.

12. (i) $x = 10$. (ii) $x = (a + b)c/a$, $y = (a + b)c/b$.

13. $x = a$, $y = b$. The equations are linear in $1/x$ and $1/y$.

14. (i) $x = 18$. (ii) $x = a/(a - b)$, $y = b/(a - b)$. (iii) $x = 7$ or $24/13$.

15. (i) $x = 8$. (ii) $x = a/2$. (iii) $x = 8$, $y = 2$; or $x = 1$, $y = -1/3$.

16. (i) $x = 6/7 = 0{\cdot}86$, $y = 2/5 = 0{\cdot}4$. (ii) $x = \pm\sqrt{7}a = \pm 2{\cdot}6a$. (iii) x is equal to 6, or to 17/7 which is 2·4.

17. (i) $x = 3$, $y = -2{\cdot}5$ or $x = -5$, $y = 1{\cdot}5$. (ii) $x = a$ or $-2a$.

18. (i) $x=1$ or $(a-b)/(b-c)$; the first root may be found by inspection. (ii) $x=3$, $y=2$, $z=7$ or $x=27/7$, $y=22/7$, $z=5$; the first and third equations give $x+y$ and z.

19. (i) $x=2$, $y=1$ or $x=12/13$, $y=-15/13$. (ii) $x=a$.

20. $a=-0{\cdot}68$; second value of x is $-0{\cdot}2$. These values are exact.

21. $a=1/2$ or -2.

22. $a=-10$, $b=11$; second solution $x=3{\cdot}15$, $y=4{\cdot}7$. These values are exact.

23. $x=q^2/p$ or $-p^2/q$.

24. $x=\{6-5a\pm\sqrt{(a^2-16a+16)}\}/(2-2a)$. Both roots are equal to $-4{\cdot}46$ when $a=1{\cdot}07$, and to $2{\cdot}46$ when $a=14{\cdot}93$; these two values of a being given by the equation $a^2-16a+16=0$.

Section 38.

2. The locus consists of two arcs whose centres are the vertices of the equilateral triangles standing upon AB.

3. Length of chord 2·3 inches.

4. Draw the perpendicular bisector of AB, and mark on it two points at a distance of 3·2 in. from AB. Through these points draw two lines 4·55 in. long, parallel to AB, and bisected at these points. The vertex lies on one of this pair of lines.

5. Each angle of the heptagon measures $128^{\circ}\,34'$.
The angles of ACF are $77^{\circ}\,9'$, $51^{\circ}\,26'$, $51^{\circ}\,26'$.
The angles of the other triangle are $12^{\circ}\,51'$, $25^{\circ}\,43'$, $141^{\circ}\,26'$.

6. The diagonals of the parallelogram bisect one another, and are therefore diameters.

7. The common chord subtends a greater angle at the centre of the smaller circle, which is nearer. So that the major segment of the smaller circle contains a greater angle than the major segment of the larger circle; and the minor segment of the smaller circle a less angle than the minor segment of the larger circle.

8. The circle touches BC at its middle point O. ADOE is a rhombus, and being inscribed in a circle is a square.

9. The vertex is given by the intersection of a semi-circle and a straight line; lengths of sides 3·2 and 1·3 inches. For an area of 3 sq. in. the two loci do not meet.

11. If the lines through D and F meet in K, those through D and E in M, triangle KDF is isosceles and KD equal to a side of the pentagon. MDAE is a parallelogram, so that MD is equal to a side of the pentagon. So that K and M coincide.

12. Call the ends A and B, and any other point P. To draw the tangent at P take two points Q R on the arc equidistant from P; the tangent is parallel to QR. To draw the tangent at A join AB BP PA; the tangent makes with AP an angle equal to ABP.

13. The angle between QAR and the tangent at A is equal to APQ and ASR.

14. First take for AB a chord equal to the radius. AB = 5 cm. BC = 5·7 cm. CD = 8·2 cm. DA = 8·7 cm.

15. The middle point of the required chord lies on the circle which has for diameter the radius through A. Possible when this circle meets the given chord.

16. The locus is an arc of a circle, for $\angle AQB = 90° + \angle APB/2$.

17. The other two angles of the triangle BPQ are constant, being angles in segments of circles.

18. The angle between the tangent and AC is equal to $\angle ABC$. Add $\angle CAX$ to the former and $\angle BAX$ to the latter.

19. A lies on the segment standing on PQ which contains the angle BAC. AD always passes through the point R of this circle so chosen that $\angle QAR = \angle CAD$.

20. $\angle ABC = 57{\cdot}5°$ $\angle ACB = 57{\cdot}5°$ AC = 1·5 in. BC = 1·6 in.

21. If the other end of the chord is called Q, $\angle ARC$ is equal to $\angle AQB$, which is constant.

22. Circle on BL passes through M because LMB is a right angle. MN is a tangent because $\angle LMN = \angle LBM$.

23. CA touches, for $\angle CAD = \angle DAE = \angle DEA$. Radius is equal to DA, for centre is the common vertex of two equilateral triangles on DA and DE.

24. The angle subtended by MN at R is half that at B, that at Q is half that at A; whence $\angle MRN = 22{\cdot}5°$, and the other results.

25. Suppose the diameter meets CA in P and CB in Q. Then angles QCP and BOP are equal, each being equal to the angle in the major segment.

26. $\angle CED = \angle CBD = \angle ABD = \angle BDE$. In some figures we must, for the proof, replace an angle by its supplement.

27. The radii to the points of contact make angles of 115° and 139° with one another.

28. The semicircle on the hypotenuse and the parallel to the hypotenuse at a distance of 2·5 cm. intersect in the vertex.

29. $\angle APB = 21°$. The angle is half ACB, which is constant.

30. B C D E being concyclic, angles EDB and MLB are each equal to ECB.

31. Suppose angle B less than angle C. The arc LM subtends at the centre the angle A minus the angle 2CAM, i.e. $A - 180 + 2C$ or $C - B$.

32. P, the point in which the join of A and its new position meets the crease, is an intersection of the semicircle on AD with the line bisecting AB and CD. Another crease passes through P. The two remaining creases pass through Q, the other intersection of the two loci. Consider the figure with AB folded on DC and with BC folded on AD, and we see that the diagonals of the parallelogram bisect one another at right angles.

33. The angle subtended by BC at the centre together with P is 180°, so that $P = 110°$; and similarly $Q = 40°$ $R = 30°$. Triangle PQR being obtuse angled, we find for triangle XYZ the angles 40° 60° 80°.

34. The vertex is found as an intersection of an arc standing on the base and a parallel to the base. The area is greatest when the triangle is isosceles, and is 170 sq. yd. (168).

35. Greatest area is 1·6 sq. in. For half that area the sides are 2·7 in. and 0·6 in. (2·73 and 0·61).

36. The middle point of the chord lies on the circle on OP as diameter. Its length is 4·2 in. (exactly).

39. (i) A = 50° B = 80° C = 130° D = 100°. (ii) BD = 1·3 in. (1·32).

40. Triangles BLK and CMK are congruent.

41. When one pair of sides are equal, the angles which they subtend (alternate angles for the other pair) are equal.

42. AD = $\{6 \cdot 3^2 + 5 \cdot 85^2 - 7 \cdot 9^2\}^{1/2}$ cm. = 3·39 cm.

43. Each angle is right, since it is equal to its opposite, and makes 180° with its opposite.

44. ∠GAH = ∠CAD = ∠CBD = ∠EBF. Add arc FG to the equal arcs GH and EF.

45. The angles T and B together make 180°, for ∠TPQ = ∠PBA and ∠TQP = ∠QBA.

46. (i) ∠ECD = ∠CDB, as they stand on equal arcs. (ii) Triangles BCD and DEB are congruent, having BC = ED, BD common, and an angle of each right.

48. Draw radii making with one another angles supplementary to 100°, etc., and draw tangents at their extremities.

49. Arc AD = arc BE = arc CF, so that ∠ACD = ∠FAC.

50. Express two opposite angles of the inner quadrilateral in terms of the angles of the original quadrilateral.

51. The other two angles of the triangle PBQ are constant.

52. Suppose A and B the given middle points. Produce AB both ways to C and D, so that CA = BD = AB. One vertex lies on the circle on AB as diameter; so does the intersection of the diagonals, which is given by drawing perpendiculars to the sides at A and B. Two adjacent vertices lie on the circles on CA and BD as diameters. The fourth lies on the circle on CD as diameter.

53. Let EG and FK be diameters parallel to AC and BD respectively, and suppose these diameters to move parallel to themselves into coincidence with the chords. The sum of arcs AD and BC remains constant (one gaining what the other loses) and equal to EK + FG or the semi-circumference. The angle between the tangents at A and D is the supplement of the angle subtended by arc AD at the centre; and so with arc BC.

54. Let EG and FK be diameters parallel to PR and QS. The quadrilateral being cyclic, the sum of arcs PS and QR is the semi-circumference. The sum is also (as in Qn. 53) equal to EK + FG or 2EK, so that EK is a quadrant.

55. $\angle BLN = \angle BPN = \angle 90° - \angle NBP = 90° - \angle ACP = \angle CPM = \angle CLM$.

56. Proof as given for Qn. 55.

57. Let perpendiculars BE CF meet in O. Join AO and produce to meet BC in D. $\angle BCF = \angle BEF = \angle OAF$, so that C A D F are concyclic and $\angle CDA$ is a right angle.

58. $\angle ADE = \angle FCE = \angle FBE = \angle FDA$.

59. Draw three radii, making with one another angles which are twice those of the given triangle (one angle is greater than 180°). Joining the extremities gives the triangle, which has an area of 23 sq. cm.

60. If circles AQR and BPR meet in O, $\angle ROP + \angle ROQ = 360° - A - B$, so that $\angle POQ = A + B = 180° - C$; and P O Q C are concyclic.

61. (i) Draw three radii, making with one another the three angles $180° - A$, $180° - B$, $180° - C$. The tangents at their extremities form the triangle. (ii) A being the smallest angle, the angles between the radii must be $180 + A$ B C.

62. BC being the base, the line in question is the bisector of the angle A, and it passes through the middle point of the arc BC of the circumcircle.

63. (i) PA = 7·8 in. PB = 5·8 in. (ii) On AB draw a triangle ABC having AC = 2 in. $\angle ACB = 110°$; P lies on AC produced.

64. Angles CBD and CDB are each 45°, so that $BD^2 = 2CD^2$.

65. A R O P are concyclic, so that $\angle OPR = \angle OAR = 60°$.

66. $\angle BAC + \angle BAD = \angle BPQ + \angle BQP = 180° - \angle EBF$.

67. Suppose ABCD the quadrilateral, O the intersection of the diagonals, MON perpendicular to CD and meeting AB in N. Angle NOA = $\angle COM = \angle CDO = \angle CAB$; so that NA = NO = NB.

68. (i) The perpendicular bisector of the base meets the circumcircle in a point of the bisector of the vertical angle. This and the given point enable us to draw the bisector of the vertical angle, and so the triangle. (ii) $\angle B = 27°$ (27·4) $\angle C = 78°$ (77·6).

69. We denote centres by X and Y. XP and YQ being parallel, triangles XAP YCQ are equiangular. $\angle SPQ + \angle SQP = \angle XAP + \angle CDA = \angle DCQ + \angle CDQ = 90°$, so that $\angle PRQ = 90°$; and $\angle RPS = 90°$, so that PRQS is a rectangle. Angle between RS and AD $= \angle SRP + \angle XAP = \angle RPQ + \angle BPQ = 90°$. Sides 6·7 cm. and 10·0 cm.; area 66 sq. cm.

70. Angles RBQ is constant, being the sum of constant angles BPA and BQA.

71. Length AR is constant, being diameter of the circle in which PQ subtends an angle BAC.

72. (i) The circles are equal because AP subtends equal angles in them. (ii) Triangles DAB EAC being congruent, angles DBA and DCA are equal.

73. Locus of P is the perpendicular bisector of AB; for, the circles being equal, angles PAC and PBC are equal.

74. Angle POQ is constant, being the sum of the constant angles BQA and PAQ.

75. Suppose AB less than CD. By marking all the angles equal to BAC and all equal to BCA, we find that XDC and YAC are equiangular isosceles triangles. Hence C A X Y are concyclic, and $\angle YXC = \angle YAC = \angle XCD$.

Section 39.

1. $x^2 + x + 5$. **2.** -4. **3.** $2x^2 + xy - 3y^2$.

4. i. $6x^5 + 9x^4 + 3x^3 - 4x^2 - 3x + 5 + (8x - 3)/(x^2 - x + 1)$.
ii. $x^8 - 3x + 2$.

5. i. $2x^5 - 6x^4 + 13x^2 - 3x - 15$. ii. $x^2 + 4x + 11$.

6. $x^6 + 4x^5 + 13x^4 + 31x^3 + 36x^2 + 33x + 12$. The work is lighter if we begin by dividing $x - 1$ into the first biquadratic and $x - 3$ into the other.

7. Product $(x+4)(x+7)(x-5)(x-6)$ or $x^4 - 63x^2 + 22x + 840$; quotient $(x+7)(x-6)$ or $x^2 + x - 42$.

8. $c = 2$.

9. i. Remainder $(30{\cdot}5 - a)x + 39 - b$. ii. $a = 30{\cdot}5$, $b = 39$.

10. i. Remainder $(a-4)x+b-15$. ii. $a=4$, $b=15$.

11. i. $13+28x+41x^2-2x^3-3x^4$. ii. Values are 109.

12. i. The difference between the two expressions is

$$10x^3+5x^4+x^5.$$

The first of these terms is the most important, and when $x=0{\cdot}005$ it is $1{\cdot}25\times10^{-6}$, which affects only the sixth place. The second and third terms are then respectively $2{\cdot}5/10^3$ and $2{\cdot}5/10^6$ of the first term, and will affect only later places.

ii. $(0{\cdot}998)^5=(1-0{\cdot}002)^5=1-0{\cdot}01+0{\cdot}00004=0{\cdot}99004$.

13. i. The two expressions must be composed of the same factors, and $a=-3$ $b=2$. ii. $(x-3)(x+3)(x-9)$; $(x^2+3x+5)(x^2-3x+5)$; $(x+1)(x-2)^2$.

14. i. $(9+4x^2)(2+x)(2-x)$. ii. $(x+p)(3x-3p-2)$.

15. i. $(x-1)(x+2)$. ii. $(x-1)(2x+1)$.
iii. $(x-1)\{a(2x+1)+b(x+2)\}$. iv. $(x-1)(201x+102)$.

16. i. $(2x+3)(x-7)$. ii. $(x-a)(x+a-1)$. iii. $x=7$ or $-1{\cdot}5$.
iv. $x=a$ or $1-a$.

17. i. $(x+14y)(14x-3y)$. ii. $(p^2+q^2)(x^2+y^2)$.

18. $(x+p)(x-p)(y+q)$. Each side of the identity can be multiplied out; or we can divide out by x^4, put y for x^2+1/x^2, and throw each side into the form $(y+2)^2$.

19. i. $a^2+9ab-10b^2$ or $(a+10b)(a-b)$.
ii. $0{\cdot}027$.

20. i. $(3x+16)(8x-3)$. ii. $(x^2+x+1)(x^2-x+1)$.
iii. $(ax+b)(bx+a+c)$. iv. $x-1$ and $x-7$.

21. $(2x^2-2xy+y^2)(2x^2+2xy+y^2)(x^2-2xy+2y^2)(x^2+2xy+2y^2)$. Put $x=10$ $y=1$ and the four factors become 181 221 82 122; the even factors may have the 2 factored out; and 221 is the product of 13 and 17.

22. The expression is $26(x^6-1)$ or

$$26(x+1)(x-1)(x^2+x+1)(x^2-x+1).$$

23. i. The remainder $(b-1-ac+c^2)x+1-a+c$ must vanish, so that $a=b=c+1$; and the quotient is $x+1$.

ii. The expression is $(x-2)(x^2-x-6)$ or

$$(x-2)(x+2)(x-3),$$

and the quotient when divided by $x+2$ is $(x-2)(x-3)$ or x^2-5x+6.

24. i. $(a+b+c)(a-b-c)$. ii. $(2x+1)(2x-5)$.

25. i. $(x+2)(x-15)$. ii. $(x-y+z)(x+y-z)$.
iii. $(x+4y)(3x-2y)$.
iv. $62357^2 - 47643^2 = 110{,}000 \times 14{,}714 = 1{,}618{,}540{,}000$.

26. i. $(x+a+b)(y-b)$. ii. $(px+q^3)(qx-p^3)$.

27. $(3x+4y-5)(2x-y+3)$.

28. The substitution of $b\ c\ a$ or of $c\ a\ b$ for $a\ b\ c$ respectively leaves the expression unchanged, so that as a function of $c-a$ or of $a-b$ it has the same properties which it has as a function of $b-c$.

29. i. $(3b-c-4)(3b-c+4)$. ii. $(a-b)(a+b+x)(a+b-x)$.
iii. $2(4xy-3)(3xy+4)$. iv. $-(b-c)(c-a)(a-b)$.

30. i. $(ac-b)(bc-a)$. ii. $k=13$; the other factor $9x-13$.

31. i. $(a-b)(c-d)(c+d)$.
ii. $\{x-(p+1)(q+1)\}\{x+(p-1)(q-1)\}$.
iii. 1 must be added.

32. i. $a=174$; remaining factors $8x+7$ and $15x-4$.
ii. $2x^2-x+1$.

33. i. $m=56$. ii. $2x^3+7x+1/3$.

34. i. $2x^2+2/x^2$.
ii. $729+\frac{1}{729}+2=(27+\frac{1}{27})^2$ and $256+\frac{1}{256}-2=(16-\frac{1}{16})^2$, so that roots are $27\frac{1}{27}$ and $15\frac{15}{16}$.

35. i. The last remainder $x-2$ will vanish and the expression be a perfect square when $x=2$. ii. $x=15$ or $55/9$.
iii. $x=y=9$.

36. The quotients of the three expressions are $x+3$ $3x+2$ x^2-2, so that they vanish simultaneously only when x^2-3x+2 vanishes, i.e. when $x=1$ or 2.

37. L.C.M. is $(x^4+x^3-x^2+5x+6)(x-4)$ or
$x^5-3x^4-5x^3+9x^2-14x-24$.

38. Factors being $(x-2)(x-3)$ $(x-2)(3x+2)$ $(x-3)(3x+2)$, the L.C.M. is $(x-2)(x-3)(3x+2)$.

39. i. $(7x+8)(8x-9)$ $(7x-2)(8x-9)$ $(7x+8)(9x-2)$.
ii. L.C.M. $(7x+8)(8x-9)(7x-2)(9x-2)$.

40. H.C.F. is $x+4$; L.C.M. is $(x-2)(x+4)^2(x^2-4x+16)$.

41. i. H.C.F. is $x+1$. ii. L.C.M. is
$(x-y)(x+y)(x^2+xy+y^2)(x^2-xy+y^2)$ or x^6-y^6.

42. i. $(1-x-y)(1+x+y+x^2-xy+y^2)$. ii. $(a-2b)(b-2c)$. iii. $2x-7$.

44. G.C.M. is $x-3$. The difference of the two expressions is $x^4(x-3)$, so that $x-3$ is the only possible common measure, and it is a measure of each.

Section 40.

1. Acute angled, since $6{\cdot}5^2+3{\cdot}6^2 > 7{\cdot}2^2$.

2. Obtuse angled. Area 280 sq. yd. (281).

3. The triangles 3–4–6 and 3–5–6 are obtuse angled.

4. The projection of AB on AC is equal to half AB.

5. Suppose $\angle$ ARQ acute, and drop the perpendicular AN on PQ. $AP^2 = AR^2+RP^2+2PR.RN$. Add to this a similar relation for AQ^2.

6. (i) When $AC^2 > AD^2+DC^2$, the angle ADC is obtuse, ADB acute, and $AB^2 < AD^2+DB^2$. (ii) As in Qn. 5.

7. If AM is dropped perpendicular to BC,

$$AB^2-AD^2 = MB^2-MD^2 = BD.DC.$$

8. (i) As in Qn. 5. (ii) The excess is $2(FA^2-FK^2)$ or $2AK.KB$.

9. The successive lengths are, in cm.: 2 3·1 5·1 7·2 9·1 10·8 12·2 13·2 13·8 14·0.

10. $OB=1{\cdot}41$ cm.

11. (i) As in Qn. 5. (ii) P is at a fixed distance from the middle point of AB.

12. (i) $AC=118{\cdot}2$ in. (ii) Suppose CM to be let fall perpendicular to AD in the strained gate. Then

$$AC^2 = AD^2+DC^2+2AD.DM,$$

which gives $DM=2{\cdot}5$ in. (2·47).

13. (i) Length is 104 mm. (ii) Length is 95 mm. exactly.

14. If M is the foot of the perpendicular from B on AC,

$$BC^2-BA^2 = MC^2-MA^2 = \{(x+y)/2\}^2-\{(x-y)/2\}^2.$$

15. Angle AOB may be acute, right, or obtuse. In the first case $AB^2+CD^2 < AO^2+BO^2+CO^2+DO^2 < AD^2+BC^2$; and so on.

16. (i) The triangle is acute if $n^2+(n-1)^2-(n+1)^2$ or $n(n-4)$ is positive. (ii) When $n=1$, one of the lengths is zero, and there is no question of the existence of a triangle. When $n=2$, one side is equal to the sum of the others, and there is no triangle. When $n=3$ the triangle is obtuse. When $n=4$ the triangle is right angled.

17. Denote by O the centre of the circle, by N the foot of the perpendicular from A on BC, and by M the foot of that from P on AN. Then $PB^2+PC^2 = 2PN^2+2NC^2$. Express PA^2 and PN^2 in terms of PO OA OM ON; remember that ON is half the radius, and the result follows.

18. Express the sum of the squares of the distances of O from the ends of each side in terms of the side and the distance of O from its middle point; and add up.

Section 41.

1. 3. **3.** $(x^2+11xy+24y^2)/(x^2-10xy+24y^2)$.

4. $(3x^2+5)/\{(1-x)(2-x)(3-x)\}$. **5.** 4.

6. (i) 1. (ii) May be expressed as

$$(6x+9-x^2)/(x^2-9) \quad \text{or} \quad 6x/(x^2-9)-1.$$

It is equal to 2/3 when $x=5{\cdot}30$ or $-1{\cdot}70$.

7. (i) $(ax+by)(bx+ay)$.

(ii) Expression $= (x-a/2)^6 \div (x-a/2)^4 = (x-a/2)^2$.

(iii) Values of x, 0 2·52 −2·52.

8. $5/(x^2+4x+16)$.

9. (i) $z=\{(1-a)x+2a\}/(2x+a-1)$. (ii) $a=-2$.

10. $2(a^3-a^2+a+1)\div(a-1)(a-2)$.

11. (i) $x=(y-c)/m$. (ii) $x=(2-3y)/(3+y)$. (iii) $x=3y^2+2y+1$

12. $y=(x-16)/(19x+18)$ $x=(18y+16)/(1-19y)$.

13. The value is $\{ab(a+b)(b-a)\}\div\{(2a+b)(2b+a)\}$.

14. (i) $(5x+14)(5x-3)$. When x is a multiple of 7 the first factor is a multiple of 7. If x is even the first factor is even; if x is odd the second factor is even.

(ii) $x=a+1$ or $(a-1)^2$; $x=-4/3$ or $-5/2$.

15. *Either*: $\Sigma(x+a)^2 - \Sigma(y+a)(z+a)$
$= \Sigma(a^2+2ax+x^2) - \Sigma\{a^2+a(y+z)+yz\}$
$= 3a^2+2a\Sigma x+\Sigma x^2 - 3a^2 - 2a\Sigma x - \Sigma yz = \Sigma x^2 - \Sigma yz.$

Or: Throw the expression into the form
$$\{(x-y)^2+(y-z)^2+(z-x)^2\}/2.$$

16. Arrange the left member in powers of x.

20. *Either*: Arrange each expression in symmetric functions of $x\ y\ z$. *Or*: $a^3+b^3+c^3-3abc = \frac{1}{2}.\Sigma a.\Sigma(b-c)^2$; and $\Sigma a = \frac{1}{2}\Sigma(y-z)^2$, while $b-c=(y-z)\Sigma x$, etc.; whence result.

21. $2bc(1+Q)=(b+c+a)(b+c-a)$, etc.

22. The expression has the value 6 when $x=2{\cdot}3$. Then the graph shows it to be greater than 6 when x lies between 2 and 2·3 or is greater than 3.

23. A B C are given in succession as 1 16 7 by equating the two expressions with 1 2 0 substituted for x. Or they can be found by multiplying out and equating coefficients.

24. $(x-3)^3+2(x-3)^2+3(x-3)+4$. Can be done by multiplying out and equating coefficients; or by putting $x=3+y$ and multiplying out $(3+y)^3-7(3+y)^2+18(3+y)-14$.

25. A sum of squares of real quantities is positive unless each of the quantities is zero.

26. (i) $(pa+qc)/(pb+qd)-a/b = (bc-ad)q/\{b(pb+qd)\}$ and $(pa+qc)/(pb+qd)-c/d = -(bc-ad)p/\{d(pb+qd)\}$; one of these quantities must be positive and one negative. (ii) $p/q=d/b$.

27. (i) Total number of candidates $(n-m)y/(n-x)$. (ii) If z is the number of girls entered $x=(my+nz)/(y+z)$; so that by the reasoning of Question 26, x lies between m and n. (iii) If $x=(m+n)/2$, expression for total entries becomes $2y$.

28. (i) $a^3b^2(a+b)(a^3-b^3)$. (ii) $(2a-b)/(a+b)$.

29. (i) a^2. (ii) $(a-3)/(a+1)^2$.

30. (i) $-4ab(a^2+b^2)/(a^4+a^2b^2+b^4)$. (ii) 0.

31. $2x/\{(x+a)(x+2a)\}$; numerical value 1/3.

32. (i) $1/(x-y)$. (ii) $8abc$; the expression is symmetrical in $a\ b\ c$, so that we know all coefficients when we have calculated those for typical terms, say for $a^3\ a^2b\ abc$.

33. (i) a/b. (ii) $x+b$.

34. (i) $1/(x-1)$. (ii) $3xyz$. **35.** 0.

36. (i) $2/(x-4)$. (ii) $(x+y)/(x-y)$.

37. $1/\{(x-a)(x-b)(x-c)\}$. Arrange the numerator in powers of x; the symmetry of each coefficient is an assistance in writing them down and a check on their correctness.

Section 42.

1. Length of side = 3·9 cm. Such a square has area of 15·21 sq. cm., as nearly equal to the rectangle as we could expect.

2. Length of side 3·0 (3·03) in. Its area is 9·0 (9·18) sq. in., that of rectangle being 9·2 sq. in. to one decimal.

3. Side of square 7·3 cm.; height of triangle 10·7 cm. Respective areas, in sq. cm. to one decimal, 53·3 and 53·5.

4. Length of side 6·580 cm. **5.** 1·7 in.

6. 3·4 cm. **7.** 1·7 in.

8. ABC has base 38 mm. and height also 38 mm., area 7·2 sq. cm. DEF has base 33·5 mm. and height 48 mm., area 8·0 sq. cm.

9. $\sqrt{\{4\sqrt{3}\}} = 2{\cdot}6$. **10.** 2·2 in.

11. 6·1 cm. First make a triangle equal to the quadrilateral.

12. (ii) Call the triangles ABC and DEF. Make a triangle equal to DEF with a base equal to BC. A triangle on base BC and of the sum of the heights is equal to the sum of the triangles.

13. OS = OP.OQ/OR = 2·93 in.

14. CD = $\sqrt{(\text{CA.CB})}$ = 1 in. The locus of D is a circle of centre C and radius 1 in.

15. (i) The converse that "if two straight lines AOB COD intersect in O and OA.OB = OC.OD, the points A B C D are concyclic" is easily proved true. A false converse can easily be found, e.g. by substituting for the consequence "a circle through A and B will also pass through C and D." (ii) The most obvious converse is that "if P is in the plane of the triangle ABC and ∠BPC > ∠BAC, then P is inside the triangle." This is untrue, for ∠BPC is greater than ∠BAC if P is anywhere inside the circum-circle of the triangle, or inside the image of this circle in BC.

16. (ii) On half the given line, i.e. on the semi-perimeter of the rectangle, draw a semicircle. Parallel to the given line at a distance equal to the side of the square draw another line, cutting the semicircle in P. The perpendicular from P on the half of the given line divides it into parts equal to the length and breadth of the rectangle. If the side of the square is greater than a quarter of the given line, the parallel and the semicircle do not meet, and the construction is impossible.

17. Construction as in Question 16. AC is 4 in. or 1 in.

18. (i) Construction as in Question 16. (ii) b must not be less than a. (iii) If x is the length of one side of the rectangle, $2b - x$ is the length of another, and $x(2b - x) = a^2$.

19. A B D E are concyclic.

20. $x(2r - x) = a^2$ or $2r = x + a^2/x$. The second root is OQ.

21. (ii) Make angle EDG equal to angle EFD. Then ED is tangent to circle GDF.

22. Express the relation $CD^2 = CE.CF$ in terms of x and y. When $CE = 3$, $y^2 + 3y = 4$; and $y = 1$ is the only positive root.

23. Denote by M the middle point of PQ.
$OS.OR = OS^2 + OS.SR = OS^2 + PS.SQ = OS^2 + MP^2 - MS^2 = OP^2$.

24. $DB.DC = DA^2$ and $LB.LC = AL^2$; whence the result.

25. Width 32·3 ft.; height 10·5 ft. For the formula apply Pythagoras to the triangle formed by A C and the centre of the circle. Radius of arch, 18 ft.

26. Draw BC cutting OX in C and making the angle OBC equal to OAB. Then ∠OBA = ∠OCB, so that the circle ABC touches OY.

27. Call the points A and B, and the line K. On AB describe a semicircle. Let AB produced meet K in C. From C draw a tangent to the semicircle to touch at D. Take points E and F on K distant CD from C. The circles ABE ABF satisfy the conditions.

28. The angle BED being right, DE is a tangent to AEC, and $DE^2 = DA.DC$. So that if the circle with centre D meets in F a circle through A and C, DF^2 is equal to DA.DC, and DF is a tangent, i.e. the circles cut at right angles in F.

29. Sum of squares of tangents $= PO.PQ + QO.QP = PQ^2$. PQ is greatest when it passes through the centres.

30. (i) On AB as diameter draw a circle. From P the intersection of AB and XY draw a tangent PQ. On XY mark off PR and PS equal to PQ. The required circles are those through A B R and A B S, and have radii of about 68 or 69 mm. and 18 or 19 mm. (ii) Suppose T a point in PR or PR produced, and let AT meet the arc ARB in U. Then $\angle ATB$ is less than $\angle AUB$, which is equal to $\angle ARB$.

31. He must stand in such position that the circle passing through the top and bottom of steeple and his eye is as small as possible. The horizontal through his eye must be a tangent. Distance from tower is 36 ft.

32. From D the middle point of BC let fall a perpendicular to the bisector AZ of angle A; on it take E the image of D in AZ; and let DE meet AC in F. To a circle on DE as diameter draw a tangent FG; and on AC mark off FH and FK equal to FG. The circles required are those through D E H and D E K.

33. The locus is the remainder of the straight line of which the common chord is part.

34. Equate $CT^2 - CP^2$ and $CT^2 + CP^2 - 2CT.CN$, which are both equal to TP^2.

35. (i) $BA = AC = AQ$, so that $BP.BC = BQ^2 = 4AB^2$.
(ii) $BC.CP = BC.BP - BC^2 = 4BA^2 - 4BD^2 = 4DA^2$.

36. (ii) The diameter being ACNB, the centre C, and the half-chord PN, we have $PN^2 = CP^2 - CN^2 = CA^2 - CN^2 = AN.NB$. (iii) If a is the distance between the supports and b the depression at the middle, we are given $b = ka^3$, where k is a constant. If R is the radius of the circle, $(a/2)^2$ is equal to $b(2R - b)$, and if we neglect b^2, to $2Rb$. Thus $R = a^2 \div 8b = a^2 \div 8ka^3 = 1/8ka$, i.e. R is inversely proportional to a.

Section 43.

1. The value of x is $1{\cdot}8$. For $1 \quad 1{\cdot}8 \quad 2$ the values of the expression are $-9 \quad -0{\cdot}168 \quad 3$.

2. The roots are $-1{\cdot}44 \quad 0{\cdot}87 \quad 3{\cdot}07$.

3. The numbers are $7 \quad 14 \quad 28$. [This is the natural set of numbers to give, but the sets $0 \quad 0 \quad 0$ and $-7 \quad -14 \quad -28$ also satisfy the conditions.]

4. (i) $x+1/x$ is equal to $13/6$ or $5/2$; whence x has one of the values $1/2 \quad 2/3 \quad 3/2 \quad 2$. (ii) Solving for y/x gives the values $5/6$ or $-11/6$; whence $x=6 \quad y=5$ or $x=6/\sqrt[3]{17}=2{\cdot}33 \quad y=-11/\sqrt[3]{17}=-4{\cdot}28$.

5. (i) $x=4$ or $-13/7$. (ii) $x=10$, $y=2$; or $x=-4$, $y=-5$.

6. (i) $x=0{\cdot}43$ or $-3{\cdot}10$. (ii) We first solve for y/x and obtain the values $4/13$ and $-3/2$; whence the four sets (the decimals being approximations):

x	2	−2	4·9	−4·9
y	−3	3	1·5	−1·5

We could also solve the equations by using the second equation to express x in terms of y, and substituting this value in the first equation.

7. We eliminate x, solve for y and find $y=0{\cdot}4$ or $-3{\cdot}8/3$. The former gives $x=0{\cdot}5$, $y=0{\cdot}4$, and $x=-0{\cdot}5$, $y=0{\cdot}4$; there is no intersection corresponding to the latter.

8. The second curve consists of two straight lines $y=(\sqrt{3}-2)x$ and $y=-(\sqrt{3}+2)x$. Only the latter meets the first curve, the points of intersection being $x=0{\cdot}197 \quad y=-0{\cdot}734$ and $x=-0{\cdot}197 \quad y=0{\cdot}734$.

9. (i) $x=-a-b$. (ii) $x=10/3$. (iii) Eliminating x gives a quadratic in y^2 with roots $y^2=20/3$ or $28/3$; whence the four solutions:

x	1·94	−1·94	2·29	−2·29
y	2·58	−2·58	−3·06	3·06

[The last entry has to be calculated carefully to make sure that 3·06 is closer than 3·05.]

10. (i) The only real root is $x=1$, and we are not concerned with anything else. (ii) The values of x are $0 \quad 3a \quad -3a$. (iii) Solutions are $x=1/2 \quad y=3/2$ and $x=3/2 \quad y=1/2$.

11. The horizontal projection of BD, plus half DE, is equal to half BC; this expressed in terms of x is the equation. The solution is

$x=\surd 85 \doteq 9{\cdot}22$. $AB=17{\cdot}89$ ft. $DE=14{\cdot}00$ ft. $FD=2{\cdot}24$ ft.

12. If the distances of the feet of the ladders from the wall are x and y feet, $y^2-x^2=17^2-10^2$ and $x+y=21$. From these $x=6$; and the height of the wall is 8 ft. (The thickness of the wall is not given, and must therefore be neglected.)

13. The triangle is acute-angled; the foot of the perpendicular is within AB and $1{\cdot}9$ cm. from A.

14. Note the unusual use of a symbol; in general $x76$ would mean $x\times 76$, but here it means $100x+76$. The square of the number is $10000x^2+15200x+5776$ or

$$(100x+76)+(10000x^2+15100x+5700).$$

As the square is to end with the digits $x \quad 7 \quad 6$, the quantity in the second bracket must be a multiple of 1000, and $151x+57$ must be a multiple of 10. By trial we see that this requires x to be 3; and the number is 376. *Or*: proceed with the ordinary arithmetical process for squaring a number.

15. There were 7 shillings, 4 florins, and 6 half-crowns.

16. $y=12-(4x-1)/5$. We can make $4x-1$ divisible by 5 by putting $x=4$; and any value $4+5m$ (m being integral) will also make $(4x-1)/5$ integral. Thus $x=4+5m$ and $y=9-4m$ give any number of solutions; those in which x and y are both positive are $x=4 \quad y=9$, $x=9 \quad y=5$, $x=14 \quad y=1$.

17. Let him make for a point x miles below the nearest point N. The boat will be there $x/6$ hrs. after passing N. So that his time in hours from house to river may be expressed as $x/6+3/4$ and as $\surd(x^2+16)/4$. The point asked for is 3 miles below N, and the second possibility $4{\cdot}2$ miles below N.

18. (i) $b^2x + a^2y = 0$. (ii) If x and y are both positive $b^2x + a^2y$ is positive; if both negative it is negative. Therefore one is positive and one negative.

19. (i) We could put the whole expression in terms of ab and $a+b$. Or more quickly thus: since a is a root of the equation, $a^2 + 7a = 9$ and $a^2(a+7)^2 = 81$. Also
$a^2 + b^2 = (a+b)^2 - 2ab = 67$; so that $ab(a^2 + b^2 - 10) = -513$;
and the given expression $= 2 \times 81 - 513 = -351$. (ii) The factor common to the two quadratic expressions is $x - 17$ and the common root is 17.

20. (i) $(x + a - 3{\cdot}7)(x - a - 1{\cdot}3)$. (ii) The value is 0. Express it as a single fraction with denominator $(x-1)(x-2)(x-3)$, and the numerator vanishes. (iii) $x + y = 3{\cdot}12$, $x - y = 26$. These come at once by adding and subtracting the given equations.

21. With $v = 100$ the body is at 100 ft. at time $1\frac{1}{4}$ sec., and again at 5 sec.; it never reaches 200 ft., its greatest height being $156\frac{1}{4}$ ft.; it is 100 ft. below the origin at time 7·1 sec. The roots of $16t^2 - vt + h = 0$ are imaginary if $h > v^2/64$, the body never reaching such a height; they are real and equal if $h = v^2/64$, this height being that at which the body turns and begins to fall; they are unequal and positive if $0 < h < v^2/64$, the body being once at that height on the upward journey and once on the return; one root is negative if $h < 0$, and if the body was in motion before the moment from which the time is reckoned, it would be at this height (or rather depth) before coming to the original position.

22. (i) The solution is $x = 0$. [If the second solution of the equation considered as a quadratic is wanted, it is $x = \infty$.] (ii) $x = 10$ $y = 12$. (iii) $x = 6$ or -8. (iv) $k = 1/6$.

23. $x + y = 117$ $xy = 47$.

24.
$$m^3 + n^3 = (m+n)(m^2 + n^2 - mn) = (m+n)^3 - 3mn(m+n)$$
$$= (3abc - b^3)/a^3.$$

25. $x^3 + y^3 = a^3 - 3ab$ (cf. Question 24).
$$x^4 + y^4 = (x^2 + y^2)^2 - 2x^2y^2 = a^4 - 4a^2b + 2b^2.$$

26. The data show that the expression vanishes if x is equal to 10 or 1·5, so that the expression is $(x - 10)(x - 1{\cdot}5)$.

27. The equation is $2y^2 - 15{\cdot}7y + 29{\cdot}61 = 0$. Its roots are greater by 1 than those of the original equation, and are $4{\cdot}7$ and $3{\cdot}15$.

28. The second equation is $q^2x^2 + (2q - p^2)x + 1 = 0$. For the numerical values of p and q, the first equation $x^2 - 5x + 6 = 0$ has roots 2 and 3, and the second equation

$$36x^2 - 13x + 1 = 0$$

has roots 1/4 and 1/9.

29. (i) $ax^2 + (2b - a^2 - 2a)x + 2a^2 - 4b = 0$. (ii) Either coefficient will give the second root, which is $3{\cdot}1415$.

30. There are two points on the graph at which $b = 21$, and at them p has the values $1{\cdot}7$ and $19{\cdot}3$: these are the roots of the equation $x^2 - 21x + 32 = 0$. The equation has equal roots when the graph gives two equal values of p, which is when $b = 11{\cdot}3$.

31. (ii) $x = 0$ or $-17/28$.

32. Roots of second equation $a/b - 1$ and $b/a - 1$.

33. (i) Other factors $3x - 7$ and $5x + 2$. (ii) The square root is $a^2 - b - 4$ or $b + 4 - a^2$.

34. Remainder is 212.

35. $f(0) = 5/2$ $f(-2) = 1/20$.

36. $a = -16$. Other values of x are 1 and 5/2.

37. $a = -4$. Other factors $x + 2$ and $x + 4$.

38. $a = 120$. Other roots 5 and -8.

39. A quadratic expression which vanishes for more than two values of x vanishes for all values. The given expression vanishes for all values of x; and if we calculate the coefficients of x^2 and x and the term independent of x, we find them all to be zero.

40. The most general function of a and b which is homogeneous, of the third degree, and symmetrical is

$$P(a^3 + b^3) + Q(a^2b + ab^2).$$

This must vanish when $a = 2b$; which gives $Q = -3P/2$. The expression is therefore a multiple of

$$2(a^3 + b^3) - 3(a^2b + ab^2).$$

Section 44.

1. Sides have lengths 7·07 7·07 4·47. The rectangle has area 35, and the three triangles by which it exceeds the given triangle have area 20, leaving 15 as area of given triangle.

2. The other corners are $(-1, 1)$ and $(0, -4)$.

3. Use the rectangle property of the circle to find the distance of the point of contact from the point in which the join of the two given points meets the given line.

4. The first curve consists of the two straight lines $x - 0{\cdot}8 = 2y$ and $y = 2(x - 0{\cdot}8)$. The solutions are $x = 2{\cdot}74$ $y = 0{\cdot}97$ and $x = 1{\cdot}60$ $y = 1{\cdot}60$.

5. The two points are $(5, 0)$ and $(-5, 0)$.

6. We may begin with any circular arc which passes through the first two points. The centre of the next arc must be collinear with the second point and the first centre, and also on the perpendicular bisector of the line joining the second and third points.

7. (i) The two axes and the lines $x = 7$ $y = 7$ form a square circumscribing the given square. By considering the four triangles by which it exceeds the inner square, we see that the other two corners are at $(7, 5)$ and $(2, 7)$. (ii) Length of side is 5·385, area 29. (iii) The rectangle is the circumscribing square with which we started.

8. (i) The centres of the two circles are at the intersections of the perpendicular at O to DE with the perpendicular bisectors of BO and OC. (ii) Bisect the exterior angles at Q and S by lines meeting in T. The two arcs have their centres at the intersections of the perpendicular from T to QS with the perpendicular bisectors of QT and ST.

9. For the master's use it may be stated that the path of P is $y = 2/(x + 1)$.

10. The three lines, as well as those for other values of k, pass through the point $(-1, 2)$.

11. A is at (1·9, 1·1), B at (0·5 or 0·6, 3·0), C at (6·0, 2·2). The equations are

BC $x/21 + y/3{\cdot}1 = 1$ (or $x/20$ for first term),

CA $-x/1{\cdot}9 + y/0{\cdot}5 = 1$ (or $-x/2{\cdot}0$ for first term),

AB $x/2{\cdot}6 + y/3{\cdot}8 = 1$.

Section 45.

1. One lap measures $2a + \pi b$ yards along the dotted line, $2a + \pi b - \pi w$ along the inner boundary, and $2a + \pi b + \pi w$ along the outer; numerical values 754 716 792. While the slower boy goes 716 yards, the faster goes about 788 yards, which is less than the length of his lap: so that the slower one wins.

2. The straight pieces 7·3 inches, and the curves π inches; total 10·4 in.

3. Suppose the radius of a corrugation to be r. The length of the chord ab (the side of the inscribed square) is $\sqrt{2}r$, and the length of the corrugation is $\frac{1}{2}\pi r$ or $\pi/2\sqrt{2}$ times the chord. The corrugated metal smoothed out would measure $100\pi/2\sqrt{2}$ sq. ft. and weighs $273\pi/2\sqrt{2}$ or 303 lb.

4. The three semicircular areas are, in sq. cm., $\frac{1}{2}\pi r^2$ $\frac{1}{2}\pi x^2$ $\frac{1}{2}\pi (r-x)^2$, whence the expression. The unshaded area is zero when C is at A or B, and has its maximum value (which is half that of the semicircle AKB) when C is midway between A and B.

5. (i) is correct. (ii) is wrong, for the square root must lie between 0·6 and 0·7. (iii) is wrong, for the area stated is less than the square on the radius. (iv) is wrong; a regular inscribed hexagon has a side 81·9 ft. long, a 12-sided figure has consequently a side of more than $\frac{1}{2} \times 81{\cdot}9$ ft., and a 7-sided figure has a side greater than the 12-sided figure.

6. For one revolution of A, a point on its rim travels 40 cm., a point on b's rim travels 40 cm., b and B make 4 revolutions, a point in B's rim travels 120 cm., a point on c's rim travels 120 cm., c and C make 12 revolutions. B turns 4 times as fast as A, and C 12 times. If A makes 1/3 revolution per second, B makes 4/3 and C 4 per second.

7. (i) In one revolution of the shaft E falls 100π cm., G rises 22π cm., and F falls 18π cm. D rises 11π cm. because of the rise of G and falls 9π cm. because of the fall of F, so that it rises 2π cm. for one revolution of the shaft. Hence when E falls 1 cm., D will rise $2\pi/100\pi$ cm. or 0·02 cm. (ii) Similar reasoning with diameters p q r cm. gives for D a rise of $(q-p)h \div 2r$ cm. when E falls h cm.

8. Let CP, drawn parallel to DA, meet BE produced in P. From the parallelism of the two lines it follows that BG = GP and GE = EP; whence GE = $\frac{1}{3}$BE and GH = $\frac{1}{3}$BC = $\frac{2}{3}$DC. Volume of cone is 14000 (or 14500) c. cm. in the first case, and $\frac{1}{3}\pi a^2 b$ in the second.

9. The perimeter of the base being 7π in., its radius is 3·5 in. and its area $\pi 3{\cdot}5^2$ sq. in. The height of the cone is therefore $3{\cdot}5\sqrt{3}$ in., and the volume 78 c. in. (77·8).

10. (i) The greatest length of the rod is 30 cm., the least length 22 cm. (22·4). (ii) Each volume is half that of the greater cylinder less half that of the lesser, and therefore 13000 c. cm. (12566).

Section 46.

1. AH = 3·7 cm. HB = 2·3 cm.

2. Denote AH by c, HB by x. Then $x(x+c) = c^2$ or $x = a(\sqrt{5}-1)/2$, whence the following construction. Draw AP at right angles to AH and equal to half of it. Join PH, and from it cut off PQ equal to PA. On AH produced mark off HB equal to HQ; which gives B.

3. Draw BP at right angles to BA and equal to half of it. Join PA, and from it cut off PQ equal to PB. From AB cut off AC equal to AQ. Then

$$(AD + AC)^2 = (PQ + AQ)^2 = AP^2 = AB^2 + AD^2.$$

4. Draw BC at right angles to BA and equal to half of it. Join BC, and produce it. Mark off CD on CB, and CE on CB produced, each equal to CA. In AB take Q so that BQ = BD, and on AB produced take P so that BP = BE; this gives P and Q. Then BP - BQ = DE = AB.

5. Denote AD by p and BC by q. The Golden Section gives $BD = 0{\cdot}5(\sqrt{5} - 1)p$ and $AB = 0{\cdot}5(\sqrt{5} + 1)p$. Then Pythagoras gives $(p+q)^2 = \{0{\cdot}5(\sqrt{5}+1)p\}^2 + q^2$ or

$$q = 0{\cdot}25(\sqrt{5}+1)p = 0{\cdot}5AB.$$

6. Take any line AB. Draw AP at right angles to AB and equal to half AB. Join PB, and from it mark off PQ equal to PA. With centres A and B and radii AB and BQ strike arcs meeting in C. ABC is the required triangle. This is Euclid's construction a little condensed.

7. Each base angle in Question 6 measures 72°.

8. The vertical angle of Question 6 is 36°; by bisecting it we obtain 18°. An angle of 15° is given by the bisection of an angle of 30°. An angle of 3° is obtained as the difference between 18° and 15°.

9. (i) Each side must measure 0·8 inch, and consecutive sides are inclined at 108°. (ii) Draw the inscribed circle, and draw tangents at the extremities of the radii to the vertices.

PART VI.

Section 47.

1. Breadth of river 142 ft. **2.** AP = 153 metres.

3. Height above A is 2600 (2636) ft.

4. B is 3030 yd. from O, 61° north of east. AB makes 54° with the east. A is 543 yd. east and 1340 (or 1350) yd. north of O; B is 940 yd. east and 1290 yd. north of A, and 1480 yd. east and 2640 yd. north of O.

5. Area 13000 (12540) sq. miles.

6. The sixteen readings are, in feet, 200 380 460 530 670 890 880 820 970 1120 1200 1000 680 520 410 300. Many of them are uncertain as to 10 ft. or so.

8. Distances of first and second points 3·8 (3·76) miles and 2·6 (2·57) miles. From the second point he must go 0·9 (0·88) mile further to the nearest point.

9. Least and greatest values for AB are 8·5 and 19·4 miles.

10. Grasmoor is 7·3 miles from Skiddaw, bearing S. 46° W.

11. 87 miles.

12. A scale of 1·1″ to 10′ would give side margins of 2″ and end margins a little wider. The nearest simple scale, such as is generally used, is 1″ to 10′.

13. BC is 1580 metres. Height CD is 330 metres (328).

Section 48.

1. 0·49979.

2. The area is 29 hectares or 72 acres.

3. Distance of horizon 154·95 miles. [The radius of the earth is given to one significant figure, and the answer ought to be asked for to a corresponding degree of accuracy. This question is included in its present form because such questions are frequently set.]

4. Volume is 15·64 c. in. **5.** Fifth root is 3·579.

6. When the breadth is 4 ft. and the depth 3 ft., 4650 cubic feet pass per minute. If the depth is reduced to 2 ft., the flow is reduced to 2740 or 2750 c. ft., 59 per cent. of the former quantity.

7. (i) $\log 10{\cdot}39 = 1{\cdot}0165$, $\log 0{\cdot}0912 = \bar{2}{\cdot}9600$,

antilog $\bar{1}{\cdot}3 = 0{\cdot}1995$.

(ii) Value is $2{\cdot}856 + 1{\cdot}226$ or $4{\cdot}08$.

8. Diameter 14·7 inches. **9.** 1·844.

10. (i) 963·4. (ii) 0·01481.

11. $\log_5 1250 = \log_5 (5^3 \times 10) = 3 + \log_5 10 = 3 + 1/x$.

12. £380. [More accurate tables give £377·0.]

13. £590. [More accurate tables give £586·1.]

14. (i) £64. [More accurately £64·3. It happens that the four-figure tables give the first decimal place correctly, but they cannot be relied on to give it, and we cannot by their help alone state the result to more than the nearest integer.] (ii) $n = 20$.

15. £424 to the elder and £376 to the younger.

16. £7700. (More accurately £7683. It is by chance that the four-figure tables give three significant figures correctly.)

17. 0·6934.

18. (i) 13·6 (13·63). (ii) $a^{5/7}$.

19. (i) Difference of height in feet $= 60360 \log 1{\cdot}5$

$= 60360 \times 0{\cdot}1761 = 10600$.

(ii) $\log (15/p_2) = 15775/60360 = 0{\cdot}2614$, so that

$\log p_2 = \log 15 - 0{\cdot}2614 = 0{\cdot}9147$ and $p_2 = \text{antilog } 0{\cdot}9147 = 8{\cdot}2$.

The pressure at the top is 8 lb. per sq. in.

20. $\text{Log}_e 2 = \log_{10} 2/\log_{10} 2{\cdot}718 = 0{\cdot}3010/0{\cdot}4343$, and the ratio is 0·85.

21. By calculating the values of the two expressions for a number of values of x, we find that the former is greater when $x = 0{\cdot}3$ and less when $x = 0{\cdot}4$, so that a root lies between these values. To the required degree of accuracy the root is 0·35. The approximate equation has a root which is also to two significant figures 0·35.

22. $\log_2 1{\cdot}7 = 0{\cdot}76$ or $0{\cdot}77$, $\log_2 5{\cdot}4 = 2{\cdot}43$; the product of 1·7 and 5·4 is 9·2.

23. (i) One of the results comes by adding the given equations, the other by squaring and adding.
(ii) $c = (k^2 - z^2)/2z = 52{\cdot}5$; $e^{x/c} = (k + z + c)/c = 7/3$, so that $x = 52{\cdot}5 \log_{10} 2{\cdot}333/\log_{10} 2{\cdot}718 = 44{\cdot}5$, and the span is 89 feet.

Section 49.

1. True value of each ratio 1·528.

2. The angles of the new triangle are equal to the old, the sides are doubled, and the area quadrupled.

3. The angles of the triangle are 84° 60° 36°. The length of the bisector is 3·1 (3·14) cm.

4. The speed of the car is 13 (13·1) yd. per sec. or 27 (26·8) miles per hour.

5. The quotients are 11·0 and 40·1.

6. P describes a straight line. When $CF = a/n$ ft., the distance BP is $(1 - 1/n)\surd(a^2 + b^2)$ ft.

7. Find expressions for the distances of M and N from AB, and then for MN, which is also equal to $d + a - b$. Numerical value for H 54 ft. (54·4).

8. If the longest side of the second triangle is 5 units, the other two sides are 3·7 and 4·5 to two significant figures.

9. The triangles are not similar. For similarity each side of one triangle must be paired with a side of the other, and the ratio of any two sides of one triangle must be equal to the ratio of the *corresponding* sides of the other.

10. The greatest of the four triangles is twice each of the adjacent ones and four times the remaining one.

11. PQ = 2·5 in. AP/AQ = PT/QU = 1/2 and BQ/BP = QS/PR = 1/2, so that AB = 3PQ = 7·5 in.

12. The two portions of the quadrilateral are in the ratio of their heights, that is, in the ratio of the portions of the diagonal.

13. (i) Let CD and AY meet in O. Equality of areas shows that OA.OD = OY.OC, whence triangles OAC OYD are similar and AC is parallel to YD. (ii) The perpendicular from O on DY = OD.OY/DY = 2·4 in., the perpendicular from O on AC = OA.OC/AC = 1·2 in.; and their sum is 3·6 in.

14. Draw PQ parallel to OX to meet OY in Q. On OY measure off OB equal to 3 times OQ.

15. Construct as for Question 14. There is one line for which D is the point of trisection next to AB, and one in which it is the point of trisection further from AB.

16. (i) The image moves away in a straight line to the right at the rate of 20 inches per second; distance from O in m seconds is $20(m+1)$ inches. (ii) The path is a semi-circle whose diameter is the line joining O to the original position of its image.

17. Denote the mid points of BC and AD by R and Q.

$$\triangle ABD = \triangle OQD - \triangle OPB + \triangle PQA$$
$$= \triangle OQA - \triangle OPC + \triangle PQD = \triangle ACD.$$

Hence BC and AD are parallel and cut CPD and BPA in the same ratio.

18. Suppose that AP meets BQ in K and meets CR in L. Then $KA/KP = AB/PQ = AC/PR = LA/LP$; so that K and L coincide.

19. (i) For the given values of a and c, b is 3·46 in. (ii) To prove the relation equate AO/OB to the corresponding ratio in a similar triangle.

20. (i) The first part as in Question 19. (ii) Triangle APQ, when isosceles, is twice the square ABCD.

21. (i) Drop the perpendicular AD upon BC. Draw AE parallel to BC and equal to AD. Join BE, cutting AC in F, which is a vertex of the square. Let FG parallel to BC meet AB in G, and GH parallel to AD meet BC in H. Then $GH/AD = GB/AB = GF/AE$; so that $GH = GF$. (ii) To prove the relation, equate the ratio of BC to the height of triangle FBC and the corresponding ratio for FAE.

22. Since the field is equally divided $xy = 735$; and a variation of the relation of Question 19 gives $15/x + 12/y = 1$. The solutions are $x = 35 \quad y = 21$ and $x = 26\frac{1}{4} \quad y = 28$.

23. (i) Capacity 780 c. cm., the value of h being $\sqrt{96}$ cm. (ii) When a slant height of x cm. is covered, the depth is $x\sqrt{0{\cdot}96}$ cm. and the radius of the surface $4 + 0{\cdot}2x$ cm.; whence the expression required. (iii) Distances of graduation marks from bottom of slant side, in cm., 0 1·5 2·8 3·9 5·0 6·0 6·9 7·7 8·5 9·3 10.

24. (i) The locus of B is a circle on ON as diameter, for the angle OBN is always right. (ii) Produce ON to N_1 so that $ON_1 = 2ON$; the locus of B when $OA.OB = ON.ON_1$ is a circle on ON_1 as diameter. (iii) Bisect ON_1 in N_2; the locus when $OA.OB = ON.ON_2$ is a circle on ON_2 as diameter.

25. The relations $CB/BA = OP/PA$ $CA/BA = OA/PA$ give

$$CB = 2xy/\sqrt{(x^2+y^2)} \text{ in.} \quad CA = 2x^2/\sqrt{(x^2+y^2)} \text{ in.}$$

The area of ABC is $2x^3y/(x^2+y^2)$ sq. in.

Section 50.

1. The tangent is 1·12, the angle being 48°.

2. The sine is 0·87 and the cosine 0·50.

3. The angle is 17·5°, the cosine 0·95, the tangent 0·31.

4. The angle is 53°, the sine 0·80, the tangent 1·33.

5. The angle is 71·5°, the sine 0·95, the cosine 0·32.

6. Sine 42° = 0·67, tan 42° = 0·90. The tables give the values to four figures, while drawing will not give more than two and not those with certainty.

7. The four angles are 11·5° 23·5° 37° 53°.

8. (i) The angle is 66·5° and the sine 0·92. (ii) The angle is 44·5° and the sine 0·70.

9. The cosine of an acute angle has a value between 0 and 1, the sine also has a value between 0 and 1, while the tangent may have any positive value.

10. In the following table the column headed SIN/COS gives, to two decimal places, the ratio of the approximate values of the sine and cosine:

	SINE.	COSINE.	TANGENT.	SIN/COS.
20°	0·34	0·94	0·36	0·36
40°	0·64	0·77	0·84	0·83
60°	0·87	0·50	1·73	1·74
80°	0·98	0·17	5·67	5·76

11. Distance from base 211 feet.

12. Breadth 33 yd. (32·7).

13. Length of chord 3·584 in.

14. The tangent = 2, the angle 63·5°, the sine 0·89, the cosine 0·45.

15. The top descends 2·0 ft.

16. (i) New length of string 24 in.; inclination of rod 19·5°. (ii) Shortest length for string 16 in.; inclination of rod 53°.

17. (i) Depth of C 24·4 in. (ii) Angle 28° (27·8°). (iii) Mean of depths of two opposite corners is depth of centre.

18. Latitude of Alexandria 31°, of Carthage 37°.

19. (i) $a^2 = p^2 + m^2$, $b^2 = p^2 + n^2$, $(m+n)^2 = a^2 + b^2$. (ii) Four sets of three points, one which gives no circle because the points are collinear. Relations,

$$a^2 = m(m+n), \quad b^2 = n(m+n), \quad p^2 = mn.$$

(iii) Two of the five quantities being known, our equations give the remaining three.

20. At intervals of 30 seconds we have for the coordinates,

x	0	0·5	0·87	1	0·87	0·5	0	−0·5	−0·87	−1	−0·87	−0·5	0
y	0	0·87	0·87	0	−0·87	−0·87	0	0·87	0·87	0	−0·87	−0·87	0

The figure is a figure-of-eight symmetrical about both coordinate axes; to see the shape, points should be plotted at intervals of 15 seconds. The curve is completely described in 6 minutes.

21. (i) Triangle OAC is isosceles and obtuse angled.
(ii) AC = 23 (23·3) cm. OM = 7 (7·0) cm. Area of OAC is 82 sq. cm.
(iii) The condition specified occurs when $\angle OCD = \angle OCM$ and each is 30°; AB/BC = 0·58, BC/AB = 1·73.

22. The radius is 1·93 in.

23. Length of pole 30 ft. (29·7).

24. Height of flagstaff 45 (44·8) ft.

25. Height of tower 99 (99·3) ft.

26. Air space 2500 cubic metres (2470 or 2480 according to selection of data). Inclination of diagonal of end wall, calculated from remaining data, lies between 44° and 44·1°, but nearer to 44°.

27. PR = $s/(\tan\beta - \tan\alpha)$, RQ = $h - b + s\tan\alpha/(\tan\beta - \tan\alpha)$. Numerical values PR = 198 ft. RQ = 24 ft.

28. Height of retaining wall 15 (14·7) ft.

29. Arc subtends 90·5° at centre; radius is 84·5 ft. Span will be increased by an inch by a rise of temperature of 69° C. (69·4).

30. The tangents at B and C meet on the perpendicular bisector of the chord BC, that is, at E. The right-angled triangle OBE gives $BE = r \cot(\theta/2)$.

31. The area within which trees may be planted is a square 642 ft. in the side. One side will take a row of 33 trees, 20 feet apart. The rows must be $\sqrt{300}$ feet apart, which gives 38 rows. There are 19 rows of 33 trees and 19 rows of 32 trees, in all 1235 trees.

32. The slant height of a face being, in cm., $2 \tan 70°$, the angle of slope is 69° (68·7).

33. Slant height of face in feet $= \sqrt{(481^2 + 378^2)} - 612$. Length of edge $= \sqrt{(481^2 + 378^2 \times 2)} - 719$. Inclination of face $= \text{arc tan}\,(481/378) = 52°$ (51·8).

34. $AB = AD \tan 43°$ and $AD = CD/\cos 52°$, whence $AB = 76$ yd.

Section 51.

1. Length of rim of cone being $28\sqrt{2}\pi$ cm., the angle of the plane sector is $360° \times 28\sqrt{2}\pi \div 56\pi$ or 255°(254·6). The weight of the cone is 3·7 kilograms (3·72).

3.

	Surfaces.	Plane angles.	Edges.	Plane surfaces.
Sphere	1	0	0	0
Cone	2	0	1	1
Cylinder	3	0	2	2
Pyramid	5	16	8	5
Cube	6	24	12	6

4. Consider the vertical section which contains a diagonal of the upper surface of the cube. The equilateral triangle must be such that a rectangle $\sqrt{2}$ in. broad and 1 in. high may be inscribed. Height of cone 2·225 in.

6. The section of the sphere by the plane of the wires is the inscribed circle of the triangle, and has radius $\sqrt{3}$ in. The height of the centre above the wires is therefore 1 in., and of the top 3 in.

7. If the centre O is x in. above E, $(1{\cdot}8 - x)^2 = x^2 + 0{\cdot}75^2$ or $x = 0{\cdot}744$, and the radius OF is 1·056 in.

8. (i) Distance of centre of triangle from foot of tripod is $70/\sqrt{3}$ or 40·4 cm. Height of tripod is $\sqrt{8367}$ or 91·5 cm. (ii) The values of y are 100 97·3 88·7 72·1 38·3. The value of x that makes y zero is 173. (iii) The relation is obtained by applying Pythagoras to the right-angled triangle whose hypotenuse is the tripod-leg, one side the height of the tripod, and the other side the distance of the centre of the equilateral triangle from a foot of the tripod.

9. The length of edge AE is the hypotenuse of the right-angled triangle whose base is the plan of AE and whose height is the height of E above A; it is 26 ft. For the slated area we first obtain the slant heights of ABE and EFCB, which are about 19·8 or 19·9 ft.; the area ABE is about 341 sq. ft., the area EFCB about 750 sq. ft., and the whole roof about 2180 sq. ft.

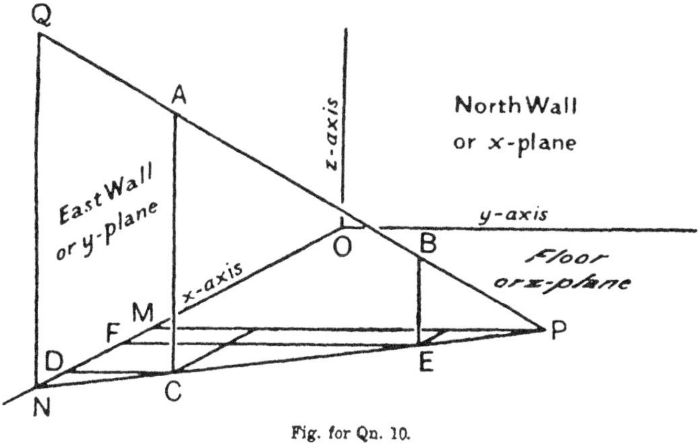

Fig. for Qn. 10.

10. The length of the cord is 4·72 ft., its prolongation meets the floor at (3·5 7·25 0) and the east wall at (5·57 0 6·21). In the figure A B are the two ends of the cord; C the projection of A on the floor, D the projection of C on the axis OX; E and F corresponding projections

of B. P is the intersection of AB and CE, M its projection on OX. Q is the intersection of AB with the east wall, N its projection on OX, N lying of course on CE produced. Now DM/FM = CP/EP = CA/EB, whence OM = 3·5; and (MP − CD)/(MP − FE) = CP/EP = CA/EB, whence MP = 7·25; thus the co-ordinates of P. Also NF/ND = EF/CD, whence ON = 39/7; and (QN − BE)/(QN − AC) = NE/NC = EF/CD, whence QN = 87/14; thus the co-ordinates of Q.

11. The slope of the line of intersection of the roofs is that of the diagonal of a cuboid, the diagonals of whose vertical faces slope at 45° and 60°. The height of the cuboid being unity, its other dimensions are 1 and $1/\sqrt{3}$, and the slope arc tan $\sqrt{0\cdot75}$ or 41°.

12. Inclination = arc cos (2·5/3) = 33·6°. One end is higher than the other by $\sqrt{(3^2 - 2\cdot5^2)}$ or 1·66 in.

13. If M is the foot of the perpendicular, BM is also perpendicular to AC, and is $5\sqrt{2}$ cm. long. Therefore

$$DM = \sqrt{(2\cdot5^2 + 50)} \text{ cm.} = 7\cdot5 \text{ cm.}$$

14. (i) The areas are: ABC 32 sq. cm., ACD 38 sq. cm., ABD 42 sq. cm. (ii) The area of BCD being 33·1 sq. cm., the volume of the pyramid is 88 c. cm. (iii) The tangent of the inclination of ABC is 80 mm. divided by the perpendicular distance of the plan of A from BC; the inclination is 79° (78·7).

15. The line of intersection of two bevels is the diagonal of a cube whose side measures $\sqrt{0\cdot5}$ inch.

16. One diagonal is horizontal. The inclination of the figure is arc cos (2/3·5) or 55°; the side of the square is 2·47 in.

17. We denote the points of the bed below A B C D by P Q R S respectively. Reasoning by similar triangles shows that the line PR meets a north-south vertical plane through D in a point T due east of A at a depth of 361·5 ft. Like reasoning shows that PQ meets the plane at a point U 320 yd. north of A at a depth of 419 ft. From the points T U we find that S (in line with them and vertically below D) is at a depth of 397 ft., or in round numbers 400 ft.

18. A B C being the points of which the shafts are sunk, BA meets the line of outcrop in F such that AF = 4AB = 4 miles;

CA meets it in E such that AE = 2 miles, CB in D such that BD = 5 miles. DEF is perpendicular to CAE, 2 miles from A, 3 miles from C, and $2\frac{1}{2}$ miles from B.

19. Let A-BCD be the given tetrahedron. Let parallels through B and C to AD meet a plane through A parallel to BCD in E and F. Consider the tetrahedra A-BEF A-BCF, which with the given one make up a prism. Compare the tetrahedra F-ABC (which is the same as A-BCF) and D-ABC (the given one); they have the same base ABC; their heights, the distances of F and D from ABC, are equal, for FD meets the plane ABC on the line AC and is bisected by it; thus the two tetrahedra are equal. Also compare the tetrahedra A-BCF and A-BEF; their bases BCF BEF are equal; their height is the same; so that these two tetrahedra are equal. Thus the original tetrahedron and two others of equal volume make up the prism.

20. Suppose a plane K drawn through P parallel to ABCD. Triangles OPQ OAB being similar, PQ is parallel to AB and lies in the plane K; and by like reasoning QR RS SO are parallel to the corresponding sides BC CD DA and lie in the plane K. Also the ratios PQ/AB QR/BC RS/CD SO/DA are all equal, being equal to OP/OA, etc.

21. Let P Q R S denotes the mid-point of AB CD CA BD. PR is parallel to BC and equal to half of it, so is QS; thus PRQS is a parallelogram and its diagonals PQ RS bisect one another. It is shown by like reasoning that the join of the mid-points of BC and AD bisects PQ and RS.

Section 52.

1. The length is 3·5 in. (3·46).

2. Along one line through O mark off OA equal to 1 unit and OB equal to 3·9, along another mark off OC equal to 1·7. A parallel to AC through B meets OC in D such that OD is the required product. OD = 6·6 (6·63).

3. Draw a straight line ABC making AB = 3·75 units and BC = 1·75. On AC describe a semicircle and erect a perpendicular at B to meet it in D. BD is the square root and is 2·6 (2·56).

4. BF/BD = BE/BC = BD/BA.

5. If EM and RH meet in T, RT is 5/16 in. If FL and PK meet in U, PU is 11/16 in.

6. Each ratio is 2·1 (2·095).

7. AD and BE bisect the angles at A and B. Therefore CF bisects the angle C, so that

$$AF = \tfrac{6}{11}AB = 2{\cdot}18 \text{ cm.}, \quad BF = \tfrac{5}{11}AB = 1{\cdot}82 \text{ cm.}$$

8. Express the relation BD/DC = BE/EC in terms of OB OD OE.

9. AE/EB = AD/DB = AD/DC = AF/FC.

10. On the given line as base make a triangle whose sides are in the ratio 4 : 3, and bisect the vertical angle.

11. The triangles POA PQB are similar and QB = 2OA. All such points lie on a circle, centre Q, radius 2 in.

12. Let O be the centre, Q the mid-point of AP, B the mid-point of AO. Triangles APO AQB are similar and BQ is constant.

13. CF/FD = CE/AD = constant. AE/AF = BE/BC = constant. If A is held at rest and F traces out a figure, E will trace a similar figure.

14. On the arms OA OB of the sector take two points C D equidistant from O, and on CD draw a square CDEF. OE and OF cut the arc of the sector in two corners of the inscribed square.

15. (i) $\angle adc = 68^\circ\ 12'$ $\angle feg = 20^\circ\ 33'$, $cd = 5{\cdot}385$ in., $eg = 8{\cdot}544$ in.

(ii) One angle is $1^\circ\ 15'$; area is 1·002 sq. in. (the last figure quite unreliable).

16. The area increases by 3·2 sq. cm. (3·16). For similarity the increments of sides would have to be proportional to the sides.

17. The quadrilateral is equal to triangle ADE, and therefore to triangle AXE, which has base AX and height 2 units. Numerical case 2·4 sq. in. (2·41).

18. Frontages on AC are 58 ft. 24 ft. 18 ft. (57·74 23·91 18·35). Length of fencing 69 ft. (69·69) and price to nearest shilling £9. 6s.

19. (i) Area about 0·91 sq. ft. or 132 sq. in. (ii) For breadth a ft., the area is about a^2 sq. ft., the coefficient differing from 1 by something like 0·003. (iii) For breadth $3a$ ft., area is $9{\cdot}0a^2$ sq. ft.

20. The ratio of the areas is that of the squares upon the sides, viz. 2.

21. The ratio of each side of the second triangle to the corresponding side of the first is $\sqrt{2}$, and the ratio of the areas 2.

22. Let O be the vertex of the isosceles triangle, PQ the base, and A its mid-point. Draw OR at 30° to OA meeting AQ in R. The mean proportional between AQ and AR is half the side of the required equilateral triangle. The length of the side is 8·5 cm. (8·54).

23. $OB = OC$ and $OA = OD$, since $OA.OB = OC.OD$ and $OA.OC = OB.OD$.

24. Let the triangle be ABC, the bisector of angle A meeting the base in D and the circum-circle in E. $AB/AD = AE/AC$, so that $AB.AC = AD(AD + DE) = AD^2 + BD.DC$.

Section 53.

1. The expression is equal to $3^7(9 - 54 + 144 - 224)$ or $-270{,}000$.

2. (i) $(a + z)n/2$. (ii) 588 feet.

3. He saved £460 in 1908 and £5320 in all.

4. (i) 10·8 sq. in. (ii) 11·88 sq. in. (iii) The area, in square inches, is $\{1 + 2 + 3 + \ldots + (n - 1)\}24/n^2$ or $12(n - 1)/n$. The three differences are, in square inches, 1·2 0·12 $12/n$.

5. 87·25.

6. The series is $11 + 16 + 21 + \ldots$, and the sum to n terms $n(5n + 17)/2$.

7. Let there be n houses on each side of the square. Suppose house a to be on the first side, house b on the third; then $n - a = b - 2n - 1$ or $n = (a + b - 1)/3$. Suppose houses a and b to be on the second and fourth sides of the square, and we find $n = (a + b - 1)/5$. If $a + b - 1$ is a multiple of 3 but not of 5, there are $(a + b - 1)4/3$ houses round the square; if a multiple of 5 but not of 3, there are $(a+b-1)4/5$; if a multiple of 15, both expressions give possible solutions, and we cannot decide between them.

8. Total payment £133,875.

9. The growth in the year $1901+n$ is $18 \times 0{\cdot}9^n$ inches, which is greater than 0·1 in. if $n=49$, and less if $n=50$; it will first be less than 0·1 in. in the year 1951. The height at the beginning of 1951 is, in feet,

$$10{\cdot}5 + 1{\cdot}5(1 + 0{\cdot}9 + 0{\cdot}9^2 + \ldots + 0{\cdot}9^{49}) \quad \text{or} \quad 25{\cdot}42 \text{ ft.}$$

The maximum height is 25·5 ft. [Questions like this and No. 11 should not be encouraged; trees and snowballs do not grow in the way described.]

10. The 28th rod has 29 feet immersed. The total length of rod below high water is 2400 ft., and the cost of sheathing £300.

11. Whole weight in lb. is $4 \times 1{\cdot}2^2(1 + 1{\cdot}2 + 1{\cdot}2^2 + \ldots + 1{\cdot}2^{19})$ or $28{\cdot}8(6{\cdot}19^2 - 1)$ or 1100 lb. (1070).

12. The number of years is $\log(100/50{\cdot}375) \div \log 1{\cdot}025$ or 28.

13. £424. 13s. [Four-figure tables are not sufficiently accurate for this.]

14. The distance of M_n from the centre is $r(1/2)^{n/2}$. The length of the spiral is $r/(\sqrt{2}-1)$, which is less than $2{\cdot}42r$.

15. If n terms must be taken, 2^n must be greater than 1000, so that $n=10$.

16. The sum is $(1-0{\cdot}8^8)/1{\cdot}8$ or 0·462. The limiting value is $1/1{\cdot}8$ or 0·556.

17. (i) The sum being 1·25, the error is 0·002. (ii) The sum is $1/(1-\sqrt{0{\cdot}2})$ or 1·809. (iii) $1/(1+\sqrt{0{\cdot}2})$ or 0·691.

18. If the decimal recurred we should have a vulgar fraction (namely the square of the root) equal to 2.

19. Twelve terms; common ratio y/x. Sum of six terms,

$$x^6(x^6 - y^6)/(x-y).$$

20. (i) $576a$. (ii) Sum to 8 terms $0{\cdot}6\{1-(2/3)^8\}$ or 0·5766; sum to infinity 0·6.

21. He saved £16 in the first year.

22 Used for 16 years.

23. (i) The nth term is $2a-b-n(a-b)$, and the sum of n terms $(3a-b)n/2-(a-b)n^2/2$. (ii) The nth term is zero if $n=1+a/(a-b)$. (iii) The nth term is the sum of n terms less the sum of $n-1$ terms, and is $10n+1$. (iv) The nth term is $2(n+1)$ if $n>1$, and the series is $5 \quad 6 \quad 8 \quad 10 \ldots$.

24. The sums are $a^2/(a-b)$ and $a^3/(a^2-b^2)$.

25. The common ratio may be $3/4$ and the sum $2{\cdot}5822$, or the common ratio may be $-3/4$ and the sum $0{\cdot}3689$.

26. If n is the number of terms $(1{\cdot}5)^n$ must be greater than 1620; n must be greater than $\log 1620/\log 1{\cdot}5$, and is 19.

27. Thirteenth term is $29{\cdot}4^2/7{\cdot}5$ or 115.

28. (i) The A.M. being a^2+b^2, the G.M. a^2-b^2, the first ratio is a^2/b^2; the second ratio is a/b. (ii) There are 6 terms, the common difference is $1{\cdot}2$, and the second term $4{\cdot}2$.

29. (i) $(a-b)^2$ being positive, $a^2+2ab+b^2>4ab$ or $(a+b)/2>\sqrt{(ab)}$.

(ii) If a is the first term and r the common ratio, we have the equations $a(1+r+r^2+r^3)=60$ and $a(1+r^3)=36$, whence $(1+r^2)/(1-r+r^2)=60/36$, and the series is $4 \quad 8 \quad 16 \quad 32$.

30. (i) If a and b are the first term and the common difference, $(a+4b)^2=a(a+10b)$ and $b=a/8$; the common ratio is $3/2$. (ii) The coefficient of a in every term of the A.P. is a fraction with 8 (or a lower power of 2) as denominator. The first four terms of the G.P. have the same property (the fourth term being the 20th of the A.P.), but all later terms have a higher power of 2 as denominator.

31. (i) The harmonic series is $1 \quad 3/2 \quad 3 \quad \infty \quad -3 \quad -3/2 \quad -1 \quad -3/4 \quad -3/5 \ldots$, and the nth term $3/(4-n)$. (ii) Consider the corresponding arithmetical progression $1/a+k \quad 1/b+k \quad 1/c+k$, and the effect on it of giving k different values.

32. From $x+z=2y$ and $xy=z^2$ we have either $x=y=z$ or $y=x/4 \quad z=-x/2$; and in either case x is the H.M. between y and z.

33. $2a=1+A$, $a^2=G$, $2/a=1+1/H$, whence $4G=(1+A)^2$ and $H=(1+A)/(3-A)$.

34. (i) The G.M. of the A.M. and H.M. is $\sqrt{\left\{\frac{a+b}{2}\cdot\frac{2ab}{a+b}\right\}}$ or $\sqrt{(ab)}$.
(ii) $a+b=91$, $ab/(a+b)=252/13$; the numbers are 28 and 63.

35. (i) Proof that $(a+b)/2 \nless \sqrt{(ab)}$ and $2ab/(a+b) \ngtr \sqrt{(ab)}$ follows Question 29 above. (ii) The harmonic mean is 16/15.

36. $x=(ba+bc-2ac)/(a+c-2b)$.

37. If a b c are the numbers,

$$a+b+c=37 \qquad 1/a+1/b+1/c=1/4 \qquad 1/a+1/c=2/b.$$

The second and third equations give $b=12$; whence $a+c=25$ and $ac=150$, so that a and c are 10 and 15. The three numbers are thus 10 12 15.

38. Suppose $1/h$ the common difference of the series $1/a_1$ $1/a_2$ .. $1/a_n$ .... Then $a_1a_2=h(a_1-a_2)$ $a_2a_3=h(a_2-a_3)$ .. $a_{n-1}a_n=h(a_{n-1}-a_n)$. By adding these equations we have $a_1a_2+a_2a_3+\ldots+a_{n-1}a_n=h(a_1-a_n)$; whence the required relation.

39. (i) The 117th term is midway between the 101st and the 133rd. The sum of the 33 terms is therefore $3\times 33=99$. (iii) Put $q=kp$ $r=k^2p$ $s=k^3p$, and the given relation becomes an identity in k.

PART VII.

ANSWERS TO TYPICAL QUESTION PAPERS.

Section 55. • London Matriculation.

ARITHMETIC AND ALGEBRA.

1. If the length, width, and depth are $f\ \ g\ \ h$ inches to the nearest hundredth, the limits for the number of cubic inches in the volume are $(f \pm 0{\cdot}005)(g \pm 0{\cdot}005)(h \pm 0{\cdot}005)$, or approximately $fgh \pm 0{\cdot}005(gh + hf + fg)$; so that the error is approximately $\pm 0{\cdot}005(gh + hf + fg)$. The volume is 210 c. in., only two figures being certain.
2. He loses 31 % of the value at the time of sale.
3. The yield from Rentes is greater, being 2s. 8d. more on £100 invested.
4. The cost is $0{\cdot}36x^3$ francs.
5. x must be greater than $-1{\cdot}75$ and less than $1{\cdot}5$.
6. $2x^4 - 5x^3 + 3x^2 + 17x + 7$. Numerical value 45.
7. For values of y between -10 and $+6$ there are two positive values of x.
8. (i) $x = b + a \quad y = b - a$. (ii) $x = +3$ or -3.
9. The nth term is $4n - 1$, and the sum to n terms $n(2n + 1)$. The 250th term is 999, differing by 1 from 1000.
10. $x^2 + 2x - 2a$ is positive when $(x + 1)^2 > 2a + 1$ or

$$x > \sqrt{(2a + 1)} - 1.$$

GEOMETRY.

1. For Q, draw a parallel to AB one inch away, and find Q equidistant from this and AC. P is 1·9 (1·87) in. from AB and AC; Q is 2·5 (2·50) in. from AB and 1·5 (1·50) in. from AC.
2. $B = 110^\circ \quad C = 130^\circ \quad D = 60^\circ \quad E = 170^\circ$.

3. Triangles ABC ADE are equal, because triangles BDC and EDC are equal. Given triangle ABC, lay the given length AD along AB; the third vertex E is the intersection of AC by a parallel through B to DC.

4. (i) The triangles BCG and FCA are congruent, and they are respectively equal to the half of the square and the half of the rectangle. (ii) Each right-angled isosceles triangle is a quarter of the square upon the same base.

6. If M is the foot of the perpendicular from A on BC, the required rectangle may have sides BC and 2BM, or BM and 2BC.

7. The common centre of the two circles is at the intersection of the perpendicular bisectors of the two lines. When the bisectors coincide any point on them serves; when they are parallel the construction is impossible.

8. T being the point of contact of the tangent from P, produce TP to R so that PR = PT. Concentric with the original circle, draw another passing through R and cutting the original line in Q, the point required.

9. The perimeter of the hexagon is 6 times the radius, that of the square 8 times.

10. Take F on AC so that $\angle ADF = \angle ADB$. Then

$$\angle DFA = \angle B = \angle DEC,$$

so that DE = DF. Also ADB and ADF being congruent DF = DB. Whence the result.

Section 56. London University School Examination.

ARITHMETIC.

1. 180. **2.** £176,285 15s. 9d.

3. £78. 15s [We must assume that 10s. 6d. is the published price. Who would not be a bookseller on these terms?]

4. 1016 kilograms is 2 ounces less than 1 ton.

5. 50 gm. of the alloy occupies 33/8·93 + 17/7·12 c. cm., so that 1 c. cm. weighs 8·2 gm.

6. £506 3s. 9d. **7** 9·7 francs (9·74).

8. 27 yards.

9. 0·856. The third figure requires care because 0·855 is almost as close.

10. Suppose $a\ b\ c\ \ldots$ to be the unit, ten, hundred, ... figures. The difference under consideration is

$$a + b10 + c10^2 + d10^3 + \ldots - (a + b + c + d + \ldots)$$

or $9b + 99c + 999d + \ldots$; every term of which is divisible by 9.

ALGEBRA.

1. If $1 - x$ is substituted for y, the given relation becomes an identity in x.

2. (i) With $a = -195$, the factors are $x - 15$ and $x + 13$. (ii) $(a - 3)(p^2 - a)$.

3. The watch indicates the passage of 5 hrs. 58 min. in 6 clock-hours or $6 \times 288 \div 289$ true hours. It loses one minute and becomes correct in $6 \times 288 \div 218$ hrs., *i.e.* at 4.56 p.m. to the nearest minute. In a day it loses $218 \div 72$ min. *i.e.* 3 min. pretty accurately.

4. $x = (a + b)/2$ $y = (a - b)/2$. The working is simplified by solving first for $x + y$ and $x - y$.

5. $p = 10{\cdot}75$. The required values of x are $2{\cdot}51$ and $-7{\cdot}51$.

6. Required values of x $2{\cdot}3$ and $-1{\cdot}3$ ($2{\cdot}303$ and $-1{\cdot}303$).

7. $x = 2{\cdot}5$.

ELEMENTARY MATHEMATICS: GEOMETRY.

1. In the quadrilateral the triangle DAC is isosceles, whence angles BAC BCA are equal, and BA is equal to BC.

2. Complete the parallelograms AOCM and PORN; let ST cut CR in O′, and draw A′O′P′ parallel to AOP. The parallelograms O′P′NR and O′AMC being equal, and OPNR and OAMC also equal, O and O′ must coincide.

3. On half the line describe a square. Half its diagonal is the distance of the dividing point from the middle of the line.

4. *Either*—O being the intersection of PR and QS, we have

$\angle POS = \angle PQS + \angle RPQ = \angle PRQ + \angle RQS = \angle SOR.$

Or—Give chord QS a parallel motion till arcs PQ QR become equal. PQ + RS remains constant, one gaining what the other loses. When PQ = QR we have also PS = SR, since PQ + RS = PS + QR; that is, QS is perpendicular to PR.

5. The centre is the intersection of the perpendicular bisector of AP and the tangent at P. The radius is 2·2 (2·19) in.

6. (i) The point of contact is at a constant distance from C, viz. $\sqrt{(CA.CB)}$. (ii) The portion of the locus is a little more than a semicircle, being bounded at one end by AC produced, and at the other by the circle on AB as diameter.

7. When the triangle is equilateral the centres of the escribed circles are at the vertices of equilateral triangles described upon its sides. The inscribed circle of the triangle formed by the three centres touches the sides of its triangle at the middle points, and is therefore the circum-circle of the original triangle.

Section 57. Board of Education Preliminary.

ARITHMETIC.

1. 4497·78 francs. [The "nearest centime" is at the same time the exact answer.]
2. 2·0873.
3. There are 19 tiles on each diagonal or side, and 361 in all.
4. Average 19·23.
5. It must be multiplied by 21, and the perfect square is $3^4 2^6 7^2$.
6. 97 % of spirit is left (97·0299).
7. By 4 lb.
8. 0·1 (0·098) of table uncovered; 0·3 (0·324) of cloth hangs over.
9. 1600 square inches are used; 13 sheets are bought; 740 sq. in. are wasted.

10. 39,208 pupils.

11. (i) £13. 18s. 4d. Express 1s. 8d. as £$\frac{1}{12}$. (ii) 140. Put the two numbers under one root symbol, and factorize. (iii) The first fraction is greater, since its numerator is greater and each term of its denominator is less than the corresponding term of the second denominator. (iv) $70\frac{1}{2}$ miles. Express the time taken as $6 - \frac{1}{8}$ hours.

MATHEMATICS I.

1. 16s. 4d. **2.** 3 farthings a mile.

3. Length 314·2 mm., and with one link cut off 309·4 mm.

4. (i) $a = 7$. (ii) $a = 7 - b/6$.

5. Prices are: 20 shillings at the pit, 25 shillings in the town.

6. (i) $(2x+19)(3x-5)$. (ii) $(a-b)(a+2b+1)$. (iii) 27; the expression is $\{(p+2)-(p-1)\}^3$.

7. (i) The equations should be solved for $x-1$ and $y-1$ as variables; the solutions are:

x	2	2	0	0
y	3	−1	3	−1.

(ii) $x = 3{\cdot}126$.

(iii) The greatest product of two numbers whose sum is 80, is 40×40 or 1600; and 2000 is impossible.

8. He reduces by z pence the cost of z lb., so that his profit is increased by 1d. per lb.

9. Suppose he walked for t hours at x miles per hour. We have the equations $(x+1/2)t4/5 = (x-1/2)(t+5/2) = xt$. The distance is 15 miles.

10. The roots are 6·0 and −20 (5·98 and −1·98).

MATHEMATICS II.

1. By placing the sides 9 cm. and 16 cm. at the same end of the 12 cm. sides, we unite the two right-angled triangles into another; whose area is 150 sq. cm.

2. Take AB 4·5 cm. long and BC perpendicular to it and 7 cm. long. Draw AD so that $\angle CAD = \angle ACD$ and D lies on BC. BD is half the base of the required triangle. Area 9·2 (9·243) sq. cm.

3. Draw a circle through B and cutting BA again in C. Through C draw the diameter CD. BD is the required perpendicular.

4. The perpendicular bisectors of the sides meet at the centre of the circumscribing circle. The radius OA is $1\frac{3}{4}$ in., for triangles OBC OCD are equilateral.

5. Let the perpendiculars at B and C meet in D, and those at C and A in E. The quadrilaterals PABF PBCD are cyclic, and a consideration of their angles shows the angles DPF DEF either equal or supplementary.

6. The radius is 3·8 (3·82) cm. Triangle KAJ is isosceles because the value of the angles at A and J is $(B + C)/2$; and so for LBJ. Since $KJ = KA$ and $LJ = LB$, $CK + CL - KL$ is equal to $CA + CB$ or 12 cm.

7. The length BP is constant, being equal to BA.

8. Volume 1810 c. in. ; height $21\frac{1}{3}$ in. ; area of curved surface 650 sq. in.

Section 58. Board of Education Certificate.

MATHEMATICS—PART I.

1. The product of the numbers is the product of the L.C.M. and H.C.F. or 9,923,837. The two numbers may be 359 and 27,643 or 2513 and 3949.

2. 1563·3. 3. 42·8 inches.

4. The true length and breadth being k b, the calculated area lies between $kb\,1{\cdot}05^2$ and $kb\,0{\cdot}95^2$. The former differs more from kb, and the percentage difference is 10·25.

5. Saving of distance 2800 ft. (the actual distance being 4798 ft. and the alternative 1996).

6. Rental £41.

7. The prices of a given quantity of coffee are as 1·05 to 1, it being 5 % dearer in the small tin.

8. (i) $13(m - 6)$ pence were contributed. (ii) Since $2m$ and $3m$ are integral, their difference m is integral ; and m must be at least 7 to make $3m - 20$ positive. (iii) The number of pence is a multiple of 13, and the sum is a multiple of

1s. 1d. (iv) To make the sum £1. 15s. 9d., m is 39, and the number of girls 69.

9. (i) $3a+b$. (ii) $x=1$. (iii) $x=-7/22$. (iv) The factor $215-157$ is 58.

10. If the distance is a miles and the time available t hours, $a=(t+1/6)2{\cdot}5=(t-1/4)4$; whence the distance is $2\frac{7}{8}$ miles.

11. (i) Area 308 sq. ft.

12. Length of ride 53 (52·6) miles. Area of property 91 sq. miles.

13. Difference of length of bases 9·6 cm.

14. Make at the centre angles of 100° and 160°.

MATHEMATICS—PART II.

1. (i) The roots are $1/p$ $1/q$. (ii) The equation is

$$ax^2+2(a+b)x+a+2b+4c=0.$$

2. £3. 5s. **3.** $x=5$.

4. (i) $x=0,\ 2{\cdot}83,\ -2{\cdot}83$; $x=5$ $y=5$, $x=1$ $y=9$, $x=9$ $y=1$. (ii) The two relations when put in integral form are seen to be identical. (iii) $(x-6)(15x+2)$.

5. Boys 60 %, girls 40 %.

6. (i) The latter is greater, the two values being 1·145 and 1·201. (ii) If x is the number $\log\{(x-10)/x\}=-0{\cdot}5$, so that $(x-10)/x=0{\cdot}3162$, and $x=14{\cdot}6$.

7. *Either*—Put $a=b-d$ $c=b+d$, and the relation becomes an identity. *Or*—Throw the given expression into the form $(b-c)(c-a)(a-b)$.

8. (i) $n(n-1)/2$ games are played. (iii) Apart from the game between B and C each of them may win $n-3$ games, and together they may win $2n-5$ games.

9. (i) $r(C_{r-1}+C_r)=C_{r-1}\left(r+\dfrac{n-r+1}{r}\times r\right)=(n+1)C_{r-1}$. (ii) The left side is the expansion of $(1-x^2)^n$, the right side is $(1+x)^n$ times the expansion of $(1-x)^n$.

10. Let M N denote the positions of A B when at the same level. Let A rise to P and B sink to Q, and let QR parallel to MP meet MN in R. QR = QN = MP, so that the middle point of PQ is always on MN.

11. He is travelling 9·5 west of south.

12. Through A draw any line, and through B and C lines suitably inclined to it.

13. (ii) $c = b + d$ because each side of the equation gives twice the distance of the centre of the parallelogram from XAY. If XAY falls within the angle DAB, perpendiculars in one direction must be reckoned positive, in the other negative.

14. (ii) If the tangent at A meets BC in T, $TB = TC = TA$, and $\angle BAC$ is right.

15. River is 28 yards broad.

16. Let fall the perpendicular RM to BC. $CM/MR = CB/BA = 6/5$ and $QM/MR = QB/BP = 4/3$, so that $QC/MR = 4/3 - 6/5$, and $MR = 15$ in. $= 5BP$. Hence $QR = 25$ in.

17. (i) Each side of BCD is $6\sqrt{2}$ in., and the area is 31 (31·2) sq. in. (ii) Remaining surface 840 (841) sq. in.

18. (i) Length of shadow of post $= 11·9$ (11·92) ft. (ii) Area of shadow of wall $25 \times 11·92 \times \sqrt{0·5}$ or 211 sq. ft.

Section 59. Cambridge Senior Locals.

ARITHMETIC.

1. G.C.M. 221.

2. (i) $14\frac{5}{8}$. (ii) 13·66; the last figure needs care, as 13·65 is almost as close.

3. £23. 11s. 1d.

4. 12 per cent.

5. £40. 10s. 7d.

6. 12·2 chains.

7. £1030. 18s. 7d.

8. Difference becomes 6 min. in 12 hr., and $16\frac{1}{2}$ min. in 33 hr. In that period the first clock has gained 11 min., and shows 9.11 p.m.; the other has lost $5\frac{1}{2}$ min., and shows $8.54\frac{1}{2}$ p.m.

9. £12500.

10. Suppose the distance a miles, the time available t hours. Then $a = (t + 1/4)3 = (t - 1/4)4$; and the distance is 6 miles.

GEOMETRY.

A 1. Let PS and QR meet in T. TPQ and TRS are isosceles triangles, so that PS = QR.

A 2. Let CK parallel to AD meet AB in K, and first construct the triangle CBK. The area is 53 (53·3) sq. cm.

A 3. (ii) The sum of the angles made by the bisectors with PQ is 90°.

A 4. (ii) When PR and XZ do not intersect,

$$\angle PXY = \angle RQY = 180^\circ - \angle RZY.$$

Slight change of proof if they do intersect.

B 6. QS/QR = QR/QP, and triangle RQS, being similar to PQR, is isosceles.

B 7. (i) Suppose L to lie on BC produced beyond B, M and N to lie within CA and AB respectively. Then

$$\angle LNB = LPB = 90^\circ - LBP = 90^\circ - CAP = MPA = MNA;$$

whence LN and NM are in line.

(ii) Produce opposite sides of the quadrilateral to meet, thus forming two triangles. About these two triangles draw circles. The circles intersect in the point from which the perpendiculars are to be drawn.

B 8. Let OQ be the intersection of the two planes A and B, O the point in which it meets PMN. Draw OC parallel to PM and OD parallel to PN, and so lying in the plane PMN. OC is perpendicular to A, and therefore to OQ, which lies in A. Similarly OD is perpendicular to OQ. So that OQ, being perpendicular to OC and OD, is perpendicular to the plane PMN, in which they lie.

B ix. Show triangles PMZ PNY to be congruent, and then XMY and XNZ.

B x. $PQ^2 = PR^2 + RQ^2 + 2NR.RQ$; so that NR = RQ.

B xi. Length between points of contact 3·8 (3·753) in. or 2·9 (2·946) in.

ALGEBRA.

1. (i) $12(1-x)(1+x)(2+x)$ or $-12x^3 - 24x^2 + 12x + 24$. (ii) $x = 1$.

2. Other factors $3x + 2$ and $4x - 5$.

3. (i) $x = -2$ or -3. (ii) $x = a + b$. (iii) $x = -b$, $y = -a$; $x = a - 1$, $y = b - 1$.
4. 18 pence a dozen, and 9 eggs a shilling.
5. Add 350 grammes of copper and 450 of tin.
6. $a^3x^2 + 4a(ac - b^2)x + 4c(ac - b^2) = 0$.
7. The entry (1, 3·9) is wrong. The values $a = -0{\cdot}1$ $b = 1{\cdot}7$ $c = 1{\cdot}5$ are found from three of the remaining entries, and the fourth entry then satisfies the equation.
8. The nth term is the sum to $n+1$ terms less the sum to n terms or $2nb - b + c$. The $(n+1)$th term less the nth is $2b$, which is constant. First term $b + c$, common difference $2b$.
9. There are 56 sets containing one a or none, 35 sets containing two a's; total 91.
10. First four terms: 1 0·8 0·3 0·07. Ratio of $(r+1)$th to rth $(17 - r) \div 20r$. Interest £118. 6s.
11. True value of $e^{0{\cdot}25}$ to four decimal places is 1·2840. Four figure tables give only three decimal places, viz. 1·284.
12. The product of the three roots being -1, the third is $1/(1 - a)$.

A. $(2x^2 + x + 3)(2x^2 - x + 3)$.

B. (i) $x = a$ $y = b$. (ii) $x = 11/7$ or $2/5$.

C. £700.

D. The point required is $x = y = 1{\cdot}37$ $(1{\cdot}371)$.

Section 60. Oxford Senior Locals.

ARITHMETIC.

1. £2 must be added.
2. 0·128125.
3. (i) Side of lawn 45 yards. (ii) He must go 10 times round.
4. Former depth 40 cm.
5. 3·65 per cent.
6. A boy 2s. 4½d., a girl 1s. 7d.
7. He takes on 400 more men.
8. The car's speed is $(135 + 27)/27$ times the man's, or 21 miles an hour.
9. Population was 50,000.

GEOMETRY.

2. (ii) Draw parallels 5 cm. and 7 cm. from the given lines. Eight of their intersections (two in each angle) satisfy the conditions.

3. On the base draw a circle one of whose segments contains 60°. A line through the middle point of the minor arc and a point of the base $1\frac{1}{2}$ inch from its end meets the major arc in the vertex of the triangle.

4. Radius is 63 or 64 mm.

ALGEBRA.

1. The second factor is $a - x + y - z$. 2. 50.

3. (i) $5b^2 \div (a - b)(a + 2b)(a + 3b)$. (ii) x.

4. (i) $x = a$ or $1/a$. (ii) $x = 4$, $y = 3$; $x = -3/2$, $y = -8$.

5. Length 22 ft. 6 in., breadth 15 ft. 4 in., height 12 ft.

6. The equation is $x^2 - (a^2 + 2b)x + b^2 = 0$.

7. There are either 8 or 19 terms. The 9th to 19th terms are 65 52 39 ... −39 −52 −65, and their sum is zero.

8. It is the number of ways in which the remaining five boys may be selected from eight, viz. 56.

9. (ii) The coefficient of x is 10. **10.** The factors are 4 and 4.

Section 61. Northern Universities Matriculation.

ARITHMETIC AND ALGEBRA.

1. 6080 ft. (6076). 2. (i) £8272. 10s. (ii) £7747. 8s.

3. $(3x + 2y)(2x - 3y)$ $(x^2 + 3xy + 4y^2)(x^2 - 3xy + 4y^2)$
$$(x^2 + 3x - 5)(x^2 - 3x + 5).$$

4. (i) $abc/(bc + ca + ab)$ days. (ii) Ratio $m^2/(m^2 - n^2)$.

5. $x^2 - 2x - 1$.

6. 7 voted for B and C, 8 for C and A, 10 for A and B.

7. (i) The difference is $x^2 - 2x + 3$ or $(x - 1)^2 + 2$. The value is least when $x = 1$, and is then 2; for all other values of x the difference is greater than 2, and cannot vanish.
(ii) $x = 1$, $y = -2$; $x = 2$, $y = -0{\cdot}6$.

8. If he goes a miles the first day, and $a+b$ the second, then

$$(2a+4b)5/2=100 \qquad (2a+6b)7/2=150$$
$$\text{or} \quad a=120/7 \qquad b=10/7.$$

To go 300 miles he takes 12 days.

9. The series $1-x/(1-x)+\{x/(1-x)\}^2 \ldots$ is convergent if $1 > x/(1-x) > -1$. The condition is satisfied by all negative values of x and by positive values less than 1/2.

GEOMETRY.

1. If P′ is the image of P in KL, R lies upon QP′ produced.
2. The lines run to appropriate points of trisection of two sides of the parallelogram.
3. BD = 0·784 cm. DC = 9·216 cm. AD = 2·688 cm.
4. Suppose AK the greater section of AB. Then AB^2+BK^2 is equal to $(AB-BK)^2+2AB.BK$, the first term of which is AK^2 and the second term $2AK^2$.
5. The first locus is a diameter; the second a concentric circle; the third a circle whose diameter is the line joining the centre of the original circle to the point of concurrence of the chords.
7. AI/ID = AB/BD = (BA + AC)/BC.
8. AF/AE = AC/AB, for triangles AFC AEB are similar. When A = 60°, △AEF is a quarter of ABC.

Section 62. Central Welsh Board.

ARITHMETIC.

1. 1·754.
2. 7·389 20·085 1·649.
3. Cost price 16s. 8d. per ton.
4. 139 days' supply.
5. 487 lb.
6. Diameter 0·6 inch.
7. 320 additional workmen.
8. £538. 5s. 9d.

ELEMENTARY MATHEMATICS: GEOMETRY.

1. (i) First make a triangle ACK, having AC of given length, CK = BD, AK = AB + CD. This gives A and C; B is in AK at the given distance from A. (ii) BC = 3·0 (2·95) in., AD = 2·8 (2·77) in.

3. Suppose OAB the position of the triangle OPQ in which PQ lies along the given line. The locus is the line through B parallel to OA; for triangles OAP OBQ are congruent. By taking the triangle on the other side of OP we get a second line as part of the locus.

4. The sum of the squares or two sides is equal to twice the square of the concurrent median plus half the square of the opposite side. The addition of the three expressions of this kind gives: twice the sum of the squares of the medians is equal to $1\frac{1}{2}$ times the sum of the squares on the sides, which, for a right-angled triangle, is three times the square on the hypotenuse.

5. On half the shortest chord through P describe a square. The diagonal of this square is the greater segment of the chord. Length of chord 3·2 (3·17) in.

6. $180^\circ - FAD = FED = ABC = 180^\circ - ADC$.

7. The point of contact is given by the rectangle property of the circle.

8. On BC, 9 cm. long, draw a circular segment containing an angle of 49°. Join the middle point of the remaining segment to a point of trisection of BC; this line cuts the circle again in A, the vertex of the triangle. The in-radius is 1·9 cm. (1·94).

ELEMENTARY MATHEMATICS: ALGEBRA.

1. (i) $(x-3)(x-4)(x-8)(x+1)$. (ii) Remainder is ah^2+bh+c.

2. (i) $x^4-6x^2+1=(x^2+2x-1)(x^2-2x-1)$. (ii) The left member is $\{2(x+y)^2+2\}/\{(x+y)^2-1\}$, or, multiplying out and writing 1 for x^2+y^2, it is $(4+4xy)/2xy=2(1+1/xy)$.

3. (i) $x=19/7$ or $-23/13$. (ii) $x=2a-b$, $y=2b-a$.

4. The expression is less than 2 when x lies between 1 and 5.

5. The fractions are 0·85 and 0·15.

6. The first term is 548/209, the common difference $-45/209$, and the sum to n terms $n(1141-45n)/418$.

7. (i) Put $b=ap$ $c=ap^2$ and the relation becomes an identity.
(ii) Cleared of fractions, the relation factorizes into

$$(ab-cd)(ad-bc)=0,$$

whence a/c must be equal to b/d or to d/b.

8. Percentage profit was $\{(100+c)a^2-100b^2\}/b^2$.

Section 63. Scotch Education Department.

MATHEMATICS. LOWER GRADE: FIRST PAPER.

1. Show triangles LMN PQR congruent, and then triangles LMO PQS.

2. Draw two parallels 2·8 cm. apart. On one take AB = 5 cm. With centre A, radius 6 cm., strike an arc meeting the parallel line in D. Complete the parallelogram ABCD.

3. Their distances are 6 cm. and 5·2 cm.

4*a*. Produce XC to meet AB in M. Then

$$\angle ABC = \angle XFC = \angle XCF = 90^\circ - BCM,$$

so that CM is perpendicular to AB.

4*b*. Draw DQ and PM perpendicular to AB, and PN perpendicular to AC. Draw QR parallel to DP to meet PM in R. Triangles DPN QRM being congruent, PN = RM. The difference of the perpendiculars is thus PR or DQ.

5*a*. If L M are the mid-points of AB AC, triangles ADE AML are congruent and $\angle AED = \angle ALM = \angle ABC$. Hence BCED is a cyclic quadrilateral.

5*b*. L = 56° M = 65° N = 59°.

MATHEMATICS. LOWER GRADE: SECOND PAPER.

1. $(5x+3)(3x-5)$ $(x-a)(x+b)$.

2. $3-x-5x^2+2x^3$. When $x=1$, dividend is 6, divisor −6, and quotient −1.

3. (i) $x=17/15$. (ii) $x=(c+d)/2a$, $y=(d-c)/2b$. (iii) $x=1/2$ or $-1/3$.

4. (i) A had £72, B had £28. (ii) $x=8$.

5*a*. a being 1/2 and b 1, the value of the expression is 17·5 when $x=5$.

5*b*. (i) $16a^3x/(x^4-16a^4)$. (ii) $2\div(x-1)(x-2)$.

6*a*. $x=4$, $y=1$; $x=4/3$, $y=3$.

6*b*. Area 150 sq. ft., AC = 30·0 ft., BC = 16·2 ft., AN = 18·6 ft.

MATHEMATICS. LOWER GRADE: THIRD PAPER.

1. 16 inches.
2. £1600, £1813. 6s. 8d., £2186. 13s. 4d.
3. 2·00312 cubic metres or 2003·12 litres.
4. First class $17\frac{3}{4}$d., third class $8\frac{1}{4}$d.
5. £4550.
6. Slowest from A to B, fastest from G to H.

MATHEMATICS. HIGHER GRADE: FIRST PAPER.

1. If O is the mid-point of BC $AD^2 - AB^2 = OD^2 - OB^2 = BD.CD$.
2. Let ST be produced to K. Angle RTK is constant, being equal to PQR. For the tangent draw from P a line making with PR the angle PQR.
3. OE and OF are tangents because their squares are equal to OA.OB.

4*a*. BD $= 9\frac{5}{11}$ in. BE $= 52$ in.

4*b*. O being the centre of the given circle, draw QR parallel to OP to meet AO in R. AR $=$ QR $= \frac{3}{8}$OA, so that R is a fixed point and Q at a fixed distance from it.

5*a*. The locus is the intersection of two planes, the first being midway between the parallel planes and the second the perpendicular bisector of the line joining the given points.

5*b*. Volume of OABC $= 1/6$ c. in. When four such tetrahedra are removed from the cube, the remaining figure has a volume of $1/3$ c. in. and an edge of length 1·414 in.

MATHEMATICS. HIGHER GRADE: SECOND PAPER.

1. £53. 6s. 8d. has been deducted, the income being £1200.
2. (i) $(6x + 5y)(3x - 4y)$. (ii) $(a + b)(a - b)(a^2 + ab + b^2)$.
   (iii) $(p^2 + s^2)(q^2 + r^2)$.
3. (i) $x = c$ or $b - c$. (ii) $x = 2/3$. [The value $x = -1$ must be rejected, as it would require the first radical to be taken with a negative sign.]

   (iii)

$x =$	2	-2	2·121	$-2·121$
$y =$	1	-1	0·707	$-0·707$

4. Point of intersection $x = 2·2$ (2·19), $y = 1·0$ (0·96).

5a. If the first travels at x miles an hour, their distances (in miles) t hours after 9 a.m. are

$$xt \quad (x+3)(t-0{\cdot}5) \quad (x+8)(t-1).$$

The time in question was noon and the speeds 15 18 23 miles an hour.

5b. Produce AB to K so that AK = 6 in. Make the angle AKC equal to 45°, and its second leg cuts the semicircle in the possible point C. AD = 4 or 4·5 in., CD = 2 or 1·5 in.

6a. (i) Suppose k added. $\frac{a+k}{b+k}-\frac{a}{b}=\frac{(b-a)k}{b(b+k)}$, so that the fraction is increased. Also $1-\frac{a+k}{b+k}=\frac{b-a}{b+k}$, so that the fraction remains less than 1. (ii) Let the line $x=1$ cut OX OP OQ OR in D A B C; then AD BD CD are respectively x/y x'/y' 1.

6b. Each of the ratios BC/BA CD/BC etc. is 12/13, which is the required common ratio.

Section 64. Intermediate Education Board.

ARITHMETIC.

1. The ratio was 3 to 5.

2. Difference £1. 10s. Total dividend £145. 10s.

3. He gained 17·6 %. **4.** 2·2 lb. **5.** £10,928. 4s.

6. If the $6\frac{1}{2}$ % is reckoned all through on the present value, the value after 7 years is £2182. 10s. If it is reckoned each year on the value at the beginning of the year, the value is £2331. 0s. to the nearest shilling. Both interpretations are reasonable.

7. The height is 12 ft., the length 24 ft., and the breadth 16 ft.

PASS GEOMETRY.

6. Length of tangent 7·2 cm. (7·23).

7. Longest side 12·7 cm.

8. If BP is the diameter through B, $\angle BPC = 30°$ $\angle BCP = 90°$ $\angle CBP = 60°$.

9. If O is the mid-point of BC, each expression is equal to $OB^2 - OM^2$.

10. They are the three escribed circles.

HONOURS GEOMETRY.

1. Make ∠RPQ′ equal to ∠CAB, Q′ lying in RQ.

$$PQ'/PR = AB/AC = PQ/PR,$$

so that $PQ' = PQ$. Either Q and Q′ coincide and the angles at B and Q are equal, or PQQ′ is an isosceles triangle and the angles at B and Q are supplementary.

4. Let the perpendicular from A meet BC in D and the circum-circle in P. Let PQ parallel to BC meet the circumcircle again in Q. Let R be the centre, RS the perpendicular on BC. Consider the right-angled triangle OPQ. RS bisects PQ, to which it is perpendicular. BC bisects OP (for ∠OBD = ∠PAC = ∠PBC). Therefore S lies on OQ at the middle point. ARQ is a straight line, for ∠APQ is right. Whence AO = 2RS.

5. ABCD being the quadrilateral, E F G H the feet of perpendiculars upon AB BC CD DA from P on the circumcircle, triangles PEF PHG are similar (for ∠PEF = ∠PBF = ∠CDP = ∠GHP, and so for angles PFE and PGH).

6. Think of APQR as a variable line always passing through A. The shapes of triangles BPQ and BQR are fixed, for the angles BPA BQA BRA are the angles in fixed segments. Therefore PQ/QR is constant.

7. If the bisectors BE CF meet in O, angle EOF is $90° + A/2$, so that O is given as the intersection of the bisector of angle A with a circular segment on EF.

8. Angle CPM = ∠CAM = ∠CBA = ∠CQM. Whence PM = MQ.

9. Through one straight line P draw a plane parallel to the other Q. From any point A on Q let fall AB perpendicular to the plane. Through B draw BC parallel to Q to meet P in C. The perpendicular at C to the plane is the line required.

10. Through A B draw AE BF parallel to XY, and draw EYF through Y parallel to AB to meet them in E and F. Through P draw PG parallel to CF and meeting BF in G; through Q draw QH parallel to DE and meeting AE in H. Let HG and XY (which are coplanar) meet in O. DE is parallel to CF; PG and HQ are parallel to one another; triangles OHQ OGP are similar, and POQ is a straight line meeting XY in O.

PASS ALGEBRA.

1. Quotient $3(a^2+b^2+c^2)$ **2.** 2·44 and −0·27.

3. $x=3 \quad y=-2 \quad z=5$.

4. Father's age 32 years, son's 8 years.

5. It is the square of $2(x^2+xy+y^2)$, as is seen by arranging according to powers of x and extracting the root.

6. $1/(3x-1)$. **7.** $x=2a+b$ or $2b+a$.

8. The fraction is 9/10.

10. The points of intersection are $x=3{\cdot}1\ (3{\cdot}09)$ $y=1{\cdot}5\ (1{\cdot}46)$ and $x=-2{\cdot}6\ (-2{\cdot}59)$ $y=4{\cdot}3\ (4{\cdot}29)$.

HONOURS ALGEBRA.

1. $x=-(b^2c^2+c^2a^2+a^2b^2+a^2bc+ab^2c+abc^2)/(bc+ca+ab)$
$y=(b+c)(c+a)(a+b)/(bc+ca+ab)$
$z=a^2b^2c^2/(bc+ca+ab)$.

First obtain two equations free from x, and then from them an equation in z only.

2. If he walks a miles an hour and runs b miles an hour, we have the equations

$$4/a+3/b=3/2 \qquad 3/(a+1)+2/(b+1)=5/6,$$

whence a quadratic of which only one root is available. He runs 8 miles an hour and walks 3 miles an hour.

3. Use the given relations to express y and z in terms of x, and substitute in the given equation.

5. When the values of $x\ y\ z$ in terms of $a\ b\ c$ are substituted, the left side of the expression has factors b^3-c^3 c^3-a^3 a^3-b^3 in numerator and denominator. These being divided out, the required result follows.

6. (i) When A must immediately precede B, the number of orders is as if A was not there, viz. $(n-1)!$ (ii) If A is to precede B anyhow, we first place B, say in the rth place. We have then a choice of $r-1$ places for A, and the remainder may be placed in $(n-2)!$ ways. Summing up, we have for the total number of ways

$$(n-2)!\,(1+2+3+\ldots+n-1) \quad \text{or} \quad n!/2.$$

7. The expression is equal to $-3(1+5x)^{-1}+\frac{2}{3}(1-\frac{2}{3}x)^{-1}$ and the required coefficient is $-3(-5)^r+\frac{2}{3}(\frac{2}{3})^r$.

8. $\text{Log}_e(5/4)=0{\cdot}22314$ $\log_e(128/125)=0{\cdot}02372$. The former being equal to $\log_e 10-3\log_e 2$ and the latter to $10\log_e 2-3\log_e 10$, we can calculate $\log_e 10$ and $\log_e 2$, and so obtain $\log_{10}2=\log_e 2/\log_e 10=0{\cdot}3010$.

GLASGOW: PRINTED AT THE UNIVERSITY PRESS BY ROBERT MACLEHOSE AND CO. LTD.

	0	1	2	3	4	5	6	7	8	9	1	2	3	4	5	6	7	8	9
10	0000	0043	0086	0128	0170						4	9	13	17	21	26	30	34	38
						0212	0253	0294	0334	0374	4	8	12	16	20	24	28	32	36
11	0414	0453	0492	0531	0569						4	8	12	15	19	23	27	31	35
						0607	0645	0682	0719	0755	4	7	11	15	19	22	26	30	33
12	0792	0828	0864	0899	0934						3	7	11	14	18	21	25	28	32
						0969	1004	1038	1072	1106	3	7	10	14	17	20	24	27	31
13	1139	1173	1206	1239	1271						3	7	10	13	16	20	23	26	30
						1303	1335	1367	1399	1430	3	7	10	13	16	19	22	25	29
14	1461	1492	1523	1553	1584						3	6	9	12	15	19	22	25	28
						1614	1644	1673	1703	1732	3	6	9	12	15	17	20	23	26
15	1761	1790	1818	1847	1875						3	6	9	11	14	17	20	23	26
						1903	1931	1959	1987	2014	3	6	8	11	14	17	19	22	25
16	2041	2068	2095	2122	2148						3	5	8	11	14	16	19	22	24
						2175	2201	2227	2253	2279	3	5	8	10	13	16	18	21	23
17	2304	2330	2355	2380	2405						3	5	8	10	13	15	18	20	23
						2430	2455	2480	2504	2529	2	5	7	10	12	15	17	20	22
18	2553	2577	2601	2625	2648						2	5	7	9	12	14	16	19	21
						2672	2695	2718	2742	2765	2	5	7	9	11	14	16	18	21
19	2788	2810	2833	2856	2878						2	4	7	9	11	13	16	18	20
						2900	2923	2945	2967	2989	2	4	6	8	11	13	15	17	19
20	3010	3032	3054	3075	3096	3118	3139	3160	3181	3201	2	4	6	8	11	13	15	17	19
21	3222	3243	3263	3284	3304	3324	3345	3365	3385	3404	2	4	6	8	10	12	14	16	18
22	3424	3444	3464	3483	3502	3522	3541	3560	3579	3598	2	4	6	8	10	12	14	15	17
23	3617	3636	3655	3674	3692	3711	3729	3747	3766	3784	2	4	6	7	9	11	13	15	17
24	3802	3820	3838	3856	3874	3892	3909	3927	3945	3962	2	4	5	7	9	11	12	14	16
25	3979	3997	4014	4031	4048	4065	4082	4099	4116	4133	2	3	5	7	9	10	12	14	15
26	4150	4166	4183	4200	4216	4232	4249	4265	4281	4298	2	3	5	7	8	10	11	13	15
27	4314	4330	4346	4362	4378	4393	4409	4425	4440	4456	2	3	5	6	8	9	11	13	14
28	4472	4487	4502	4518	4533	4548	4564	4579	4594	4609	2	3	5	6	8	9	11	12	14
29	4624	4639	4654	4669	4683	4698	4713	4728	4742	4757	1	3	4	6	7	9	10	12	13
30	4771	4786	4800	4814	4829	4843	4857	4871	4886	4900	1	3	4	6	7	9	10	11	13
31	4914	4928	4942	4955	4969	4983	4997	5011	5024	5038	1	3	4	6	7	8	10	11	12
32	5051	5065	5079	5092	5105	5119	5132	5145	5159	5172	1	3	4	5	7	8	9	11	12
33	5185	5198	5211	5224	5237	5250	5263	5276	5289	5302	1	3	4	5	6	8	9	10	12
34	5315	5328	5340	5353	5366	5378	5391	5403	5416	5428	1	3	4	5	6	8	9	10	11
35	5441	5453	5465	5478	5490	5502	5514	5527	5539	5551	1	2	4	5	6	7	9	10	11
36	5563	5575	5587	5599	5611	5623	5635	5647	5658	5670	1	2	4	5	6	7	8	10	11
37	5682	5694	5705	5717	5729	5740	5752	5763	5775	5786	1	2	3	5	6	7	8	9	10
38	5798	5809	5821	5832	5843	5855	5866	5877	5888	5899	1	2	3	5	6	7	8	9	10
39	5911	5922	5933	5944	5955	5966	5977	5988	5999	6010	1	2	3	4	5	7	8	9	10
40	6021	6031	6042	6053	6064	6075	6085	6096	6107	6117	1	2	3	4	5	6	8	9	10
41	6128	6138	6149	6160	6170	6180	6191	6201	6212	6222	1	2	3	4	5	6	7	8	9
42	6232	6243	6253	6263	6274	6284	6294	6304	6314	6325	1	2	3	4	5	6	7	8	9
43	6335	6345	6355	6365	6375	6385	6395	6405	6415	6425	1	2	3	4	5	6	7	8	9
44	6435	6444	6454	6464	6474	6484	6493	6503	6513	6522	1	2	3	4	5	6	7	8	9
45	6532	6542	6551	6561	6571	6580	6590	6599	6609	6618	1	2	3	4	5	6	7	8	9
46	6628	6637	6646	6656	6665	6675	6684	6693	6702	6712	1	2	3	4	5	6	7	7	8
47	6721	6730	6739	6749	6758	6767	6776	6785	6794	6803	1	2	3	4	5	5	6	7	8
48	6812	6821	6830	6839	6848	6857	6866	6875	6884	6893	1	2	3	4	4	5	6	7	8
49	6902	6911	6920	6928	6937	6946	6955	6964	6972	6981	1	2	3	4	4	5	6	7	8

LOGARITHMS.

	0	1	2	3	4	5	6	7	8	9	1	2	3	4	5	6	7	8	9
50	6990	6998	7007	7016	7024	7033	7042	7050	7059	7067	1	2	3	3	4	5	6	7	8
51	7076	7084	7093	7101	7110	7118	7126	7135	7143	7152	1	2	3	3	4	5	6	7	8
52	7160	7168	7177	7185	7193	7202	7210	7218	7226	7235	1	2	2	3	4	5	6	7	7
53	7243	7251	7259	7267	7275	7284	7292	7300	7308	7316	1	2	2	3	4	5	6	6	7
54	7324	7332	7340	7348	7356	7364	7372	7380	7388	7396	1	2	2	3	4	5	6	6	7
55	7404	7412	7419	7427	7435	7443	7451	7459	7466	7474	1	2	2	3	4	5	5	6	7
56	7482	7490	7497	7505	7513	7520	7528	7536	7543	7551	1	2	2	3	4	5	5	6	7
57	7559	7566	7574	7582	7589	7597	7604	7612	7619	7627	1	2	2	3	4	5	5	6	7
58	7634	7642	7649	7657	7664	7672	7679	7686	7694	7701	1	1	2	3	4	4	5	6	7
59	7709	7716	7723	7731	7738	7745	7752	7760	7767	7774	1	1	2	3	4	4	5	6	7
60	7782	7789	7796	7803	7810	7818	7825	7832	7839	7846	1	1	2	3	4	4	5	6	6
61	7853	7860	7868	7875	7882	7889	7896	7903	7910	7917	1	1	2	3	4	4	5	6	6
62	7924	7931	7938	7945	7952	7959	7966	7973	7980	7987	1	1	2	3	3	4	5	6	6
63	7993	8000	8007	8014	8021	8028	8035	8041	8048	8055	1	1	2	3	3	4	5	5	6
64	8062	8069	8075	8082	8089	8096	8102	8109	8116	8122	1	1	2	3	3	4	5	5	6
65	8129	8136	8142	8149	8156	8162	8169	8176	8182	8189	1	1	2	3	3	4	5	5	6
66	8195	8202	8209	8215	8222	8228	8235	8241	8248	8254	1	1	2	3	3	4	5	5	6
67	8261	8267	8274	8280	8287	8293	8299	8306	8312	8319	1	1	2	3	3	4	5	5	6
68	8325	8331	8338	8344	8351	8357	8363	8370	8376	8382	1	1	2	3	3	4	4	5	6
69	8388	8395	8401	8407	8414	8420	8426	8432	8439	8445	1	1	2	2	3	4	4	5	6
70	8451	8457	8463	8470	8476	8482	8488	8494	8500	8506	1	1	2	2	3	4	4	5	6
71	8513	8519	8525	8531	8537	8543	8549	8555	8561	8567	1	1	2	2	3	4	4	5	5
72	8573	8579	8585	8591	8597	8603	8609	8615	8621	8627	1	1	2	2	3	4	4	5	5
73	8633	8639	8645	8651	8657	8663	8669	8675	8681	8686	1	1	2	2	3	4	4	5	5
74	8692	8698	8704	8710	8716	8722	8727	8733	8739	8745	1	1	2	2	3	4	4	5	5
75	8751	8756	8762	8768	8774	8779	8785	8791	8797	8802	1	1	2	2	3	3	4	5	5
76	8808	8814	8820	8825	8831	8837	8842	8848	8854	8859	1	1	2	2	3	3	4	5	5
77	8865	8871	8876	8882	8887	8893	8899	8904	8910	8915	1	1	2	2	3	3	4	4	5
78	8921	8927	8932	8938	8943	8949	8954	8960	8965	8971	1	1	2	2	3	3	4	4	5
79	8976	8982	8987	8993	8998	9004	9009	9015	9020	9025	1	1	2	2	3	3	4	4	5
80	9031	9036	9042	9047	9053	9058	9063	9069	9074	9079	1	1	2	2	3	3	4	4	5
81	9085	9090	9096	9101	9106	9112	9117	9122	9128	9133	1	1	2	2	3	3	4	4	5
82	9138	9143	9149	9154	9159	9165	9170	9175	9180	9186	1	1	2	2	3	3	4	4	5
83	9191	9196	9201	9206	9212	9217	9222	9227	9232	9238	1	1	2	2	3	3	4	4	5
84	9243	9248	9253	9258	9263	9269	9274	9279	9284	9289	1	1	2	2	3	3	4	4	5
85	9294	9299	9304	9309	9315	9320	9325	9330	9335	9340	1	1	2	2	3	3	4	4	5
86	9345	9350	9355	9360	9365	9370	9375	9380	9385	9390	1	1	2	2	3	3	4	4	5
87	9395	9400	9405	9410	9415	9420	9425	9430	9435	9440	0	1	1	2	2	3	3	4	4
88	9445	9450	9455	9460	9465	9469	9474	9479	9484	9489	0	1	1	2	2	3	3	4	4
89	9494	9499	9504	9509	9513	9518	9523	9528	9533	9538	0	1	1	2	2	3	3	4	4
90	9542	9547	9552	9557	9562	9566	9571	9576	9581	9586	0	1	1	2	2	3	3	4	4
91	9590	9595	9600	9605	9609	9614	9619	9624	9628	9633	0	1	1	2	2	3	3	4	4
92	9638	9643	9647	9652	9657	9661	9666	9671	9675	9680	0	1	1	2	2	3	3	4	4
93	9685	9689	9694	9699	9703	9708	9713	9717	9722	9727	0	1	1	2	2	3	3	4	4
94	9731	9736	9741	9745	9750	9754	9759	9763	9768	9773	0	1	1	2	2	3	3	4	4
95	9777	9782	9786	9791	9795	9800	9805	9809	9814	9818	0	1	1	2	2	3	3	4	4
96	9823	9827	9832	9836	9841	9845	9850	9854	9859	9863	0	1	1	2	2	3	3	4	4
97	9868	9872	9877	9881	9886	9890	9894	9899	9903	9908	0	1	1	2	2	3	3	4	4
98	9912	9917	9921	9926	9930	9934	9939	9943	9948	9952	0	1	1	2	2	3	3	4	4
99	9956	9961	9965	9969	9974	9978	9983	9987	9991	9996	0	1	1	2	2	3	3	3	4